Springer Monographs in Mathematics

Springer-Verlag Berlin Heidelberg GmbH

Kazuaki Taira

Semigroups, Boundary Value Problems and Markov Processes

Springer

Kazuaki Taira
University of Tsukuba
Institute of Mathematics
305-8571 Tsukuba, Ibaraki
Japan
e-mail: taira@math.tsukuba.ac.jp

Cataloging-in-Publication Data applied for

A catalog record for this book is available from the Library of Congress.

Bibliographic information published by Die Deutsche Bibliothek
Die Deutsche Bibliothek lists this publication in the Deutsche Nationalbibliografie;
detailed bibliographic data is available in the Internet at http://dnb.ddb.de

Mathematics Subject Classification (2000): 60J, 47D, 35J, 35B

DOI 10.1007/978-3-662-09857-8

http://www.springer.de

Originally published by Springer-Verlag Berlin Heidelberg New York in 2004
MyCopy version of the original edition 2004

Typesetting by the author using a Springer TEX macro package
Cover production: Erich Kirchner, Heidelberg

Printed on acid-free paper 41/3142db - 5 4 3 2 1 0
www.springer.com/mycopy

To my mother
Yasue Taira
and to the memory of my father
Yasunori Taira

Preface

The purpose of this book is to provide a careful and accessible account along modern lines of the subject which the title deals, as well as to discuss problems of current interest in the field. Unlike many other books on Markov processes, this book focuses on the relationship between Markov processes and elliptic boundary value problems, with emphasis on the study of analytic semigroups. More precisely, this book is devoted to the functional analytic approach to a class of *degenerate* boundary value problems for second-order elliptic integro-differential operators, called Waldenfels operators, which includes as particular cases the Dirichlet and Robin problems. We prove that this class of boundary value problems provides a new example of analytic semigroups both in the L^p topology and in the topology of uniform convergence. As an application, we construct a strong Markov process corresponding to such a physical phenomenon that a Markovian particle moves both by jumps and continuously in the state space until it "dies" at the time when it reaches the set where the particle is definitely absorbed.

The approach here is distinguished by the extensive use of the techniques characteristic of recent developments in the theory of partial differential equations. The main technique used is the calculus of pseudo-differential operators which may be considered as a modern theory of potentials. Several recent developments in the theory of partial differential equations have made possible further progress in the study of boundary value problems and hence the study of Markov processes. The presentation of these new results is the main purpose of this book. We confined ourselves to the simple but important boundary condition. This makes it possible to develop our basic machinery with a minimum of bother and the principal ideas can be presented concretely and explicitly.

This monograph is an expanded and revised version of a set of lecture notes for graduate courses given by the author at Hiroshima University in 1995–1997 and at the University of Tsukuba in 1998–2000. In 1988–1990 I gave a course in functional analysis and partial differential equations at the University of Tsukuba, the notes for which were published in the Springer Lecture Notes in Mathematics series under the title *Boundary Value Problems and Markov Processes* in 1991. These notes were found useful by a number of people, but they went out of print after a few years. Moreover, in the ten years since the lecture notes appeared there have been a number of much more comprehensive treatments of the material.

Out of all this has emerged the present book. This new edition has been revised to streamline some of the analysis and to give better coverage of important examples and applications. It is addressed to advanced undergraduates or beginning-graduate students and also mathematicians with an interest in functional analysis, partial differential equations and probability. For the former, it may serve as an effective introduction to these three interrelated fields of mathematics. For the latter, it provides a method for the analysis of elliptic boundary value problems, a powerful method clearly capable of extensive further development. I have revised and updated the bibliography, but I have preferred to give references to expository books and articles rather than to research papers.

It is possible to regard the present book as the first volume of a three-volume series by the author, the others being *Analytic Semigroups and Semilinear Initial Boundary Value Problems* (Cambridge University Press, 1995) and *Brownian Motion and Index Formulas for the de Rham Complex* (Wiley-VCH, 1998). I hope that this monograph will lead to a better insight into the study of three interrelated subjects in analysis: semigroups, elliptic boundary value problems and Markov processes.

I would like to express my hearty thanks to the two referees whose comments and corrigenda concerning portions of various preliminary drafts of this book have resulted in a number of improvements. I am grateful to my colleagues and students at Tsukuba and Hiroshima who have provided the stimulating environment in which ideas germinate and flourish.

My special thanks go to the editorial staffs of Springer-Verlag Tokyo and Heidelberg for their unfailing helpfulness and cooperation during the production of the book.

This research was partially supported by Grant-in-Aid for General Scientific Research (No. 13440041), Ministry of Education, Culture, Sports, Science and Technology, Japan.

Last, but not least, I owe a great debt of gratitude to my wife, Naomi, who gave me moral support during the preparation of this book.

Tsukuba,
April 2003,

Kazuaki Taira

Contents

1 **Introduction and Main Results** 1

2 **Theory of Semigroups** 19
2.1 Banach Space Valued Functions 19
2.2 Operator Valued Functions 21
2.3 Exponential Functions 22
2.4 Contraction Semigroups 23
2.5 Analytic Semigroups 35

3 **Markov Processes and Semigroups** 47
3.1 Markov Processes 47
3.2 Transition Functions and Feller Semigroups 51
3.3 Generation Theorems for Feller Semigroups 60
3.4 Borel Kernels and the Maximum Principle 70

4 **Theory of Distributions** 73
4.1 Notation 73
4.2 L^p Spaces 74
4.3 Distributions 80
4.4 The Fourier Transform 87
4.5 Operators and Kernels 92
4.6 Layer Potentials 93
4.6.1 The Jump Formula 93
4.6.2 Single and Double Layer Potentials 94
4.6.3 The Green Representation Formula 95

5 **Theory of Pseudo-Differential Operators** 97
5.1 Function Spaces 97
5.2 Fourier Integral Operators 102
5.2.1 Symbol Classes 102
5.2.2 Phase Functions 104

5.2.3 Oscillatory Integrals . . . 105
5.2.4 Fourier Integral Operators . . . 107
5.3 Pseudo-Differential Operators . . . 108
5.4 Potentials and Pseudo-Differential Operators . . . 115
5.5 The Transmission Property . . . 116
5.6 The Boutet de Monvel Calculus . . . 118

6 Elliptic Boundary Value Problems . . . 121
6.1 The Dirichlet Problem . . . 121
6.2 Formulation of a Boundary Value Problem . . . 127
6.3 Reduction to the Boundary . . . 129

7 Elliptic Boundary Value Problems and Feller Semigroups . . . 135
7.1 Formulation of a Problem . . . 135
7.2 Transversal Case . . . 140
7.2.1 Generation Theorem for Feller Semigroups . . . 141
7.2.2 Sketch of Proof of Theorem 7.1 . . . 141
7.2.3 Proof of Theorem 7.15 . . . 147
7.3 Non-Transversal Case . . . 155
7.3.1 The Space $C_0(\overline{D} \setminus M)$ . . . 156
7.3.2 Generation Theorem for Feller Semigroups . . . 157
7.3.3 Sketch of Proof of Theorem 7.20 . . . 159

8 Proof of Theorem 1.1 . . . 169
8.1 Regularity Theorem for Problem (1.1) . . . 170
8.2 Uniqueness Theorem for Problem (1.1) . . . 173
8.3 Existence Theorem for Problem (1.1) . . . 174
8.3.1 Proof of Theorem 8.8 . . . 175
8.3.2 Proof of Proposition 8.11 . . . 183

9 Proof of Theorem 1.2 . . . 187

10 A Priori Estimates . . . 191

11 Proof of Theorem 1.3 . . . 197
11.1 Proof of Part (i) of Theorem 1.3 . . . 197
11.2 Proof of Part (ii) of Theorem 1.3 . . . 201

12 Proof of Theorem 1.4, Part (i) . . . 203
12.1 Sobolev's Imbedding Theorems . . . 203
12.2 Proof of Part (i) of Theorem 1.4 . . . 204

13 Proofs of Theorem 1.5 and Theorem 1.4, Part (ii) 213
13.1 Existence Theorem for Feller Semigroups 213
13.2 Feller Semigroups with Reflecting Barrier 226
13.3 Proof of Theorem 1.5 234
13.4 Proof of Part (ii) of Theorem 1.4 243

14 Boundary Value Problems for Waldenfels Operators 245
14.1 Formulation of a Boundary Value Problem 245
14.2 Proof of Theorem 1.6 247
14.3 Proof of Theorem 1.7 253
14.4 Proof of Theorem 1.8 258
14.5 Proof of Theorem 1.9 262
14.6 Concluding Remarks 270

A Boundedness of Pseudo-Differential Operators 271
A.1 The Littlewood-Paley Series 271
A.2 Definition of Sobolev and Besov Spaces 273
A.3 Non-Regular Symbols 275
A.4 The L^p Boundedness Theorem 280
A.5 Proof of Proposition A.7 281
A.6 Proof of Proposition A.8 287
A.6.1 Proof of the Case $\delta = 1$ 288
A.6.2 Proof of the Case $0 \leq \delta < 1$ 296

B Unique Solvability of Pseudo-Differential Operators 303

C The Maximum Principle 311
C.1 The Weak Maximum Principle 312
C.2 The Strong Maximum Principle 314
C.3 The Boundary Point Lemma 319

References 325

Index of Symbols 329

Index 333

1

Introduction and Main Results

This book is devoted to the study of three interrelated subjects in analysis: semigroups, elliptic boundary value problems and Markov processes. The purpose of the book provides a careful and accessible exposition of the functional analytic approach to the problem of construction of Markov processes with boundary conditions in probability theory. We construct a Feller semigroup corresponding to such a physical phenomenon that a Markovian particle moves both by jumps and continuously in the state space until it "dies" at the time when it reaches the set where the particle is definitely absorbed. Our approach is distinguished by the extensive use of the ideas and techniques characteristic of the recent developments in the theory of partial differential equations. The following diagram gives a bird's eye view of Markov processes, semigroups and boundary value problems and how these relate to each other:

Probability	Functional Analysis	Boundary Value Problems
Markov process (X_t)	Feller semigroup $T_t f(\cdot) = \int p_t(\cdot, dy) f(y)$ $p_t(x, dy)$ Markov transition function	Infinitesimal generator $\mathfrak{W}$ $T_t = \exp[t\mathfrak{W}]$
Markov property Starting afresh property	Semigroup property $T_{t+s} = T_t \cdot T_s$	Waldenfels operator Ventcel' boundary condition

This introductory chapter is devoted to the functional analytic approach to a class of *degenerate* boundary value problems for second-order elliptic differential operators which includes as particular cases the Dirichlet and Robin problems. We prove that this class of boundary value problems provides a new example of *analytic semigroups* both in the L^p topology and in the topology of uniform convergence.

Let D be a bounded domain of Euclidean space $\mathbf{R}^N$, with smooth boundary ∂D; its closure $\overline{D} = D \cup \partial D$ is an N-dimensional, compact smooth manifold with boundary.

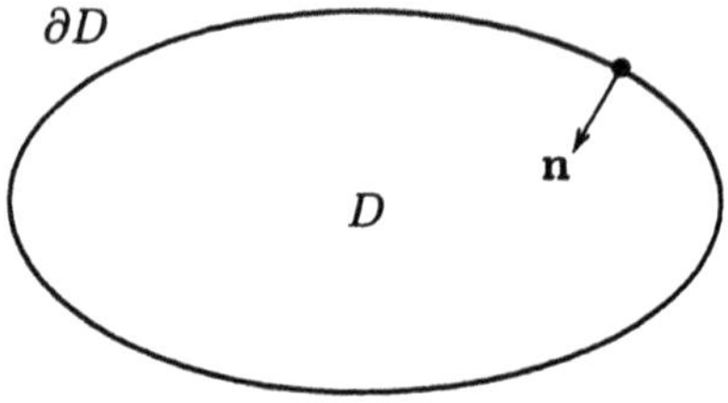

Fig. 1.1.

We let

$$Au(x) = \sum_{i,j=1}^{N} a^{ij}(x) \frac{\partial^2 u}{\partial x_i \partial x_j}(x) + \sum_{i=1}^{N} b^i(x) \frac{\partial u}{\partial x_i}(x) + c(x)u(x)$$

be a second-order, *elliptic* differential operator with real coefficients such that

(1) $a^{ij}(x) \in C^\infty(\overline{D})$, $a^{ij}(x) = a^{ji}(x)$ and there exists a constant $a_0 > 0$ such that

$$\sum_{i,j=1}^{N} a^{ij}(x)\xi_i\xi_j \geq a_0|\xi|^2, \quad x \in \overline{D},\ \xi \in \mathbf{R}^N .$$

(2) $b^i(x) \in C^\infty(\overline{D})$.
(3) $c(x) \in C^\infty(\overline{D})$ and $c(x) \leq 0$ on $\overline{D}$, but $c(x) \not\equiv 0$ in D.

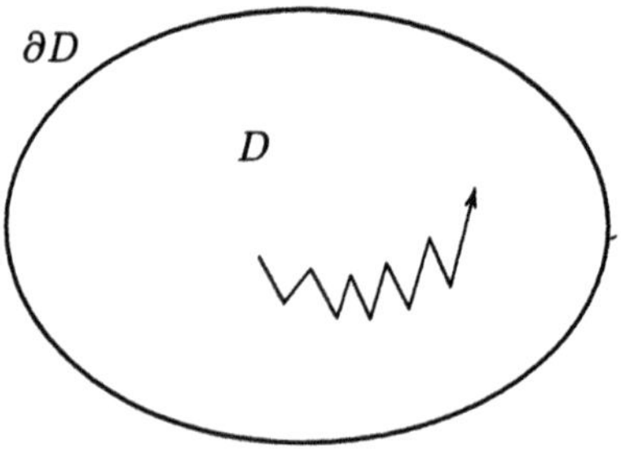

Fig. 1.2.

The functions $a^{ij}(x)$, $b^i(x)$ and $c(x)$ are called the diffusion coefficients, the drift coefficients and the termination coefficient, respectively. The differential operator A is called a *diffusion operator* which describes analytically a strong Markov process with continuous paths in the interior D such as Brownian motion (see Fig. 1.2).

Let L_0 be a first-order, boundary condition with real coefficients such that

$$L_0 u(x') = \mu(x')\frac{\partial u}{\partial \mathbf{n}}(x') + \gamma(x')u(x') .$$

Here:

(1) $\mu(x') \in C^\infty(\partial D)$ and $\mu(x') \geq 0$ on ∂D.
(2) $\gamma(x') \in C^\infty(\partial D)$ and $\gamma(x') \leq 0$ on ∂D.
(3) $\mathbf{n} = (n_1, n_2, \ldots, n_N)$ is the unit interior normal to the boundary ∂D (see Fig. 1.1).

The boundary condition L_0 is a special case of general *Ventcel' boundary conditions* (cf. [We], [BCP]). The two terms of L_0

$$\mu(x')\frac{\partial u}{\partial \mathbf{n}}(x'), \quad \gamma(x')u(x')$$

are supposed to correspond to the reflection phenomenon and the absorption phenomenon, respectively (see Fig. 1.3).

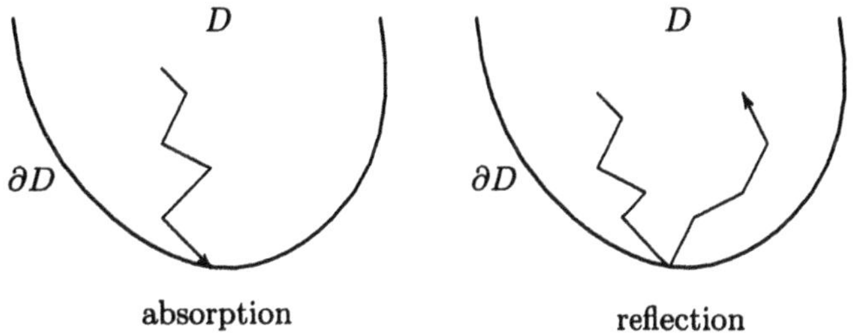

Fig. 1.3.

We consider the following boundary value problem: Given functions f and φ defined in D and on ∂D, respectively, find a function u in D such that

$$\begin{cases} Au = f & \text{in } D , \\ L_0 u = \mu(x')\dfrac{\partial u}{\partial \mathbf{n}} + \gamma(x')u = \varphi & \text{on } \partial D . \end{cases} \tag{1.1}$$

It should be noticed that if $\mu(x') \equiv 0$ and $\gamma(x') \equiv -1$ on ∂D (resp. $\mu(x') > 0$ on ∂D), then the boundary condition L_0 is the so-called Dirichlet (resp. Robin) condition. It is easy to see that problem (1.1) is non-degenerate (or coercive) if and only if either $\mu(x') > 0$ on ∂D or $\mu(x') \equiv 0$ and $\gamma(x') < 0$ on ∂D. The generation theorem for analytic semigroups is well established in the non-degenerate case both in the L^p topology (see Friedman [Fr1], Tanabe [Tn]) and in the topology of uniform convergence (see Masuda [Ma] for the Dirichlet case).

Our fundamental hypothesis is the following:
(H) $\mu(x') + |\gamma(x')| > 0$ on ∂D.
The intuitive meaning of hypothesis (H) is that either the reflection phenomenon or the absorption phenomenon occurs at each point of the boundary ∂D. More precisely, condition (H) implies that absorption phenomenon occurs at each point of the set $M = \{x' \in \partial D : \mu(x') = 0\}$, while reflection phenomenon occurs at each point of the set $\partial D \setminus M = \{x' \in \partial D : \mu(x') > 0\}$. In other words, a Markovian particle moves continuously in the space $\overline{D} \setminus M$ until it "dies" at which time it reaches the set M (see Fig. 1.4).

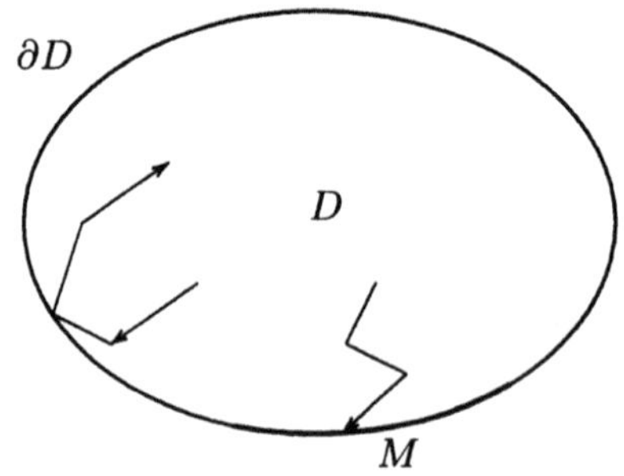

Fig. 1.4.

The first purpose of this book is to prove existence and uniqueness theorems for problem (1.1) in the framework of L^p spaces and Hölder spaces. The crucial point is how to define function subspaces in which problem (1.1) is uniquely solvable.

If k is a positive integer and $1 < p < \infty$, we define the Sobolev space of L^p style

$$\begin{aligned} H^{k,p}(D) = {} & \text{the space of (equivalence classes of) functions} \\ & u \in L^p(D) \text{ whose derivatives } D^\alpha u,\ |\alpha| \le k, \text{ in the} \\ & \text{sense of distributions are in } L^p(D), \end{aligned}$$

and the space

$$\begin{aligned} B^{k-1/p,p}(\partial D) = {} & \text{the space of the boundary values } \varphi \text{ of functions} \\ & u \in H^{k,p}(D). \end{aligned}$$

In the space $B^{k-1/p,p}(\partial D)$, we introduce a norm

$$|\varphi|_{B^{k-1/p,p}(\partial D)} = \inf\left\{\|u\|_{H^{k,p}(D)} : u \in H^{k,p}(D),\ u|_{\partial D} = \varphi\right\} .$$

The space $B^{k-1/p,p}(\partial D)$ is a Banach space with respect to this norm $|\cdot|_{B^{k-1/p,p}(\partial D)}$. More precisely, it is a Besov space (see Bergh–Löfström [BL], Taibleson [Tb], Triebel [Tr]).

We introduce a subspace of $B^{k-1/p,p}(\partial D)$ which is associated with the boundary condition L_0 in the following way: We let

$$B_{L_0}^{1-1/p,p}(\partial D) = \left\{\varphi = \mu(x')\varphi_1 - \gamma(x')\varphi_2 : \varphi_1 \in B^{1-1/p,p}(\partial D),\ \varphi_2 \in B^{2-1/p,p}(\partial D)\right\} ,$$

and define a norm

$$|\varphi|_{B_{L_0}^{1-1/p,p}(\partial D)} = \inf\left\{|\varphi_1|_{B^{1-1/p,p}(\partial D)} + |\varphi_2|_{B^{2-1/p,p}(\partial D)} : \varphi = \mu(x')\varphi_1 - \gamma(x')\varphi_2\right\} .$$

Then it is easy to verify that the space $B_{L_0}^{1-1/p,p}(\partial D)$ is a Banach space with respect to this norm $|\cdot|_{B_{L_0}^{1-1/p,p}(\partial D)}$. It should be noticed that the space $B_{L_0}^{1-1/p,p}(\partial D)$ is an "interpolation space" between the Besov spaces $B^{2-1/p,p}(\partial D)$ and $B^{1-1/p,p}(\partial D)$. More precisely, we have

$$\begin{cases} B_{L_0}^{1-1/p,p}(\partial D) = B^{2-1/p,p}(\partial D) & \text{if } \mu(x') \equiv 0 \text{ on } \partial D , \\ B_{L_0}^{1-1/p,p}(\partial D) = B^{1-1/p,p}(\partial D) & \text{if } \mu(x') > 0 \text{ on } \partial D . \end{cases}$$

Now we can state an existence and uniqueness theorem for the boundary value problem (1.1) in the framework of L^p spaces:

Theorem 1.1. *Let $1 < p < \infty$. Assume that condition (H) is satisfied:*
(H) $\mu(x') + |\gamma(x')| > 0$ on ∂D.
Then the mapping

$$(A, L_0) \colon H^{2,p}(D) \longrightarrow L^p(D) \bigoplus B_{L_0}^{1-1/p,p}(\partial D)$$

is an algebraic and topological isomorphism. *In particular, for any $f \in L^p(D)$ and any $\varphi \in B_{L_0}^{1-1/p,p}(\partial D)$, there exists a unique solution $u \in H^{2,p}(D)$ of problem (1.1).*

Furthermore, in order to study problem (1.1) in the framework of Hölder spaces, we introduce a subspace of $C^{1+\theta}(\partial D)$, $0 < \theta < 1$, which is a Hölder space version of $B_{L_0}^{1-1/p,p}(\partial D)$. We let

$$C_{L_0}^{1+\theta}(\partial D) = \left\{\varphi = \mu(x')\varphi_1 - \gamma(x')\varphi_2 : \varphi_1 \in C^{1+\theta}(\partial D),\ \varphi_2 \in C^{2+\theta}(\partial D)\right\} ,$$

and define a norm

$$|\varphi|_{C^{1+\theta}_{L_0}(\partial D)} = \inf\left\{|\varphi_1|_{C^{1+\theta}(\partial D)} + |\varphi_2|_{C^{2+\theta}(\partial D)} : \varphi = \mu(x')\varphi_1 - \gamma(x')\varphi_2\right\} .$$

Then it is easy to verify that the space $C^{1+\theta}_{L_0}(\partial D)$ is a Banach space with respect to the norm $|\cdot|_{C^{1+\theta}_{L_0}(\partial D)}$. We remark that

$$\begin{cases} C^{1+\theta}_{L_0}(\partial D) = C^{2+\theta}(\partial D) & \text{if } \mu(x') \equiv 0 \text{ on } \partial D , \\ C^{1+\theta}_{L_0}(\partial D) = C^{1+\theta}(\partial D) & \text{if } \mu(x') > 0 \text{ on } \partial D . \end{cases}$$

The next theorem is a Hölder space version of Theorem 1.1:

Theorem 1.2. *Let $0 < \theta < 1$. If condition (H) is satisfied, then the mapping*

$$(A, L_0): C^{2+\theta}(\overline{D}) \longrightarrow C^{\theta}(\overline{D}) \bigoplus C^{1+\theta}_{L_0}(\partial D)$$

is an algebraic and topological isomorphism. *In particular, for any $f \in C^{\theta}(\overline{D})$ and any $\varphi \in C^{1+\theta}_{L_0}(\partial D)$, there exists a unique solution $u \in C^{2+\theta}(\overline{D})$ of problem (1.1).*

The second purpose of this book is to study the boundary value problem (1.1) from the point of view of analytic semigroup theory in functional analysis, and is to generalize the generation theorem for analytic semigroups to the *degenerate* case.

We associate with problem (1.1) a unbounded linear operator A_p from $L^p(D)$ into itself as follows:

(a) The domain of definition $\mathcal{D}(A_p)$ of A_p is the set

$$\mathcal{D}(A_p) = \left\{u \in H^{2,p}(D) : L_0 u = \mu(x')\frac{\partial u}{\partial \mathbf{n}} + \gamma(x')u = 0 \text{ on } \partial D\right\} .$$

(b) $A_p u = Au$, $u \in \mathcal{D}(A_p)$.

Then we can prove a generation theorem for analytic semigroups in the framework of L^p spaces:

Theorem 1.3. *Let $1 < p < \infty$. If condition (H) is satisfied, then we have the following two assertions:*

(i) For every $\varepsilon > 0$, there exists a constant $r_p(\varepsilon) > 0$ such that the resolvent set of A_p contains the set

$$\Sigma_p(\varepsilon) = \left\{\lambda = r^2 e^{i\vartheta} : r \geq r_p(\varepsilon), -\pi + \varepsilon \leq \vartheta \leq \pi - \varepsilon\right\} ,$$

and that the resolvent $(A_p - \lambda I)^{-1}$ satisfies the estimate

$$\|(A_p - \lambda I)^{-1}\| \leq \frac{c_p(\varepsilon)}{|\lambda|}, \quad \lambda \in \Sigma_p(\varepsilon) , \tag{1.2}$$

where $c_p(\varepsilon) > 0$ is a constant depending on ε.

(ii) The operator A_p generates a semigroup e^{zA_p} on the space $L^p(D)$ which is analytic *in the sector $\Delta_\varepsilon = \{z = t + is : z \neq 0, |\arg z| < \pi/2 - \varepsilon\}$ for any $0 < \varepsilon < \pi/2$ (see Fig. 1.5).*

It should be noticed that Theorem 1.3 for $p = 2$ is proved by Taira [Ta1, Theorem 1].

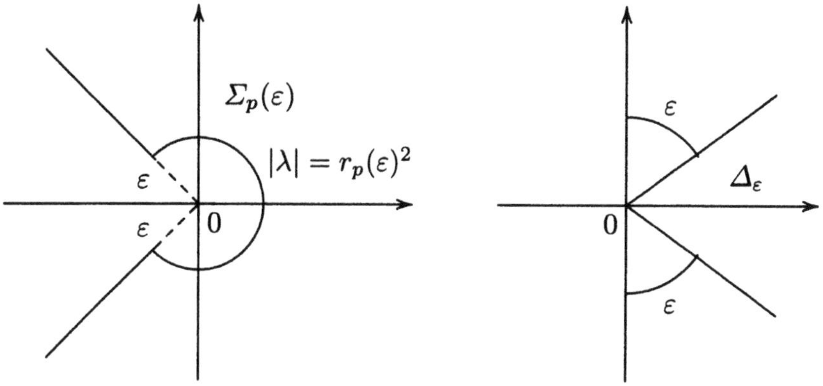

Fig. 1.5.

Secondly, we state a generation theorem for analytic semigroups in the topology of uniform convergence.

Let $C(\overline{D})$ be the space of real-valued, continuous functions f on $\overline{D}$. We equip the space $C(\overline{D})$ with the topology of uniform convergence on the whole $\overline{D}$; hence it is a Banach space with the maximum norm

$$\|f\|_\infty = \max_{x \in \overline{D}} |f(x)| .$$

We introduce a subspace of $C(\overline{D})$ which is associated with the boundary condition L_0. We remark that the boundary condition

$$L_0 u = \mu(x') \frac{\partial u}{\partial \mathbf{n}} + \gamma(x') u = 0 \quad \text{on } \partial D$$

includes the condition

$$u = 0 \quad \text{on } M = \{x' \in \partial D : \mu(x') = 0\} ,$$

if $\gamma(x') \neq 0$ on M. With this fact in mind, we let

$$C_0(\overline{D} \setminus M) = \{u \in C(\overline{D}) : u = 0 \text{ on } M\} .$$

The space $C_0(\overline{D} \setminus M)$ is a closed subspace of $C(\overline{D})$; hence it is a Banach space. Further we introduce a linear operator $\mathfrak{A}$ from $C_0(\overline{D} \setminus M)$ into itself as follows:

(a) The domain of definition $\mathcal{D}(\mathfrak{A})$ of $\mathfrak{A}$ is the set

$$\mathcal{D}(\mathfrak{A}) = \left\{u \in C_0(\overline{D} \setminus M) : Au \in C_0(\overline{D} \setminus M),\ L_0 u = 0 \text{ on } \partial D\right\} . \quad (1.3)$$

(b) $\mathfrak{A}u = Au$, $u \in \mathcal{D}(\mathfrak{A})$.

Here Au and $L_0 u$ are taken in the sense of *distributions* (see Chap. 13).

Then Theorem 1.3 remains valid with $L^p(D)$ and A_p replaced by $C_0(\overline{D}\setminus M)$ and $\mathfrak{A}$, respectively. More precisely, we can prove the following:

Theorem 1.4. *If condition (H) is satisfied, then we have the following two assertions:*

(i) For every $\varepsilon > 0$, there exists a constant $r(\varepsilon) > 0$ such that the resolvent set of $\mathfrak{A}$ contains the set

$$\Sigma(\varepsilon) = \left\{\lambda = r^2 e^{i\vartheta} : r \geq r(\varepsilon), -\pi + \varepsilon \leq \vartheta \leq \pi - \varepsilon\right\} ,$$

and that the resolvent $(\mathfrak{A} - \lambda I)^{-1}$ satisfies the estimate

$$\|(\mathfrak{A} - \lambda I)^{-1}\| \leq \frac{c(\varepsilon)}{|\lambda|}, \quad \lambda \in \Sigma(\varepsilon) , \quad (1.4)$$

where $c(\varepsilon) > 0$ is a constant depending on ε.

(ii) The operator $\mathfrak{A}$ generates a semigroup $e^{z\mathfrak{A}}$ on the space $C_0(\overline{D} \setminus M)$ which is analytic *in the sector $\Delta_\varepsilon = \{z = t + is : z \neq 0, |\arg z| < \pi/2 - \varepsilon\}$ for any $0 < \varepsilon < \pi/2$ (see Fig. 1.6).*
Moreover, the operators $\{e^{t\mathfrak{A}}\}_{t\geq 0}$ are non-negative and contractive on the space $C_0(\overline{D} \setminus M)$:

$$f \in C_0(\overline{D}\setminus M),\ 0 \leq f(x) \leq 1 \text{ on } \overline{D} \setminus M \Longrightarrow 0 \leq e^{t\mathfrak{A}} f(x) \leq 1 \text{ on } \overline{D} \setminus M .$$

Theorems 1.3 and 1.4 express a *regularizing effect* for the parabolic differential operator $\partial/\partial t - A$ with homogeneous boundary condition L_0.

A strongly continuous, non-negative and contraction semigroup $\{T_t\}_{t\geq 0}$ on the Banach space $C_0(\overline{D} \setminus M)$ is called a *Feller semigroup* on $\overline{D} \setminus M$. Thus part (ii) of Theorem 1.4 may be reformulated as follows:

Theorem 1.5. *If condition (H) is satisfied, then the operator $\mathfrak{A}$ generates a Feller semigroup $\{e^{t\mathfrak{A}}\}_{t\geq 0}$ on $\overline{D} \setminus M$.*

Theorem 1.5 is a generalization of Bony–Courrège–Priouret [BCP, Théorème XIX] to the degenerate case where $\mu(x') \geq 0$ on ∂D (see also [Ta4, Theorem 4]).

It is known (see Dynkin [Dy, Chap. III], Taira [Ta2, Chap. 9]) that if $\{T_t\}_{t\geq 0}$ is a Feller semigroup on $\overline{D} \setminus M$, then there exists a unique Markov transition function $p_t(x, \cdot)$ on $\overline{D} \setminus M$ such that

$$T_t f(x) = \int_{\overline{D}\setminus M} p_t(x, dy) f(y), \quad f \in C_0(\overline{D} \setminus M) .$$

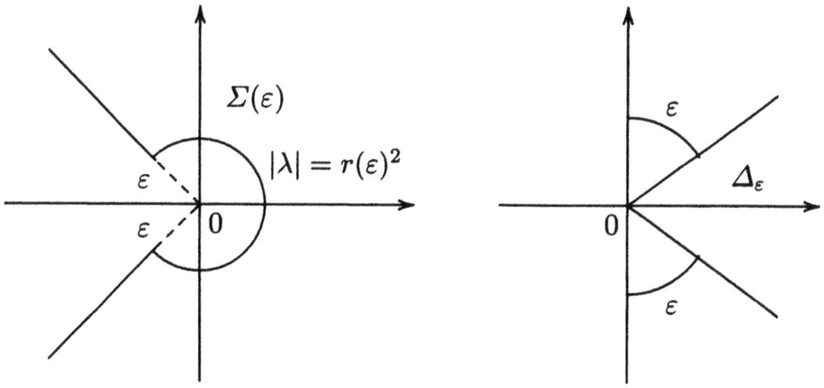

Fig. 1.6.

Furthermore, it can be shown that the function $p_t(x, \cdot)$ is the transition function of some strong *Markov process*; hence the value $p_t(x, E)$ expresses the transition probability that a Markovian particle starting at position x will be found in the set E at time t.

Rephrased, Theorem 1.5 asserts that there exists a Feller semigroup corresponding to such a physical phenomenon that a Markovian particle moves continuously in the space $\overline{D} \setminus M$ until it "dies" at which time it reaches the set M, as in Fig. 1.4.

It is worth while pointing out here that the condition

(H.1) $\mu(x') \geq 0$ and $\gamma(x') \leq 0$ on ∂D

is necessary in order that the operator $\mathfrak{A}$ be the infinitesimal generator of a Feller semigroup $\{e^{t\mathfrak{A}}\}_{t\geq 0}$ on $\overline{D} \setminus M$ (see [Ta2, Sect. 9.5]). Moreover, if condition (H.1) is satisfied, then it is easy to see that condition (H) is equivalent to the condition

(H.2) $\gamma(x') < 0$ on $M = \{x' \in \partial D : \mu(x') = 0\}$.

Furthermore, it should be emphasized that Theorem 1.2 through Theorem 1.5 may be extended to the *integro-differential operator* case. For simplicity, we assume that the domain D is . *convex.*

Let W be a second-order, elliptic integro-differential operator with real coefficients such that

$$\begin{aligned} Wu(x) &= Au(x) + Su(x) \\ &:= \left(\sum_{i,j=1}^{N} a^{ij}(x) \frac{\partial^2 u}{\partial x_i \partial x_j}(x) + \sum_{i=1}^{N} b^i(x) \frac{\partial u}{\partial x_i}(x) + c(x)u(x) \right) \\ &\quad + \int_{\mathbf{R}^N \setminus \{0\}} \left(u(x+z) - u(x) - \sum_{j=1}^{N} z_j \frac{\partial u}{\partial x_j}(x) \right) s(x,z)\, m(dz) \,. \end{aligned}$$

Here:

(1) $a^{ij}(x) \in C^\infty(\overline{D})$, $a^{ij}(x) = a^{ji}(x)$ and there exists a constant $a_0 > 0$ such that

$$\sum_{i,j=1}^{N} a^{ij}(x)\xi_i\xi_j \geq a_0|\xi|^2, \quad x \in \overline{D},\ \xi \in \mathbf{R}^N .$$

(2) $b^i(x) \in C^\infty(\overline{D})$.

(3) $c(x) \in C^\infty(\overline{D})$, and $c(x) \leq 0$ in D, but $c(x) \not\equiv 0$ in D.

(4) $s(x,z) \in C(\overline{D} \times \mathbf{R}^N)$ and $0 \leq s(x,z) \leq 1$ on $\overline{D} \times \mathbf{R}^N$, and there exist constants $C_0 > 0$ and $0 < \theta < 1$ such that

$$|s(x,z) - s(y,z)| \leq C_0|x-y|^\theta, \quad x,\ y \in \overline{D},\ z \in \mathbf{R}^N , \tag{1.5}$$

and

$$s(x,z) = 0 \quad \text{if } x+z \notin \overline{D} . \tag{1.6}$$

Condition (1.6) implies that the integral operator S may be considered as an operator acting on functions u defined on the closure $\overline{D}$ (see Garroni–Menaldi [GM, Chap. II, Remark 1.19]).

(5) The measure $m(dz)$ is a Radon measure on the space $\mathbf{R}^N \setminus \{0\}$ which satisfies the *moment condition*

$$\int_{\{|z|\leq 1\}} |z|^2\, m(dz) + \int_{\{|z|>1\}} |z|\, m(dz) < \infty . \tag{1.7}$$

Condition (1.7) is a standard condition for the measure $m(dz)$, and it implies that a Markovian particle does not move by jumps so far.

The operator W is called a second-order *Waldenfels operator* (cf. [BCP], [Wa]). The differential operator A is called a diffusion operator which describes analytically a strong Markov process with continuous paths in the interior D such as Brownian motion. The integral operator S is called a second-order *Lévy operator* which is supposed to correspond to the jump phenomenon in the closure $\overline{D}$ (see Fig. 1.7). In this context, condition (1.6) implies that any Markovian particle does not move by jumps from $x \in D$ to the outside of $\overline{D}$.

First, we consider (instead of problem (1.1)) the following boundary value problem: Given functions f and φ defined in D and on ∂D, respectively, find a function u in D such that

$$\begin{cases} Wu = f & \text{in } D , \\ L_0 u = \varphi & \text{on } \partial D . \end{cases} \tag{1.8}$$

The next theorem is a generalization of Theorem 1.2 to the integro-differential operator case:

Theorem 1.6. *Assume that condition (H) is satisfied:*
(H) $\mu(x') + |\gamma(x')| > 0$ *on* ∂D.
Then the mapping

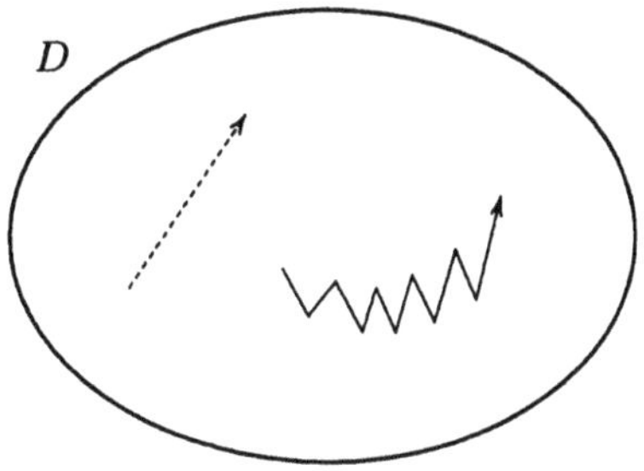

Fig. 1.7.

$$(W, L_0): C^{2+\theta}(\overline{D}) \longrightarrow C^{\theta}(\overline{D}) \bigoplus C^{1+\theta}_{L_0}(\partial D)$$

is an algebraic and topological isomorphism. *In particular, for any* $f \in C^{\theta}(\overline{D})$ *and any* $\varphi \in C^{1+\theta}_{L_0}(\partial D)$, *there exists a unique solution* $u \in C^{2+\theta}(\overline{D})$ *of problem (1.8).*

As an application of Theorem 1.6, we can construct a Feller semigroup corresponding to such a physical phenomenon that a Markovian particle moves both by jumps and continuously in the state space until it "dies" at the time when it reaches the set where the particle is definitely absorbed, generalizing Theorem 1.5.

To do this, we define a linear operator $\mathcal{W}$ from the Banach space $C_0(\overline{D} \setminus M)$ into itself as follows:

(a) The domain of definition $\mathcal{D}(\mathcal{W})$ is the set

$$\mathcal{D}(\mathcal{W}) = \{u \in C^2(\overline{D}) \cap C_0(\overline{D} \setminus M) : Wu \in C_0(\overline{D} \setminus M) , \\ L_0 u = 0 \text{ on } \partial D\} .$$

(b) $\mathcal{W}u = Wu$, $u \in \mathcal{D}(\mathcal{W})$.

The next theorem asserts that there exists a Feller semigroup on $\overline{D} \setminus M$ corresponding to such a physical phenomenon that a Markovian particle moves both by jumps and continuously in the state space $\overline{D} \setminus M$ until it "dies" at the time when it reaches the set M as in Fig. 1.4:

Theorem 1.7. *If condition (H) is satisfied, then the operator* $\mathcal{W}$ *is closable in the space* $C_0(\overline{D} \setminus M)$, *and its minimal closed extension* $\overline{\mathcal{W}}$ *is the infinitesimal generator of some Feller semigroup* $\{e^{t\overline{\mathcal{W}}}\}_{t \geq 0}$ *on* $\overline{D} \setminus M$.

Remark 1.1. For the non-degenerate case, the reader is referred to Komatsu [Ko, Theorem 5.2], Stroock [St, Theorem 2.2], Garroni–Menaldi [GM, Chap. VIII, Theorem 3.3] and also Galakhov–Skubachevskiĭ [GB, Theorem 5.1].

Secondly, we study the boundary value problem (1.8) from the point of view of analytic semigroup theory, generalizing Theorems 1.3 and 1.4.

To do this, we associate with problem (1.8) an unbounded linear operator W_p from $L^p(D)$ into itself as follows:

(a) The domain of definition $\mathcal{D}(W_p)$ is the set

$$\mathcal{D}(W_p) = \{u \in H^{2,p}(D) : L_0 u = 0 \text{ on } \partial D\} .$$

(b) $W_p u = W u$, $u \in \mathcal{D}(W_p)$.

The next theorem is a generalization of Theorem 1.3 to the integro-differential operator case:

Theorem 1.8. *Let $1 < p < \infty$. Assume that condition (H) is satisfied. Then we have the following two assertions:*

(i) For every $\varepsilon > 0$, there exists a constant $r_p(\varepsilon) > 0$ such that the resolvent set of W_p contains the set

$$\Sigma_p(\varepsilon) = \left\{\lambda = r^2 e^{i\vartheta} : r \geq r_p(\varepsilon), -\pi + \varepsilon \leq \vartheta \leq \pi - \varepsilon\right\} ,$$

and that the resolvent $(W_p - \lambda I)^{-1}$ satisfies the estimate

$$\left\|(W_p - \lambda I)^{-1}\right\| \leq \frac{c_p(\varepsilon)}{|\lambda|}, \quad \lambda \in \Sigma_p(\varepsilon) , \tag{1.9}$$

where $c_p(\varepsilon) > 0$ is a constant depending on ε.

(ii) The operator W_p generates a semigroup e^{zW_p} on the space $L^p(D)$ which is analytic *in the sector $\Delta_\varepsilon = \{z = t + is : z \neq 0, |\arg z| < \pi/2 - \varepsilon\}$ for any $0 < \varepsilon < \pi/2$.*

Moreover, we introduce a linear operator $\mathfrak{W}$ from $C_0(\overline{D} \setminus M)$ into itself as follows:

(a) The domain of definition $\mathcal{D}(\mathfrak{W})$ is the set

$$\begin{aligned} \mathcal{D}(\mathfrak{W}) = \{u \in C_0(\overline{D} \setminus M) \cap H^{2,p}(D) : Wu \in C_0(\overline{D} \setminus M) , \\ L_0 u = 0 \text{ on } \partial D\} , \qquad N < p < \infty . \end{aligned}$$

(b) $\mathfrak{W}u = Wu$, $u \in \mathcal{D}(\mathfrak{W})$.

Here it should be noticed that the domain $\mathcal{D}(\mathfrak{W})$ is *independent* of $N < p < \infty$ (see the proof of Lemma 14.10).

The next theorem is a generalization of Theorem 1.4 to the integro-differential operator case:

Theorem 1.9. *Let $N < p < \infty$. If condition (H) is satisfied, then we have the following two assertions:*

(i) For every $\varepsilon > 0$, there exists a constant $r(\varepsilon) > 0$ such that the resolvent set of $\mathfrak{W}$ contains the set

$$\Sigma(\varepsilon) = \left\{\lambda = r^2 e^{i\vartheta} : r \geq r(\varepsilon), -\pi + \varepsilon \leq \vartheta \leq \pi - \varepsilon\right\} ,$$

and that the resolvent $(\mathfrak{W} - \lambda I)^{-1}$ satisfies the estimate

$$\|(\mathfrak{W} - \lambda I)^{-1}\| \leq \frac{c(\varepsilon)}{|\lambda|}, \quad \lambda \in \Sigma(\varepsilon) , \tag{1.10}$$

where $c(\varepsilon) > 0$ is a constant depending on ε.

(ii) The operator $\mathfrak{W}$ generates a semigroup $e^{z\mathfrak{W}}$ on the space $C_0(\overline{D}\setminus M)$ which is analytic *in the sector $\Delta_\varepsilon = \{z = t + is : z \neq 0, |\arg z| < \pi/2 - \varepsilon\}$ for any $0 < \varepsilon < \pi/2$.*

Remark 1.2. By combining Theorems 1.7 and 1.9, we can prove that the operator $\mathfrak{W}$ coincides with the minimal closed extension $\overline{\mathcal{W}}$: $\mathfrak{W} = \overline{\mathcal{W}}$ (see Sect. 14.6, Theorem 14.14).

Theorems 1.8 and 1.9 express a *regularizing effect* for the parabolic integro-differential operator $\partial/\partial t - W$ with homogeneous boundary condition L_0 (cf. [GM, Chap. VIII, Theorem 3.1]).

The rest of this book is organized as follows.

The first part (Chaps. 2–5) provides the elements of semigroups, distributions and pseudo-differential operators which are used throughout the book. The material in these preparatory chapters is given for completeness, to minimize the necessity of consulting too many outside references. This makes the book fairly self-contained.

Chapter 2 is devoted to a review of standard topics from the theory of semigroups such as contraction semigroups and analytic semigroups. These topics form a necessary background for the proof of Theorems 1.3 and 1.4 (Theorem 2.11).

In Chap. 3 we introduce a class of semigroups associated with Markov processes, called Feller semigroups, and prove generation theorems for Feller semigroups (Theorems 3.3 and Theorem 3.5) which form a functional analytic background for the proof of Theorems 1.5 and 1.7. Moreover, following Bony–Courrège–Priouret [BCP] we give useful characterizations of linear operators which satisfy the positive maximum principle (Theorems 3.7 and 3.9).

In Chap. 4 we present a brief description of the basic concepts and results of the theory of distributions or generalized functions which will be used in subsequent chapters. Distribution theory has become a convenient tool in the study of partial differential equations. Many problems in partial differential equations can be formulated in terms of abstract operators acting between suitable spaces of distributions, and these operators are then analyzed by the methods of functional analysis.

Several recent developments in the theory of partial differential equations have made possible further progress in the study of boundary value problems and hence the study of Markov processes. The main technique used is the calculus of pseudo-differential operators which may be considered as a modern theory of potentials. The presentation of these new results is the main purpose of Chap. 5.

In Appendix A, by using Peetre's equivalent definition of Besov spaces and Sobolev spaces of L^p style we prove an L^p boundedness theorem for pseudo-differential operators (a global version of Theorem 5.14) as Theorem A.6 which plays a fundamental role throughout the book.

In Chap. 6, by using the L^p theory of pseudo-differential operators we study the boundary value problem (1.1) in the framework of Sobolev spaces of L^p style. We confine ourselves to the simple but important boundary condition. This makes it possible to develop our basic machinery with a minimum of bother and the principal ideas can be presented concretely and explicitly.

The idea of our approach is stated as follows. First, we consider the following *Neumann* problem:

$$\begin{cases} Av = f & \text{in } D\,, \\ \dfrac{\partial v}{\partial \mathbf{n}} = 0 & \text{on } \partial D\,. \end{cases} \tag{1.11}$$

The existence and uniqueness theorem for problem (1.11) is well established in the framework of Sobolev spaces of L^p style (see Agmon–Douglis–Nirenberg [ADN], Lions–Magenes [LM], Gilbarg–Trudinger [GT]). We let

$$v = G_N f\,.$$

The operator G_N is the Green operator for the Neumann problem. Then it follows that a function u is a solution of the problem

$$\begin{cases} Au = f & \text{in } D\,, \\ L_0 u = 0 & \text{on } \partial D \end{cases} \tag{1.12}$$

if and only if the function $w = u - v$ is a solution of the problem

$$\begin{cases} Aw = 0 & \text{in } D\,, \\ L_0 w = -L_0 v = -\mu(x')\dfrac{\partial v}{\partial \mathbf{n}} - \gamma(x')v = -\gamma(x')v & \text{on } \partial D\,. \end{cases}$$

However, we know that every solution w of the homogeneous equation

$$Aw = 0 \quad \text{in } D$$

can be expressed by means of a single layer potential as follows:

$$w = P\psi\,.$$

The operator P is the Poisson operator for the Dirichlet problem. Thus, by using the operators G_N and P we can reduce the study of problem (1.12) to that of the equation

$$T\psi := L_0 P\psi = -\gamma(x')v \quad \text{on } \partial D .$$

This is a generalization of the classical Fredholm integral equation.

It is well known (see Hörmander [Ho1], Seeley [Se2], Chazarain–Piriou [CP], Èskin [Es], Kumano-go [Ku], Taylor [Ty], Rempel–Schulze [RS]) that the operator $T = L_0 P$ is a pseudo-differential operator of first order on the boundary ∂D. We prove that the study of problem (1.12) can be reduced to that of the operator T (Theorems 6.9, 6.10 and 6.11).

In the early 1950's, W. Feller characterized completely the analytic structure of one-dimensional diffusion processes; he gave an intrinsic representation of the infinitesimal generator $\mathfrak{A}$ of a one-dimensional diffusion process and determined all possible boundary conditions which describe the domain $D(\mathfrak{A})$. The probabilistic meaning of Feller's work was clarified by E. B. Dynkin, K. Itô, H. P. McKean, Jr., D. B. Ray and others. One-dimensional diffusion processes are completely studied both from analytic and probabilistic viewpoints. The purpose of Chap. 7 is to generalize Feller's work to the multidimensional case. In 1959, A. D. Ventcel' (Wentzell) studied the problem of determining all possible boundary conditions for multidimensional diffusion processes. The main results (Theorems 7.1 and 7.20) discussed there are adapted from Bony–Courrège–Priouret [BCP], Taira [Ta2] and [Ta4]. Our proof is based on the generation theorems for Feller semigroups discussed in Chap. 3. Moreover, a unique solvability theorem for pseudo-differential operators (Theorem 7.16) plays an essential role in the construction of Feller semigroups.

In Appendix B, we give a sketch of the proof of a unique solvability theorem for pseudo-differential operators (Theorem B.1) which is an essential step for the construction of Feller semigroups in Chap. 7 (Theorem 7.16).

Chapter 8 is devoted to the proof of Theorem 1.1. We study the pseudo-differential operator T in question, and prove that if condition (H) is satisfied, then there exists a *parametrix* S for T in the Hörmander class $L^0_{1,1/2}(\partial D)$ (Lemma 8.4). Theorem 1.1 follows by applying a *Besov-space boundedness theorem* (Theorem 5.14) to the parametrix S (Theorems 8.2, 8.6 and 8.8). In the proof of surjectivity of the operator T we make use of Agmon's method (Proposition 8.11). This is a technique of treating a spectral parameter as a second-order differential operator of an extra variable and relating the old problem to a new one with the additional variable. More precisely, we introduce an auxiliary variable y of the unit circle

$$S = \mathbf{R}/2\pi\mathbf{Z} ,$$

and consider instead of problem (1.1) the following boundary value problem:

$$\begin{cases} \widetilde{A}\tilde{u} := \left(A + \dfrac{\partial^2}{\partial y^2}\right)\tilde{u} = \tilde{f} & \text{in } D \times S , \\ L_0\tilde{u} = \mu(x')\dfrac{\partial \tilde{u}}{\partial \mathbf{n}} + \gamma(x')\tilde{u} = \tilde{\varphi} & \text{on } \partial D \times S . \end{cases}$$

In Chap. 9 we study the boundary value problem (1.1) in the framework of Hölder spaces, and prove Theorem 1.2 (Theorem 9.1). Theorem 1.2 follows from Theorem 1.1 by using the Hölder space theory of pseudo-differential operators (Proposition 9.2).

In Chap. 10 we study the operator A_p, and prove estimate (1.2) for $A_p - \lambda I$ (Theorem 10.3). Once again Agmon's method plays an important role in the proof of estimate (1.2) (Proposition 10.4) just as in Chap. 8, but we replace the differential operator $A - \lambda$, $\lambda = r^2 e^{i\vartheta}$, $-\pi < \vartheta < \pi$, by the differential operator

$$\widetilde{\Lambda}(\vartheta) := A + e^{i\vartheta} \frac{\partial^2}{\partial y^2} .$$

Chapter 11 is devoted to the proof of Theorem 1.3 (Theorems 11.1 and 11.3). Just as in Chap. 8, we make use of Agmon's method to prove the surjectivity of the operator $A_p - \lambda I$ (Proposition 11.2).

Chapters 12 and 13 are devoted to the proof of Theorem 1.4 and Theorem 1.5. In Chap. 12 we prove part (i) of Theorem 1.4. Part (i) of Theorem 1.4 follows from an application of Theorem 1.3 by using Sobolev's imbedding theorems (Theorems 12.1 and 12.2) and a λ-*dependent localization* argument essentially due to Masuda [Ma] (Lemma 12.4). An essential point in the proof is how to construct a localizing function φ_0 which is adapted to the boundary condition L_0 (see the proof of Claim 12.5). It should be emphasized here that if $\mu(x') \equiv 0$ and $\gamma(x') \equiv -1$ on ∂D (the Dirichlet case), then a proof of Theorem 1.4 is stated in Pazy [Pa, Chap. 7, Theorem 3.7]. However, his proof of estimate (1.4) is incomplete (see [Pa, p. 217, estimate (3.22)]. Moreover, if $\mu(x') > 0$ on ∂D (the Robin case), then a proof of Theorem 1.4 is given by Stewart [Sw, Theorem 2]. However, his λ-dependent localization argument concerning the boundary term is not correct (see the proof of [Sw, Theorem 1]).

In Chap. 13 we prove Theorem 1.5 and part (ii) of Theorem 1.4. The main idea of our approach is essentially the same as in Chap. 7. More precisely, we study Feller semigroups with reflecting barrier (Theorem 13.14) and then, by using these Feller semigroups we construct Feller semigroups corresponding to such a physical phenomenon that either absorption or reflection phenomenon occurs at each point of the boundary (Theorem 13.18). Part (i) of Theorem 1.3, together with Theorem 1.5, proves part (ii) of Theorem 1.4.

The final Chap. 14 is devoted to the proof of Theorem 1.6 through Theorem 1.9. The essential point in the proof is to estimate the integral operator S in terms of Hölder norms (Lemmas 14.2 and 14.3). We show that the operator (W, L_0) may be considered as a perturbation of a *compact* operator to the operator (A, L_0) in the framework of Hölder spaces. Thus the proof is reduced to the differential operator case which is studied in Chaps. 8 through 13.

In Appendix C, we prove various maximum principles for second-order elliptic Waldenfels operators W such as the weak and strong maximum principles and the boundary point lemma (Theorems C.1, C.2 and C.3 and Lemma C.5) which play an important role in Chaps. 13 and 14.

2

Theory of Semigroups

This chapter is devoted to a review of standard topics from the theory of semigroups such as contraction semigroups and analytic semigroups. These topics form a necessary background for what follows. The material in this chapter is adapted from the books of Yosida [Yo] and Friedman [Fr1], and also part of Taira [Ta2]. For more leisurely treatments of semigroups, the reader is referred to Engel–Nagel [EN].

2.1 Banach Space Valued Functions

Let E be a Banach space over the real or complex number field, equipped with a norm $\|\cdot\|$. A function $u(t)$ defined on an interval I with values in E is said to be *strongly continuous* at a point t_0 of I if it satisfies the condition

$$\lim_{t \to t_0} \|u(t) - u(t_0)\| = 0 .$$

If $u(t)$ is strongly continuous at every point of I, then it is said to be strongly continuous on I. If $u(t)$ is strongly continuous on I, then the function $\|u(t)\|$ is continuous on I and also, for any f in the dual space E' of E, the function $f(u(t))$ is continuous on I.

As in the case of scalar valued functions, the following two results hold:

(1) If $u(t)$ is strongly continuous on a bounded closed interval I, then it is uniformly strongly continuous on I.
(2) If a sequence $\{u_n(t)\}$ of strongly continuous functions on I converges uniformly strongly to a function $u(t)$ on I, then the function $u(t)$ is strongly continuous on I.

If $u(t)$ is a strongly continuous function on I such that

$$\int_I \|u(t)\| \, dt < \infty , \tag{2.1}$$

then the Riemann integral

$$\int_I u(t)\,dt$$

can be defined just as in the case of scalar valued functions; we then say that the function $u(t)$ is *strongly integrable* on I. By the triangle inequality, we have

$$\left\|\int_I u(t)\,dt\right\| \leq \int_I \|u(t)\|\,dt\,.$$

Furthermore, we easily obtain the following:

Theorem 2.1. *Let $u(t)$ be a strongly continuous function defined on an interval I which satisfies condition (2.1), and let T be a bounded linear operator on E into itself. Then the function $Tu(t)$ is strongly integrable on I, and we have*

$$T\left(\int_I u(t)\,dt\right) = \int_I Tu(t)\,dt\,.$$

Similarly, we have, for any $f \in E'$,

$$f\left(\int_I u(t)\,dt\right) = \int_I f(u(t))\,dt\,.$$

As in the case of scalar valued functions, the following two results hold:

(3) If a sequence $\{u_n(t)\}$ of strongly continuous functions on a bounded closed interval I converges uniformly strongly to a function $u(t)$ on I, then the function $u(t)$ is strongly integrable on I, and we have

$$\int_I u(t)\,dt = \lim_{n\to\infty} \int_I u_n(t)\,dt\,.$$

(4) If $u(t)$ is strongly continuous in a neighborhood of a point t_0 of I, then we have

$$\lim_{h\to 0}\left\|\frac{1}{h}\int_{t_0}^{t_0+h} u(t)\,dt - u(t_0)\right\| = 0\,.$$

A function $u(t)$ defined on an open interval I is said to be *strongly differentiable* at a point t_0 of I if the limit

$$\lim_{h\to 0}\frac{u(t_0+h)-u(t_0)}{h} \tag{2.2}$$

exists in E. The value of formula (2.2) is denoted by

$$\frac{du}{dt}(t_0) \text{ or } u'(t_0)\,.$$

If $u(t)$ is strongly differentiable at every point of I, then it is said to be strongly differentiable on I. A strongly differentiable function is strongly continuous.

As in the case of scalar valued functions, the following two results hold:

(5) If $u(t)$ is strongly differentiable on I and $u'(t)$ is strongly continuous on I, then we have, for any $a, b \in I$,

$$u(b) - u(a) = \int_a^b u'(t)\, dt \,.$$

(6) If $u(t)$ is strongly continuous on I, then, for each $c \in I$, the integral $\int_c^t u(s)\, ds$ is strongly differentiable on I, and we have

$$\frac{d}{dt}\left(\int_c^t u(s)\, ds\right) = u(t) \,.$$

2.2 Operator Valued Functions

Let $L(E, E)$ the space of all bounded linear operators on a Banach space E into itself. The space $L(E, E)$ is a Banach space with the norm

$$\|T\| = \sup_{\substack{x \in E \\ \|x\| \leq 1}} \|Tx\| \,.$$

A function $T(t)$ defined on an interval I with values in the space $L(E, E)$ is said to be *strongly continuous* at a point t_0 of I if it satisfies the condition

$$\lim_{t \to t_0} \|T(t)x - T(t_0)x\| = 0, \quad x \in E \,.$$

We say that $T(t)$ is *norm continuous* at t_0 if it satisfies the condition

$$\lim_{t \to t_0} \|T(t) - T(t_0)\| = 0 \,.$$

If $T(t)$ is strongly (resp. norm) continuous at every point of I, then it is said to be strongly (resp. norm) continuous on I. A norm continuous function is strongly continuous.

The next theorem is an immediate consequence of the *principle of uniform boundedness* (see [Fr2, Theorem 4.5.1]):

Theorem 2.2. *If $T(t)$ is strongly continuous on I, then the function $\|T(t)\|$ is bounded uniformly in t over bounded closed intervals contained in I.*

A function $T(t)$ defined on an open interval I is said to be *strongly differentiable* at a point t_0 of I if there exists an operator $S(t_0)$ in $L(E, E)$ such that

$$\lim_{h \to 0} \left\| \left(\frac{T(t_0 + h) - T(t_0)}{h}\right) x - S(t_0)x \right\| = 0, \quad x \in E \,.$$

We say that $T(t)$ is *norm differentiable* at t_0 if it satisfies the condition

$$\lim_{h \to 0} \left\| \frac{T(t_0 + h) - T(t_0)}{h} - S(t_0) \right\| = 0 \,.$$

The operator $S(t_0)$ is denoted by

$$\frac{dT}{dt}(t_0) \text{ or } T'(t_0) .$$

If $T(t)$ is strongly (resp. norm) differentiable at every point of I, then it is said to be strongly (resp. norm) differentiable on I. A norm differentiable function is strongly differentiable.

It should be emphasized that the Leibniz formula can be extended to strongly or norm differentiable functions:

Theorem 2.3. *(i) If $u(t)$ and $T(t)$ are both strongly continuous (resp. differentiable) on I, then the function $T(t)u(t)$ is also strongly continuous (resp. differentiable) on I. In the differentiable case, we have*

$$\frac{d}{dt}\left(T(t)u(t)\right) = \frac{dT}{dt}(t)u(t) + T(t)\frac{du}{dt}(t) .$$

(ii) If $T(t)$ and $S(t)$ are both norm (resp. strongly) differentiable on I, then the function $S(t)T(t)$ is also norm (resp. strongly) differentiable on I, and we have

$$\frac{d}{dt}\left(S(t)T(t)\right) = \frac{dS}{dt}(t)T(t) + S(t)\frac{dT}{dt}(t) .$$

2.3 Exponential Functions

Just as in the case of numerical series, we have the following:

Theorem 2.4. *If A is a bounded linear operator on a Banach space E into itself, then the series*

$$e^{tA} = \sum_{m=0}^{\infty} \frac{t^m}{m!} A^m \quad (-\infty < t < \infty) \tag{2.3}$$

converges in the space $L(E, E)$, and enjoys the following three properties:

(a) $\|e^{tA}\| \leq e^{|t|\|A\|}$.
(b) $e^{tA}e^{sA} = e^{(t+s)A}$ $(-\infty < t, s < \infty)$.
(c) The function e^{tA} is norm differentiable on $\mathbf{R}$, and we have

$$\frac{d}{dt}(e^{tA}) = Ae^{tA} = e^{tA}A . \tag{2.4}$$

Proof. It is clear that series (2.3) converges in the space $L(E, E)$ and enjoys property (a). Property (b) is proved by rearranging the series

$$e^{(t+s)A} = \sum_{m=0}^{\infty} \frac{(t+s)^m}{m!} A^m$$

so that

$$\left(\sum_{m=0}^{\infty} \frac{t^m}{m!} A^m\right) \left(\sum_{m=0}^{\infty} \frac{s^m}{m!} A^m\right),$$

just as in the case of numerical series.

We prove property (c). Since A is a bounded linear operator, it follows that

$$Ae^{tA} = \sum_{m=0}^{\infty} \frac{t^m}{m!} A^{m+1} = e^{tA} A$$

and this series converges uniformly in t over bounded intervals of $\mathbf{R}$. Hence, the function Ae^{tA} is norm continuous on $\mathbf{R}$, and we have

$$\int_0^t Ae^{sA}\, ds = \sum_{m=0}^{\infty} \frac{t^{m+1}}{(m+1)!} A^{m+1} = e^{tA} - I\,. \tag{2.5}$$

Therefore, we find that the left-hand side of formula (2.5) and hence the function e^{tA} is norm differentiable on $\mathbf{R}$, and relation (2.4) holds. □

Theorem 2.5. *If A and B are bounded linear operators on a Banach space E into itself and if A and B commute, then we have*

$$e^{A+B} = e^A e^B = e^B e^A\,.$$

Proof. Since $AB = BA$, we can rearrange the series

$$e^{A+B} = \sum_{m=0}^{\infty} \frac{(A+B)^m}{m!}$$

so that

$$\left(\sum_{m=0}^{\infty} \frac{t^m}{m!} A^m\right) \left(\sum_{m=0}^{\infty} \frac{s^m}{m!} B^m\right),$$

just as in the case of numerical series. □

2.4 Contraction Semigroups

Let E be a Banach space. A one-parameter family $\{T_t\}_{t\geq 0}$ of bounded linear operators on E into itself is called a *contraction semigroup of class* (C_0) or simply a *contraction semigroup* if it satisfies the following three conditions:

(i) $T_{t+s} = T_t \cdot T_s$, t, $s \geq 0$.
(ii) $\lim_{t\downarrow 0} \|T_t x - x\| = 0$, $x \in E$.

(iii) $\|T_t\| \leq 1$, $t \geq 0$.

Condition (i) is called the *semigroup property*.

Remark 2.1. In view of conditions (i) and (ii), it follows that $T_0 = I$. Hence condition (ii) is equivalent to the strong continuity of $\{T_t\}_{t \geq 0}$ at $t = 0$. Moreover, it is easy to verify that a contraction semigroup $\{T_t\}_{t\geq 0}$ is strongly continuous on the interval $[0, \infty)$.

Example 2.1. Let $E = L^p(\mathbf{R}^n)$ with $1 \leq p < \infty$, and

$$K_t(x) = \frac{1}{(4\pi t)^{n/2}} \exp\left(-\frac{|x|^2}{4t}\right), \quad t > 0 .$$

Then it is easy to verify that a one-parameter family $\{T_t\}_{t\geq 0}$ of linear operators on E, defined by the formula

$$T_t f(x) = \begin{cases} f(x) & \text{for } t = 0 , \\ \int_{\mathbf{R}^n} K_t(x-y) f(y)\, dy & \text{for } t > 0 , \end{cases}$$

forms a contraction semigroup of class (C_0) (see Corollary 4.3 and Example 4.4 in Chap. 4). The function $K_t(x)$ is called the *Gaussian kernel* or *heat kernel*. Physically, the function $T_t f(x) = K_t * f(x)$ represents the temperature at position x and time t in a homogeneous isotropic medium $\mathbf{R}^n$ with unit coefficient of thermal diffusivity, given that the temperature at position x and time 0 is $f(x)$.

If $\{T_t\}_{t\geq 0}$ is a contraction semigroup of class (C_0), then we let

$$\mathcal{D} = \text{the set of all } x \in E \text{ such that the limit} \quad \lim_{h\downarrow 0} \frac{T_h x - x}{h} \quad \text{exists in } E .$$

Then we define a linear operator $\mathfrak{A}$ from E into itself as follows:

(a) The domain $\mathcal{D}(\mathfrak{A})$ of $\mathfrak{A}$ is the set $\mathcal{D}$.
(b) $\mathfrak{A}x = \lim_{h\downarrow 0} (T_h x - x)/h$, $x \in \mathcal{D}(\mathfrak{A})$.

The operator $\mathfrak{A}$ is called the *infinitesimal generator* of $\{T_t\}_{t\geq 0}$.

First, we derive a differential equation associated with the semigroup:

Proposition 2.6. *Let $\mathfrak{A}$ be the infinitesimal generator of a contraction semigroup $\{T_t\}_{t\geq 0}$. If $x \in \mathcal{D}(\mathfrak{A})$, then we have $T_t x \in \mathcal{D}(\mathfrak{A})$ for all $t > 0$, and the function $T_t x$ is strongly differentiable on the interval $(0, \infty)$ and satisfies the equation*

$$\frac{d}{dt}(T_t x) = \mathfrak{A}(T_t x) = T_t(\mathfrak{A}x), \quad t > 0 . \tag{2.6}$$

Proof. Let $h > 0$. Then we have, by the semigroup property,

$$\frac{T_h(T_t x) - T_t x}{h} = T_t \left(\frac{T_h - I}{h} x \right) .$$

However, since T_t is bounded and $x \in \mathcal{D}(\mathfrak{A})$, it follows that

$$T_t \left(\frac{T_h - I}{h} x \right) \longrightarrow T_t(\mathfrak{A}x) \quad \text{as } h \downarrow 0 .$$

This implies that

$$\begin{cases} T_t x \in \mathcal{D}(\mathfrak{A}) , \\ \mathfrak{A}(T_t x) = T_t(\mathfrak{A}x) . \end{cases}$$

Therefore, we find that $T_t x$ is strongly right-differentiable on $(0, \infty)$ and satisfies the equation

$$\frac{d}{dt}^{+} (T_t x) = \mathfrak{A}(T_t x) = T_t(\mathfrak{A}x), \quad t > 0 .$$

Similarly, we have, for each $0 < h < t$,

$$\frac{T_{t-h}x - T_t x}{-h} - T_t(\mathfrak{A}x) = T_{t-h} \left(\frac{T_h - I}{h} x - T_h(\mathfrak{A}x) \right) .$$

However, since $\|T_{t-h}\| \leq 1$ and $T_h(\mathfrak{A}x) \to \mathfrak{A}x$ as $h \downarrow 0$, we obtain that

$$\frac{T_{t-h}x - T_t x}{-h} \longrightarrow T_t(\mathfrak{A}x) \quad \text{as } h \downarrow 0 .$$

This proves that $T_t x$ is strongly left-differentiable on $(0, \infty)$ and satisfies the equation

$$\frac{d}{dt}^{-} (T_t x) = \mathfrak{A}(T_t x) = T_t(\mathfrak{A}x), \quad t > 0 .$$

Summing up, we have proved that $T_t x$ is strongly differentiable on $(0, \infty)$ and satisfies equation (2.6). □

The next proposition characterizes the infinitesimal generator $\mathfrak{A}$:

Proposition 2.7. *Let $\mathfrak{A}$ be the infinitesimal generator of a contraction semigroup $\{T_t\}_{t \geq 0}$. Then $\mathfrak{A}$ is a densely defined, closed linear operator in E.*

Proof. The proof is divided into two steps.

Step 1: First, we show that the operator $\mathfrak{A}$ is *closed.*

To do this, we assume that $x_n \in \mathcal{D}(\mathfrak{A})$, $x_n \to x_0$ and $\mathfrak{A}x_n \to y_0$ in E. Then it follows from an application of equation (2.6) that

$$T_t x_n - x_n = \int_0^t \frac{d}{ds}(T_s x_n)\, ds = \int_0^t T_s(\mathfrak{A} x_n)\, ds . \tag{2.7}$$

However, we have, as $n \to \infty$,

$$T_t x_n - x_n \longrightarrow T_t x_0 - x_0$$

and also

$$\begin{aligned} \left\| \int_0^t T_s(\mathfrak{A} x_n)\, ds - \int_0^t T_s y_0\, ds \right\| &= \left\| \int_0^t T_s(\mathfrak{A} x_n - y_0)\, ds \right\| \\ &\leq \int_0^t \| T_s(\mathfrak{A} x_n - y_0) \|\, ds \\ &\leq t\, \| \mathfrak{A} x_n - y_0 \| \longrightarrow 0 . \end{aligned}$$

Hence, by letting $n \to \infty$ in formula (2.7) we obtain that

$$T_t x_0 - x_0 = \int_0^t T_s y_0\, ds . \tag{2.8}$$

Furthermore, it follows that, as $t \downarrow 0$,

$$\frac{1}{t} \int_0^t T_s y_0\, ds \longrightarrow y_0 ,$$

since the integrand $\{T_t\}$ is strongly continuous.

Therefore, we find from formula (2.8) that, as $t \downarrow 0$,

$$\frac{T_t x_0 - x_0}{t} = \frac{1}{t} \int_0^t T_s y_0\, ds \longrightarrow y_0 .$$

This proves that

$$\begin{cases} x_0 \in \mathcal{D}(\mathfrak{A}), \\ \mathfrak{A} x_0 = y_0 . \end{cases}$$

Step 2: Secondly, we show the *density* of the domain $\mathcal{D}(\mathfrak{A})$ in E. Let x be an arbitrary element of E. For each $\delta > 0$, we let

$$x_\delta = \frac{1}{\delta} \int_0^\delta T_s x\, ds .$$

Then we have, for any $0 < h < \delta$,

$$T_h(x_\delta) = \frac{1}{\delta} \int_0^\delta T_h(T_s x)\, ds = \frac{1}{\delta} \int_0^\delta T_{h+s} x\, ds = \frac{1}{\delta} \int_h^{\delta+h} T_s x\, ds .$$

Hence it follows that

$$\begin{aligned}\left(\frac{T_h - I}{h}\right) x_\delta &= \frac{1}{\delta}\left(\frac{1}{h}\int_h^{\delta+h} T_s x\, ds - \frac{1}{h}\int_0^\delta T_s x\, ds\right) \\ &= \frac{1}{\delta}\left(\frac{1}{h}\int_\delta^{\delta+h} T_s x\, ds - \frac{1}{h}\int_0^h T_s x\, ds\right) . \end{aligned} \tag{2.9}$$

However, it follows that, as $h \downarrow 0$,

$$\frac{1}{h}\int_\delta^{\delta+h} T_s x\, ds \longrightarrow T_\delta x ,$$
$$\frac{1}{h}\int_0^h T_s x\, ds \longrightarrow x ,$$

since $\{T_t\}$ is strongly continuous.

Therefore, we find from formula (2.9) that, as $h \downarrow 0$,

$$\left(\frac{T_h - I}{h}\right) x_\delta \longrightarrow \frac{1}{\delta}(T_\delta x - x) .$$

This proves that

$$\begin{cases} x_\delta \in \mathcal{D}(\mathfrak{A}) , \\ \mathfrak{A} x_\delta = \dfrac{1}{\delta}(T_\delta x - x) . \end{cases}$$

Moreover, it follows that, as $\delta \downarrow 0$,

$$x_\delta = \frac{1}{\delta}\int_0^\delta T_s x\, ds \longrightarrow x .$$

Summing up, we have proved that $\mathcal{D}(\mathfrak{A})$ is dense in E. □

Let $\{T_t\}_{t\geq 0}$ be a contraction semigroup. Then the integral

$$\int_0^s e^{-\alpha t} T_t x\, dt, \quad x \in E , \tag{2.10}$$

is strongly integrable for all $s > 0$, since the integrand is strongly continuous on the interval $[0, \infty)$. Moreover, if $\alpha > 0$, then the limit $G_\alpha x$ of the integral (2.10) exists in E as $s \to \infty$:

$$G_\alpha x = \int_0^\infty e^{-\alpha t} T_t x\, dt, \quad x \in E,\ \alpha > 0 .$$

Thus $G_\alpha x$ is defined for all $x \in E$ if $\alpha > 0$. It is easy to see that the operator G_α is a bounded linear operator on E into itself with norm $1/\alpha$:

$$\|G_\alpha x\| \leq \frac{1}{\alpha}\|x\|, \quad x \in E . \tag{2.11}$$

The family $\{G_\alpha\}_{\alpha>0}$ of bounded linear operators is called the *resolvent* of the semigroup $\{T_t\}_{t\geq 0}$.

The next proposition characterizes the resolvent G_α:

Proposition 2.8. *Let $\{T_t\}_{t\geq 0}$ be a contraction semigroup defined on a Banach space E and $\mathfrak{A}$ the infinitesimal generator of $\{T_t\}$. For each $\alpha > 0$, the operator $(\alpha I - \mathfrak{A})$ is a bijection of $\mathcal{D}(\mathfrak{A})$ onto E, and its inverse $(\alpha I - \mathfrak{A})^{-1}$ is the resolvent G_α:*

$$(\alpha I - \mathfrak{A})^{-1} x = \int_0^\infty e^{-\alpha t} T_t x \, dt, \quad x \in E . \tag{2.12}$$

Proof. The proof is divided into three steps.

Step 1: First, we show that $(\alpha I - \mathfrak{A})$ is *surjective* for each $\alpha > 0$.

Let x be an arbitrary element of E. Then we have, for each $h > 0$,

$$\begin{aligned} T_h(G_\alpha x) &= \int_0^\infty e^{-\alpha t} T_h(T_t x) \, dt \\ &= \int_0^\infty e^{-\alpha t} T_{t+h} x \, dt \\ &= e^{\alpha h} \int_h^\infty e^{-\alpha t} T_t x \, dt . \end{aligned}$$

Hence it follows that

$$\begin{aligned} T_h(G_\alpha x) - G_\alpha x &= e^{\alpha h} \int_h^\infty e^{-\alpha t} T_t x \, dt - \int_0^\infty e^{-\alpha t} T_t x \, dt \\ &= \left(e^{\alpha h} - 1\right) \int_h^\infty e^{-\alpha t} T_t x \, dt - \int_0^h e^{-\alpha t} T_t x \, dt , \end{aligned}$$

so that

$$\begin{aligned} &\frac{T_h(G_\alpha x) - G_\alpha x}{h} \\ &= \left(\frac{e^{\alpha h} - 1}{h}\right) \int_h^\infty e^{-\alpha t} T_t x \, dt - \frac{1}{h} \int_0^h e^{-\alpha t} T_t x \, dt . \end{aligned} \tag{2.13}$$

However, we obtain that, as $h \downarrow 0$,

$$\begin{aligned} &\left(\frac{e^{\alpha h} - 1}{h}\right) \longrightarrow \alpha , \\ &\int_h^\infty e^{-\alpha t} T_t x \, dt \longrightarrow \int_0^\infty e^{-\alpha t} T_t x \, dt = G_\alpha x , \\ &\frac{1}{h} \int_0^h e^{-\alpha t} T_t x \, dt \longrightarrow e^{-\alpha t} T_t x|_{t=0} = x . \end{aligned}$$

By letting $h \downarrow 0$ in formula (2.13), we have proved that

$$\frac{T_h(G_\alpha x) - G_\alpha x}{h} \longrightarrow \alpha G_\alpha x - x \quad \text{as } h \downarrow 0 .$$

This implies that

$$\begin{cases} G_\alpha x \in \mathcal{D}(\mathfrak{A}) \,, \\ \mathfrak{A}(G_\alpha x) = \alpha G_\alpha x - x \,, \end{cases}$$

or equivalently,

$$(\alpha I - \mathfrak{A}) G_\alpha x = x, \quad x \in E \,.$$

Step 2: Secondly, we show that $(\alpha I - \mathfrak{A})$ is *injective* for each $\alpha > 0$.

Now we assume that $x \in \mathcal{D}(\mathfrak{A})$ and $(\alpha I - \mathfrak{A})x = 0$. If we introduce a function $u(t)$ by the formula

$$u(t) = e^{-\alpha t} T_t x, \quad t > 0 \,,$$

then it follows from an application of Proposition 2.6 that

$$\begin{aligned} \frac{d}{dt}(u(t)) &= -\alpha e^{-\alpha t} T_t x + e^{-\alpha t} T_t \mathfrak{A} x \\ &= -e^{-\alpha t} T_t (\alpha I - \mathfrak{A}) x \\ &= 0 \,, \end{aligned}$$

so that

$$u(t) = \text{a constant}, \quad t > 0 \,.$$

However, we have, by letting $t \downarrow 0$,

$$u(t) = x \,.$$

On the other hand, we have, by letting $t \uparrow \infty$,

$$u(t) \longrightarrow 0 \,.$$

Indeed, it suffices to note that

$$\|u(t)\| = e^{-\alpha t} \|T_t x\| \le e^{-\alpha t} \|x\| \,.$$

Hence it follows that $x = 0$.

Step 3: Summing up, we have proved that $(\alpha I - \mathfrak{A})$ is a *bijection* of $\mathcal{D}(\mathfrak{A})$ onto E and that $(\alpha I - \mathfrak{A})^{-1} = G_\alpha$. □

Now we consider when a linear operator is the infinitesimal generator of some contraction semigroup. This question is answered by the following theorem:

Theorem 2.9 (Hille–Yosida). *Let $\mathfrak{A}$ be a linear operator from a Banach space E into itself with domain $\mathcal{D}(\mathfrak{A})$. In order that $\mathfrak{A}$ be the infinitesimal generator of some contraction semigroup, it is necessary and sufficient that $\mathfrak{A}$ satisfies the following three conditions:*

(i) The operator $\mathfrak{A}$ is closed and its domain $\mathcal{D}(\mathfrak{A})$ is dense in E.

(ii) For every $\alpha > 0$ the equation

$$(\alpha I - \mathfrak{A})x = y$$

has a unique solution $x \in \mathcal{D}(\mathfrak{A})$ for any $y \in E$; we then write

$$x = (\alpha I - \mathfrak{A})^{-1} y .$$

(iii) For any $\alpha > 0$, we have

$$\|(\alpha I - \mathfrak{A})^{-1}\| \leq \frac{1}{\alpha} . \tag{2.14}$$

Proof. The necessity of conditions (i) through (iii) follows from Propositions 2.7 and 2.8 and inequality (2.11).

The sufficiency is proved in six steps.

Step 1: For each $\alpha > 0$, we define linear operators

$$J_\alpha = \alpha(\alpha I - \mathfrak{A})^{-1} ,$$

and

$$\mathfrak{A}_\alpha = \mathfrak{A} J_\alpha .$$

Then we can prove the following four assertions:

$$\|J_\alpha\| \leq 1 , \tag{2.15a}$$

$$\lim_{\alpha \to +\infty} J_\alpha x = x, \quad x \in E , \tag{2.15b}$$

and

$$\|\mathfrak{A}_\alpha\| \leq 2\alpha , \tag{2.16a}$$

$$\lim_{\alpha \to +\infty} \mathfrak{A}_\alpha x = \mathfrak{A}x, \quad x \in \mathcal{D}(\mathfrak{A}) . \tag{2.16b}$$

The operators $\mathfrak{A}_\alpha$ are called the *Yosida approximations* to $\mathfrak{A}$.

First, we remark that assertion (2.15a) is an immediate consequence of inequality (2.14). Furthermore, we have, for all $x \in \mathcal{D}(\mathfrak{A})$,

$$\begin{aligned} J_\alpha x - x &= \alpha(\alpha I - \mathfrak{A})^{-1}x - (\alpha I - \mathfrak{A})^{-1}(\alpha I - \mathfrak{A})x \\ &= (\alpha I - \mathfrak{A})^{-1}(\mathfrak{A}x) . \end{aligned}$$

Hence it follows that, as $\alpha \to +\infty$,

$$\begin{aligned} \|J_\alpha x - x\| &\leq \|(\alpha I - \mathfrak{A})^{-1}\| \, \|\mathfrak{A}x\| \\ &\leq \frac{1}{\alpha} \|\mathfrak{A}x\| \longrightarrow 0 . \end{aligned}$$

This implies assertion (2.15b), since $\|J_\alpha\| \leq 1$ and $\mathcal{D}(\mathfrak{A})$ is dense in E.

Assertion (2.16b) follows from assertion (2.15b). Indeed, we have, as $\alpha \to +\infty$,

$$\mathfrak{A}_\alpha x = \mathfrak{A} J_\alpha x = J_\alpha(\mathfrak{A}x) \longrightarrow \mathfrak{A}x, \quad x \in \mathcal{D}(\mathfrak{A}) .$$

On the other hand, it follows that

$$\mathfrak{A}_\alpha = -\alpha + \alpha J_\alpha ,$$

so that

$$\|\mathfrak{A}_\alpha\| \leq \alpha + \alpha \|J_\alpha\| \leq 2\alpha .$$

This proves assertion (2.16a).

Step 2: We define

$$T_t(\alpha) = \exp\left[t\mathfrak{A}_\alpha\right], \quad \alpha > 0 .$$

Since we have $\mathfrak{A}_\alpha = -\alpha + \alpha J_\alpha$, we obtain from an application of Theorem 2.5 that the operators

$$T_t(\alpha) = e^{-\alpha t} \exp\left[\alpha t J_\alpha\right], \quad t \geq 0 , \tag{2.17}$$

form a contraction semigroup for each $\alpha > 0$. Indeed, it suffices to note that

$$\begin{aligned}
\|T_t(\alpha)\| &= e^{-\alpha t} \left\|\exp\left[\alpha t J_\alpha\right]\right\| \\
&\leq e^{-\alpha t} \left(\sum_{n=0}^{\infty} \frac{(\alpha t)^n}{n!} \|J_\alpha^n\| \right) \\
&\leq e^{-\alpha t} e^{\alpha t} \\
&= 1 .
\end{aligned}$$

Step 3: We show that the operator $T_t(\alpha)$ has a strong limit T_t as $\alpha \to +\infty$:

$$T_t x = \lim_{\alpha \to +\infty} T_t(\alpha) x, \quad x \in E .$$

Moreover, this convergence is uniform in t over bounded intervals $[0, t_0]$, $t_0 > 0$.

If x is an arbitrary element of $\mathcal{D}(\mathfrak{A})$, then it follows that

$$\begin{aligned}
&T_t(\alpha)x - T_t(\beta)x \\
&= \int_0^t \frac{d}{ds} \left(T_{t-s}(\beta) T_s(\alpha) x\right) ds \\
&= \int_0^t \left(T_{t-s}(\beta)(-\mathfrak{A}_\beta) T_s(\alpha) x + T_{t-s}(\beta) T_s(\alpha)(\mathfrak{A}_\alpha x)\right) ds \\
&= \int_0^t T_{t-s}(\beta) T_s(\alpha) \left(\mathfrak{A}_\alpha x - \mathfrak{A}_\beta x\right) ds .
\end{aligned}$$

Hence we have, for all $0 \leq t \leq t_0$,

$$\begin{aligned}
&\|T_t(\alpha)x - T_t(\beta)x\| \\
&= \left\| \int_0^t \frac{d}{ds} (T_{t-s}(\beta)T_s(\alpha)x) \, ds \right\| \\
&\leq \int_0^t \|T_{t-s}(\beta)\| \, \|T_s(\alpha)\| \, ds \, \|\mathfrak{A}_\alpha x - \mathfrak{A}_\beta x\| \\
&\leq t \, \|\mathfrak{A}_\alpha x - \mathfrak{A}_\beta x\| \\
&\leq t_0 \, \|\mathfrak{A}_\alpha x - \mathfrak{A}_\beta x\| \, .
\end{aligned}$$

However, recall that, as $\alpha \to +\infty$,

$$\mathfrak{A}_\alpha x \longrightarrow \mathfrak{A}x \, .$$

Therefore, we obtain that, as $\alpha, \beta \to +\infty$,

$$\|T_t(\alpha)x - T_t(\beta)x\| \longrightarrow 0 \, ,$$

and that the convergence is uniform in t over the interval $[0, t_0]$.

We can define a linear operator T_t by the formula

$$T_t x = \lim_{\alpha \to +\infty} T_t(\alpha)x, \quad x \in \mathcal{D}(\mathfrak{A}) \, .$$

Furthermore, since $\|T_t(\alpha)\| \leq 1$ and $\mathcal{D}(\mathfrak{A})$ is dense in E, it follows that the operator $T_t(\alpha)$ has a strong limit T_t as $\alpha \to +\infty$:

$$T_t x = \lim_{\alpha \to +\infty} T_t(\alpha)x, \quad x \in E \, , \tag{2.18}$$

and that the convergence is uniform in t over bounded intervals $[0, t_0]$, $t_0 > 0$.

Step 4: We show that the family $\{T_t\}_{t \geq 0}$ forms a contraction semigroup of class (C_0).

First, it follows from an application of the principle of uniform boundedness that the operator T_t is bounded and satisfies the condition

$$\|T_t\| \leq \liminf_{\alpha \to +\infty} \|T_t(\alpha)\| \leq 1, \quad t \geq 0 \, .$$

Next the semigroup property of $\{T_t\}$

$$T_t(T_s x) = T_{t+s}x, \quad x \in E$$

follows from that of $\{T_t(\alpha)\}$. Indeed, we have, as $\alpha \to +\infty$,

$$\begin{aligned}
&\|T_t(T_s x) - T_t(\alpha)(T_s(\alpha)x)\| \\
&\leq \|(T_t - T_t(\alpha))T_s x\| + \|T_t(\alpha)(T_s x - T_s(\alpha)x)\| \\
&\leq \|(T_t - T_t(\alpha))T_s x\| + \|T_s x - T_s(\alpha)x\| \longrightarrow 0 \, ,
\end{aligned}$$

so that

$$\begin{aligned}T_t(T_s x) &= \lim_{\alpha\to+\infty} T_t(\alpha)(T_s(\alpha)x) \\ &= \lim_{\alpha\to+\infty} T_{t+s}(\alpha)x \\ &= T_{t+s}x, \quad x\in E\,.\end{aligned}$$

Furthermore, since the convergence of (2.18) is uniform in t over bounded sub-intervals of the interval $[0,\infty)$, it follows that the function $T_t x$, $x \in E$, is strongly continuous on the interval $[0,\infty)$. Consequently the family $\{T_t\}_{t\geq 0}$ forms a contraction semigroup.

Step 5: We show that the infinitesimal generator of the semigroup $\{T_t\}_{t\geq 0}$ thus obtained is precisely the operator $\mathfrak{A}$.

Let $\mathfrak{A}_0$ be the infinitesimal generator of $\{T_t\}_{t\geq 0}$. If x is an arbitrary element of $\mathcal{D}(\mathfrak{A})$, it follows an application of Proposition 2.6 that

$$e^{t\mathfrak{A}_\alpha}x - x = \int_0^t \frac{d}{ds}\left(e^{s\mathfrak{A}_\alpha}x\right) ds = \int_0^t e^{s\mathfrak{A}_\alpha}(\mathfrak{A}_\alpha x)\, ds\,. \tag{2.19}$$

However, we have, as $\alpha \to +\infty$,

$$e^{t\mathfrak{A}_\alpha}x - x \longrightarrow T_t x - x\,,$$

and also

$$\int_0^t e^{s\mathfrak{A}_\alpha}(\mathfrak{A}_\alpha x)\, ds \longrightarrow \int_0^t T_s(\mathfrak{A}x)\, ds\,.$$

Indeed, it suffices to note that, as $\alpha \to +\infty$,

$$\begin{aligned}\left\|\int_0^t e^{s\mathfrak{A}_\alpha}(\mathfrak{A}_\alpha x)\, ds - \int_0^t T_s(\mathfrak{A}x)\, ds\right\| &\leq \left\|\int_0^t e^{s\mathfrak{A}_\alpha}(\mathfrak{A}_\alpha x - \mathfrak{A}x)\, ds\right\| \\ &\quad + \left\|\int_0^t (e^{s\mathfrak{A}_\alpha} - T_s)(\mathfrak{A}x)\, ds\right\| \\ &\leq \int_0^t \left\|e^{s\mathfrak{A}_\alpha}\right\| ds\, \|\mathfrak{A}_\alpha x - \mathfrak{A}x\| \\ &\quad + \int_0^t \left\|e^{s\mathfrak{A}_\alpha}(\mathfrak{A}x) - T_s(\mathfrak{A}x)\right\| ds \\ &\leq t\, \|\mathfrak{A}_\alpha x - \mathfrak{A}x\| \\ &\quad + \int_0^t \left\|e^{s\mathfrak{A}_\alpha}(\mathfrak{A}x) - T_s(\mathfrak{A}x)\right\| ds \longrightarrow 0\,.\end{aligned}$$

Hence, by letting $\alpha \to +\infty$ in formula (2.19) we have, for all $x \in \mathcal{D}(\mathfrak{A})$,

$$T_t x - x = \int_0^t T_s(\mathfrak{A}x)\, ds\,.$$

Moreover, it follows that, as $t \downarrow 0$,

$$\frac{T_t x - x}{t} = \frac{1}{t} \int_0^t T_s(\mathfrak{A}x)\, ds \longrightarrow \mathfrak{A}x ,$$

since the integrand $T_t(\mathfrak{A}x)$ is strongly continuous.

Summing up, we have proved that

$$\begin{cases} x \in \mathcal{D}(\mathfrak{A}_0) , \\ \mathfrak{A}_0 x = \mathfrak{A}x . \end{cases}$$

This implies that

$$\mathfrak{A} \subset \mathfrak{A}_0 .$$

It remains to show that $\mathcal{D}(\mathfrak{A}) = \mathcal{D}(\mathfrak{A}_0)$. If y is an arbitrary element of $\mathcal{D}(\mathfrak{A}_0)$, we let

$$x = (I - \mathfrak{A})^{-1}(I - \mathfrak{A}_0)y .$$

Then we have

$$\begin{cases} x \in \mathcal{D}(\mathfrak{A}) \subset \mathcal{D}(\mathfrak{A}_0) , \\ (I - \mathfrak{A})x = (I - \mathfrak{A}_0)y , \end{cases}$$

and so

$$(I - \mathfrak{A}_0)x = (I - \mathfrak{A}_0)y ,$$

This implies that

$$y = x \in \mathcal{D}(\mathfrak{A}) ,$$

since $(I - \mathfrak{A}_0)$ is bijective.

Step 6: Finally, we show the uniqueness of the semigroup.

Let $\{U_t\}_{t \geq 0}$ be another contraction semigroup which has $\mathfrak{A}$ as its infinitesimal generator. For each $x \in \mathcal{D}(\mathfrak{A})$ and each $t > 0$, we introduce a function $w(s)$ as follows:

$$w(s) = T_{t-s}(U_s x), \quad 0 \leq s \leq t .$$

Then we obtain from an application of Proposition 2.6 that

$$\begin{aligned} \frac{dw}{ds} &= \left(\frac{d}{ds} T_{t-s}\right) U_s x + T_{t-s}\left(\frac{d}{ds} U_s x\right) \\ &= -\mathfrak{A}T_{t-s}(U_s x) + T_{t-s}(\mathfrak{A}U_s x) \\ &= -T_{t-s}(\mathfrak{A}U_s x) + T_{t-s}(\mathfrak{A}U_s x) \\ &= 0, \quad 0 < s < t , \end{aligned}$$

so that

$$w(s) = \text{a constant}, \quad 0 \leq s \leq t .$$

In particular, it follows that $w(0) = w(t)$, that is,

$$T_t x = U_t x, \quad x \in \mathcal{D}(\mathfrak{A}) .$$

This implies that $T_t = U_t$ for all $t \geq 0$, since T_t and U_t are both bounded and since $\mathcal{D}(\mathfrak{A})$ is dense in E.

Now the proof of Theorem 2.9 is complete. □

2.5 Analytic Semigroups

This section provides a review of standard topics from the theory of analytic semigroups for the proof of Theorems 1.3 and 1.4 (Theorem 2.11). For more leisurely treatments of analytic semigroups, the reader is referred to Friedman [Fr1], Pazy [Pa], Tanabe [Tn], Yosida [Yo] and also Taira [Ta5].

Let E be a Banach space over the real or complex number field, equipped with a norm $\|\cdot\|$, and let $A\colon E \to E$ be a *densely defined*, closed linear operator with domain $\mathcal{D}(A)$. We assume that the operator A satisfies the following two conditions (see Fig. 2.1 below):

(1) The resolvent set of A contains the region $\Sigma_\omega = \{\lambda \in \mathbf{C} : \lambda \neq 0, |\arg \lambda| < \pi/2 + \omega\}$, $0 < \omega < \pi/2$.
(2) For each $\varepsilon > 0$, there exists a constant $M_\varepsilon > 0$ such that the resolvent $R(\lambda) = (A - \lambda I)^{-1}$ satisfies the estimate

$$\|R(\lambda)\| \leq \frac{M_\varepsilon}{|\lambda|}, \quad \lambda \in \Sigma_\omega^\varepsilon = \left\{\lambda \in \mathbf{C} : \lambda \neq 0,\ |\arg \lambda| \leq \frac{\pi}{2} + \omega - \varepsilon\right\}. \tag{2.20}$$

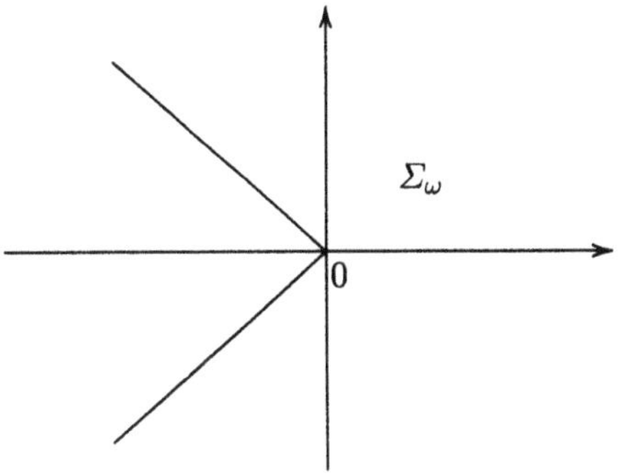

Fig. 2.1.

Then we let

$$U_t = -\frac{1}{2\pi i}\int_\Gamma e^{\lambda t} R(\lambda)\, d\lambda = -\frac{1}{2\pi i}\int_\Gamma e^{\lambda t}(A - \lambda I)^{-1}\, d\lambda\,, \tag{2.21}$$

where Γ is a path in the set $\Sigma_\omega^\varepsilon$ consisting of the following three curves $\Gamma^{(1)}$, $\Gamma^{(2)}$ and $\Gamma^{(3)}$ (see Fig. 2.2):

$$\Gamma^{(1)} = \left\{re^{-i(\pi/2+\omega-\varepsilon)} : 1 \leq r < \infty\right\},$$
$$\Gamma^{(2)} = \left\{e^{i\theta} : -\left(\frac{\pi}{2} + \omega - \varepsilon\right) \leq \theta \leq \frac{\pi}{2} + \omega - \varepsilon\right\},$$
$$\Gamma^{(3)} = \left\{re^{i(\pi/2+\omega-\varepsilon)} : 1 \leq r < \infty\right\}.$$

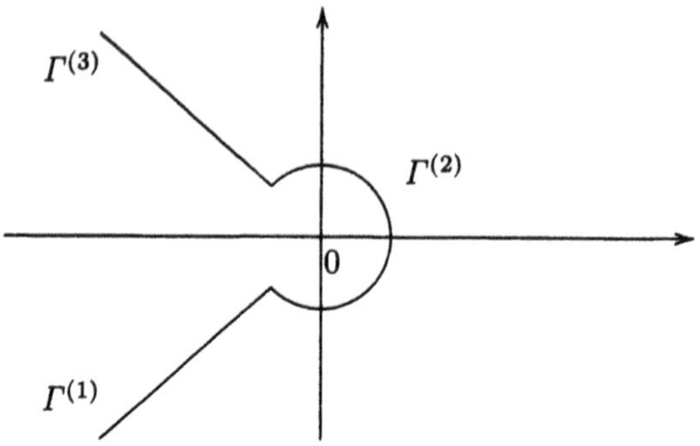

Fig. 2.2.

It is easy to see that the integral

$$U_t = -\frac{1}{2\pi i}\sum_{k=1}^{3}\int_{\Gamma^{(k)}} e^{\lambda t} R(\lambda)\, d\lambda$$

converges in the uniform operator topology for $t > 0$, and thus defines a bounded linear operator on E.

Furthermore, we have the following:

Proposition 2.10. *The operators U_t, defined by formula (2.21), form a semigroup on E, that is,*

$$U_{t+s} = U_t \cdot U_s, \quad t, s > 0\,.$$

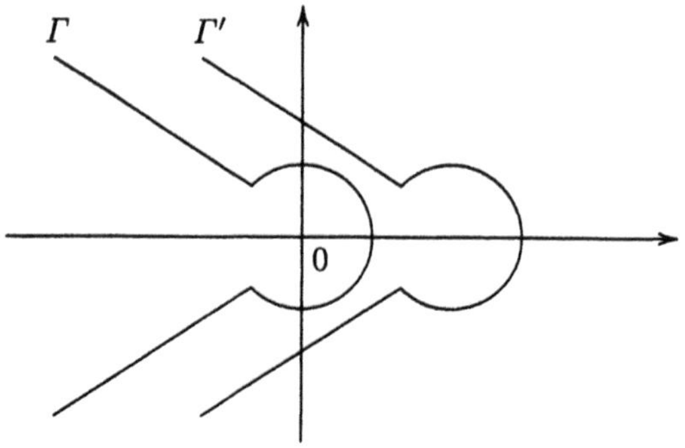

Fig. 2.3.

Proof. By Cauchy's theorem, we may assume that

$$U_s = -\frac{1}{2\pi i}\int_{\Gamma'} e^{\mu s} R(\mu)\, d\mu, \quad s > 0\,.$$

Here Γ' is a path obtained from the path Γ by translating each point of Γ to the right by a fixed small positive distance (see Fig. 2.3). Then we have, by Fubini's theorem,

$$\begin{aligned} U_t \cdot U_s &= \frac{1}{(2\pi i)^2}\int_\Gamma \int_{\Gamma'} e^{\lambda t} e^{\mu s} R(\lambda)\, R(\mu)\, d\lambda\, d\mu \\ &= \frac{1}{(2\pi i)^2}\int_\Gamma \int_{\Gamma'} e^{\lambda t} e^{\mu s} \frac{R(\lambda) - R(\mu)}{\lambda - \mu}\, d\lambda\, d\mu \\ &= \frac{1}{2\pi i}\int_\Gamma e^{\lambda t} R(\lambda) \left[\frac{1}{2\pi i}\int_{\Gamma'} \frac{e^{\mu s}}{\lambda - \mu}\, d\mu\right] d\lambda \\ &\quad - \frac{1}{2\pi i}\int_{\Gamma'} e^{\mu s} R(\mu) \left[\frac{1}{2\pi i}\int_{\Gamma} \frac{e^{\lambda t}}{\lambda - \mu}\, d\lambda\right] d\mu\,. \end{aligned}$$

We calculate the two terms in the last part.

(1) We let

$$f(\mu) = \frac{e^{\mu s}}{\lambda - \mu}\,.$$

Then, applying the residue theorem we obtain that (see Fig. 2.4)

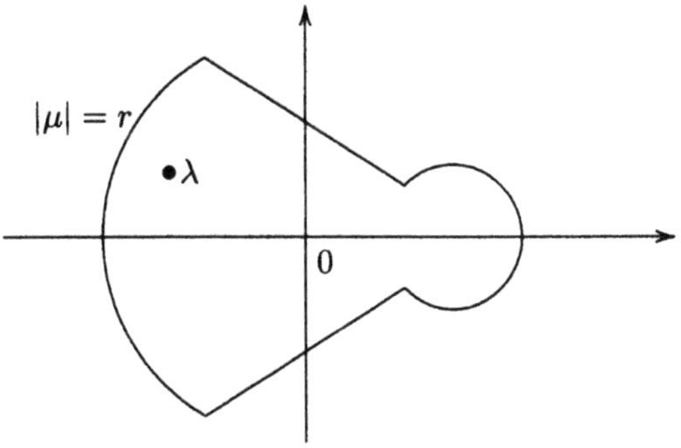

Fig. 2.4.

$$\begin{aligned} &\int_{\Gamma' \cap \{|\mu| \le r\}} f(\mu)\, d\mu + \int_{-(\pi/2+\omega-\varepsilon)}^{\pi/2+\omega-\varepsilon} f(re^{i\theta}) r i e^{i\theta}\, d\theta \\ &= \int_{\Gamma'^{(1)} \cap \{|\mu| \le r\}} f(\mu)\, d\mu + \int_{\Gamma'^{(2)}} f(\mu)\, d\mu + \int_{\Gamma'^{(3)} \cap \{|\mu| \le r\}} f(\mu)\, d\mu \end{aligned}$$

$$
\begin{aligned}
&+\int_{-(\pi/2+\omega-\varepsilon)}^{\pi/2+\omega-\varepsilon} f(re^{i\theta})rie^{i\theta}\,d\theta \\
&= 2\pi i \operatorname{Res}[f(\mu)]_{\mu=\lambda} \\
&= -2\pi i e^{\lambda s}\,.
\end{aligned}
$$

However, we have, as $r \to \infty$,

$$
\int_{\Gamma'^{(1)}\cap\{|\mu|\leq r\}} f(\mu)\,d\mu \longrightarrow \int_{\Gamma'^{(1)}} f(\mu)\,d\mu\,,
$$

$$
\int_{\Gamma'^{(3)}\cap\{|\mu|\leq r\}} f(\mu)\,d\mu \longrightarrow \int_{\Gamma'^{(3)}} f(\mu)\,d\mu\,,
$$

and also

$$
\left|\int_{-(\pi/2+\omega-\varepsilon)}^{\pi/2+\omega-\varepsilon} f(re^{i\theta})rie^{i\theta}\,d\theta\right| \leq e^{-rs\cdot\sin(\omega-\varepsilon)}\int_{-(\pi/2+\omega-\varepsilon)}^{\pi/2+\omega-\varepsilon}\frac{d\theta}{|\lambda/r-e^{i\theta}|} \longrightarrow 0\,.
$$

Therefore, we find that

$$
\begin{aligned}
\frac{1}{2\pi i}\int_{\Gamma'}\frac{e^{\mu s}}{\lambda-\mu}\,d\mu &= \frac{1}{2\pi i}\int_{\Gamma'^{(1)}+\Gamma'^{(2)}+\Gamma'^{(3)}} f(\mu)\,d\mu \\
&= -e^{\lambda s}\,.
\end{aligned}
$$

(2) Similarly, since the path Γ lies to the left of the path Γ', we find that

$$
\frac{1}{2\pi i}\int_{\Gamma}\frac{e^{\lambda t}}{\lambda-\mu}\,d\lambda = 0\,.
$$

Summing up, we obtain that

$$
U_t\cdot U_s = -\frac{1}{2\pi i}\int_{\Gamma} e^{\lambda(t+s)}R(\lambda)\,d\lambda = U_{t+s}\,.
$$

The proof of Proposition 2.10 is complete. □

The next theorem states that the semigroup U_t can be extended to an analytic semigroup in some sector containing the positive real axis.

Theorem 2.11. *The semigroup U_t can be extended to a semigroup U_z which is* analytic *in the sector $\Delta_\omega = \{z = t + is : z \neq 0, |\arg z| < \omega\}$, and enjoys the following three properties:*

(a) The operators AU_z and dU_z/dz are bounded operators on E for each $z \in \Delta_\omega$, and satisfy the relation

$$
\frac{dU_z}{dz} = AU_z\,, \quad z\in\Delta_\omega\,. \tag{2.22}
$$

(b) For each $0 < \varepsilon < \omega/2$, there exist constants $M_0(\varepsilon) > 0$ and $M_1(\varepsilon) > 0$ such that

$$\|U_z\| \leq M_0(\varepsilon), \quad z \in \Delta_\omega^{2\varepsilon}, \tag{2.23}$$

$$\|AU_z\| \leq \frac{M_1(\varepsilon)}{|z|}, \quad z \in \Delta_\omega^{2\varepsilon}, \tag{2.24}$$

where (see Fig. 2.5)

$$\Delta_\omega^{2\varepsilon} = \{z \in \mathbf{C} : z \neq 0, |\arg z| \leq \omega - 2\varepsilon\} .$$

(c) For each $x \in E$, we have, as $z \to 0$, $z \in \Delta_\omega^{2\varepsilon}$,

$$U_z x \longrightarrow x \quad \text{in } E .$$

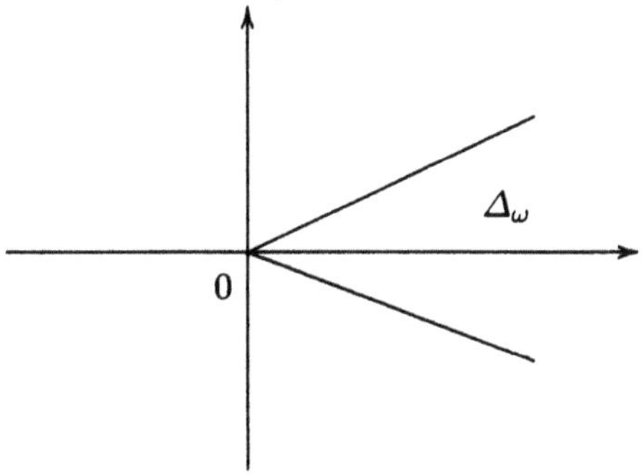

Fig. 2.5.

Proof. The proof is divided into two steps.

Step 1: First, we prove the *analyticity* of the semigroup U_z.

If $\lambda \in \Gamma^{(3)}$ and $z \in \Delta_\omega^{2\varepsilon}$, that is, if

$$\lambda = |\lambda| e^{i\theta}, \quad \theta = \frac{\pi}{2} + \omega - \varepsilon ,$$

$$z = |z| e^{i\varphi}, \quad |\varphi| \leq \omega - 2\varepsilon ,$$

then we have

$$\lambda z = |\lambda|\, |z| e^{i(\theta+\varphi)} ,$$

with

$$\frac{\pi}{2} + \varepsilon \leq \theta + \varphi \leq \frac{\pi}{2} + 2\omega - 3\varepsilon < \frac{3\pi}{2} - 3\varepsilon .$$

Note that

$$\cos(\theta + \varphi) \leq \cos\left(\frac{\pi}{2} + \varepsilon\right) = -\sin\varepsilon .$$

Hence it follows that

$$|e^{\lambda z}| \leq e^{-|\lambda|\,|z|\sin\varepsilon}, \quad \lambda \in \Gamma^{(3)}, \; z \in \Delta_\omega^{2\varepsilon} . \tag{2.25}$$

Similarly, we have

$$|e^{\lambda z}| \leq e^{-|\lambda|\,|z|\sin\varepsilon}, \quad \lambda \in \Gamma^{(1)}, \; z \in \Delta_\omega^{2\varepsilon} . \tag{2.26}$$

For each small $\varepsilon > 0$, we let

$$K_\omega^\varepsilon = \Delta_\omega^{2\varepsilon} \cap \{z \in \mathbf{C} : |z| \geq \varepsilon\} = \{z \in \mathbf{C} : |z| \geq \varepsilon, |\arg z| \leq \omega - 2\varepsilon\} .$$

Then, combining estimates (2.20), (2.25) and (2.26), we obtain that

$$\left\|e^{\lambda z} R(\lambda)\right\| \leq \frac{M_\varepsilon}{|\lambda|} e^{-\varepsilon \sin\varepsilon \cdot |\lambda|}, \quad \lambda \in \Gamma^{(1)} \cup \Gamma^{(3)}, \; z \in K_\omega^\varepsilon . \tag{2.27}$$

On the other hand, we have

$$\left\|e^{\lambda z} R(\lambda)\right\| \leq M_\varepsilon e^{|z|}, \quad \lambda \in \Gamma^{(2)}, \; z \in K_\omega^\varepsilon . \tag{2.28}$$

Therefore, we find that the integral

$$U_z = -\frac{1}{2\pi i} \int_\Gamma e^{\lambda z} R(\lambda)\, d\lambda = -\frac{1}{2\pi i} \sum_{k=1}^{3} \int_{\Gamma^{(k)}} e^{\lambda z} R(\lambda)\, d\lambda \tag{2.21$'$}$$

converges in the Banach space $L(E, E)$, uniformly in $z \in K_\omega^\varepsilon$, for every $\varepsilon > 0$. This proves that the operator U_z is analytic in the domain $\Delta_\omega = \cup_{\varepsilon>0} K_\omega^\varepsilon$.

By the analyticity of U_z, it follows that the operators U_z also enjoy the semigroup property

$$U_{z+w} = U_z \cdot U_w, \quad z, w \in \Delta_\omega .$$

Step 2: Next, we prove that the operators U_z enjoy properties (a), (b) and (c).

(i) *Property (b)*: By using Cauchy's theorem, we obtain that

$$U_z = -\frac{1}{2\pi i} \int_\Gamma e^{\lambda z} R(\lambda)\, d\lambda = -\frac{1}{2\pi i} \int_{\Gamma_{|z|}} e^{\lambda z} R(\lambda)\, d\lambda ,$$

where $\Gamma_{|z|}$ is a path consisting of the following three curves $\Gamma_{|z|}^{(1)}$, $\Gamma_{|z|}^{(2)}$ and $\Gamma_{|z|}^{(3)}$ (see Fig. 2.6):

$$\Gamma_{|z|}^{(1)} = \left\{ r e^{-i(\pi/2+\omega-\varepsilon)} : \frac{1}{|z|} \leq r < \infty \right\} ,$$

$$\Gamma_{|z|}^{(2)} = \left\{ \frac{1}{|z|} e^{i\theta} : -\left(\frac{\pi}{2} + \omega - \varepsilon\right) \leq \theta \leq \frac{\pi}{2} + \omega - \varepsilon \right\} ,$$

$$\Gamma_{|z|}^{(3)} = \left\{ r e^{i(\pi/2+\omega-\varepsilon)} : \frac{1}{|z|} \leq r < \infty \right\} .$$

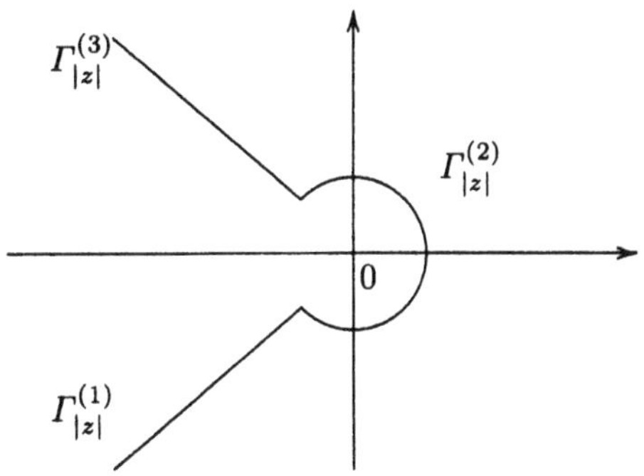

Fig. 2.6.

However, by estimates (2.20), (2.25) and (2.26) it follows that

$$\left\|e^{\lambda z}R(\lambda)\right\| \leq \frac{M_\varepsilon}{|\lambda|}e^{-|\lambda|\,|z|\sin\varepsilon}, \quad \lambda \in \Gamma^{(1)}_{|z|} \cup \Gamma^{(3)}_{|z|},\ z \in \Delta^{2\varepsilon}_{\omega}\ .$$

Hence we have, for $k = 1, 3$,

$$\begin{aligned}\int_{\Gamma^{(k)}_{|z|}} \left\|e^{\lambda z}R(\lambda)\right\| |d\lambda| &\leq M_\varepsilon \int_{1/|z|}^{\infty} e^{-\rho|z|\sin\varepsilon}\rho^{-1}\,d\rho \\ &= M_\varepsilon \int_1^\infty e^{-\sin\varepsilon\cdot s}s^{-1}\,ds\ .\end{aligned}$$

We also have, for $k = 2$,

$$\begin{aligned}\int_{\Gamma^{(2)}_{|z|}} \left\|e^{\lambda z}R(\lambda)\right\| |d\lambda| &\leq M_\varepsilon \int_{-(\pi/2+\omega-\varepsilon)}^{\pi/2+\omega-\varepsilon} e\,d\theta \\ &= 2eM_\varepsilon\left(\frac{\pi}{2}+\omega-\varepsilon\right) \\ &\leq 2\pi e M_\varepsilon\ .\end{aligned}$$

Summing up, we obtain the following estimate:

$$\begin{aligned}\|U_z\| &\leq \frac{1}{2\pi}\sum_{k=1}^{3}\int_{\Gamma^{(k)}_{|z|}} \left\|e^{\lambda z}R(\lambda)\right\| |d\lambda| \\ &\leq \frac{1}{2\pi}\left(2M_\varepsilon\int_1^\infty s^{-1}e^{-\sin\varepsilon\cdot s}\,ds + 2\pi e M_\varepsilon\right) \\ &= \frac{M_\varepsilon}{\pi}\left(\int_1^\infty s^{-1}e^{-\sin\varepsilon\cdot s}\,ds + \pi e\right)\ .\end{aligned}$$

This proves estimate (2.23), with

$$M_0(\varepsilon) = \frac{M_\varepsilon}{\pi} \left(\int_1^\infty s^{-1} e^{-\sin \varepsilon \cdot s} \, ds + \pi e \right) .$$

To prove estimate (2.24), note that

$$AR(\lambda) = (A - \lambda I + \lambda I) R(\lambda) = I + \lambda R(\lambda) ,$$

so that, by estimate (2.20),

$$\|AR(\lambda)\| \leq 1 + M_\varepsilon, \quad \lambda \in \Sigma_\omega^\varepsilon .$$

Hence, arguing as in the proof of estimate (2.23), we obtain that

$$\begin{aligned} \left\| \int_\Gamma e^{\lambda z} AR(\lambda) \, d\lambda \right\| &\leq 2 \int_{1/|z|}^\infty e^{-\rho |z| \sin \varepsilon} (1 + M_\varepsilon) \, d\rho \\ &\quad + \int_{-(\pi/2+\omega-\varepsilon)}^{\pi/2+\omega-\varepsilon} (1 + M_\varepsilon) \, e \frac{1}{|z|} \, d\theta \\ &\leq 2 (1 + M_\varepsilon) \left(\int_1^\infty e^{-\sin \varepsilon \cdot s} \, ds + \pi e \right) \frac{1}{|z|} . \end{aligned} \tag{2.29}$$

This proves that the integral $\int_\Gamma e^{\lambda z} AR(\lambda) \, d\lambda$ is convergent for every $z \in \Delta_\omega^{2\varepsilon}$. By the closedness of A, it follows that, for all $z \in \Delta_\omega^{2\varepsilon}$,

$$U_z x \in \mathcal{D}(A), \quad x \in E ,$$

and

$$AU_z x = -\frac{1}{2\pi i} \int_\Gamma e^{\lambda z} AR(\lambda) x \, d\lambda . \tag{2.30}$$

Therefore, the desired estimate (2.24) follows from estimate (2.29), with

$$M_1(\varepsilon) = \frac{1 + M_\varepsilon}{\pi} \left(\int_1^\infty e^{-\sin \varepsilon \cdot s} \, ds + \pi e \right) .$$

We find that formula (2.30) remains valid for all $z \in \Delta_\omega$, since $\Delta_\omega = \cup_{\varepsilon > 0} \Delta_\omega^{2\varepsilon}$.

(ii) *Property (a)*: By estimates (2.27) and (2.28), we can differentiate formula (2.21′) under the integral sign to obtain that

$$\frac{dU_z}{dz} = -\frac{1}{2\pi i} \int_\Gamma e^{\lambda z} \lambda R(\lambda) \, d\lambda, \quad z \in \Delta_\omega . \tag{2.31}$$

On the other hand, it follows from formula (2.30) that

$$\begin{aligned} AU_z &= -\frac{1}{2\pi i} \int_\Gamma e^{\lambda z} AR(\lambda) \, d\lambda \\ &= -\frac{1}{2\pi i} \int_\Gamma e^{\lambda z} (I + \lambda R(\lambda)) \, d\lambda \\ &= -\frac{1}{2\pi i} \int_\Gamma e^{\lambda z} \lambda R(\lambda) \, d\lambda, \quad z \in \Delta_\omega , \end{aligned} \tag{2.32}$$

since we have, by Cauchy's theorem,

$$\frac{1}{2\pi i}\int_\Gamma e^{\lambda z}\,d\lambda = 0\,,$$

just as in the proof of Proposition 2.10.

Therefore, the desired relation (2.22) follows from formulas (2.31) and (2.32).

(iii) *Property (c)*: Now let x_0 be an arbitrary element of $\mathcal{D}(A)$. First, we have

$$\lambda R(\lambda)x_0 = AR(\lambda)x_0 - x_0 = R(\lambda)Ax_0 - x_0\,,$$

and so

$$R(\lambda)x_0 + \frac{1}{\lambda}x_0 = \frac{1}{\lambda}R(\lambda)Ax_0\,.$$

Moreover, we have, by the residue theorem,

$$x_0 = \frac{1}{2\pi i}\int_\Gamma \frac{e^{\lambda z}}{\lambda}x_0\,d\lambda\,.$$

Hence it follows that

$$\begin{aligned} U_z x_0 - x_0 &= -\frac{1}{2\pi i}\int_\Gamma e^{\lambda z}\left(R(\lambda)+\frac{1}{\lambda}\right)x_0\,d\lambda \\ &= -\frac{1}{2\pi i}\int_\Gamma \frac{e^{\lambda z}}{\lambda}R(\lambda)Ax_0\,d\lambda\,. \end{aligned}$$

Here it should be noticed that, by estimate (2.20),

$$\left\|\frac{1}{\lambda}R(\lambda)\right\| \le \frac{M_\varepsilon}{|\lambda|^2}, \quad \lambda\in\Gamma\,,$$

and that

$$\begin{aligned} \left|e^{\lambda z}\right| &\le 2e^{-|\lambda||z|\sin\varepsilon} + e^{|z|} \\ &\le 2+e, \quad \lambda\in\Gamma;\ z\in\Delta_\omega^{2\varepsilon},\ |z|\le 1\,. \end{aligned}$$

Thus we obtain from an application of the Lebesgue dominated convergence theorem that, as $z\to 0$, $z\in\Delta_\omega^{2\varepsilon}$,

$$U_z x_0 - x_0 \longrightarrow -\frac{1}{2\pi i}\int_\Gamma \frac{1}{\lambda}R(\lambda)Ax_0\,d\lambda\,.$$

However, we have

$$\int_\Gamma \frac{1}{\lambda}R(\lambda)Ax_0\,d\lambda = 0\,.$$

Indeed, by Cauchy's theorem and estimate (2.20) it follows that

$$\begin{aligned} \int_\Gamma \frac{1}{\lambda}R(\lambda)Ax_0\,d\lambda &= \lim_{r\to\infty}\int_{\Gamma\cap\{|\lambda|\le r\}} \frac{1}{\lambda}R(\lambda)Ax_0\,d\lambda \\ &= -\lim_{r\to\infty}\int_{C_r}\frac{1}{\lambda}R(\lambda)Ax_0\,d\lambda \\ &= 0\,, \end{aligned}$$

where C_r is the part of the circle $\{|\lambda| = r\}$ which lies to the right of Γ shown in Fig. 2.7.

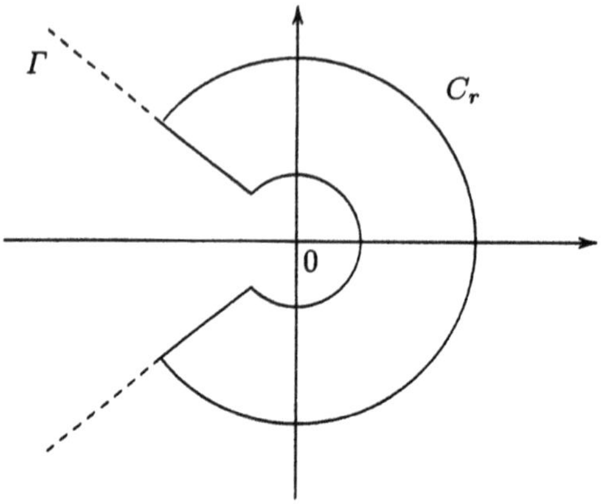

Fig. 2.7.

Summing up, we have proved that

$$U_z x_0 \longrightarrow x_0 \quad \text{as } z \to 0,\ z \in \Delta_\omega^{2\varepsilon},$$

for each $x_0 \in \mathcal{D}(A)$.

Since the domain $\mathcal{D}(A)$ is dense in E and $\|U_z\| \leq M_0(\varepsilon)$ for all $z \in \Delta_\omega^{2\varepsilon}$, it follows that, for each $x \in E$,

$$U_z x \longrightarrow x \quad \text{as } z \to 0,\ z \in \Delta_\omega^{2\varepsilon}.$$

The proof of Theorem 2.11 is now complete. □

Remark 2.2. Assume that the operator A satisfies a stronger condition than condition (2.20):

$$\|R(\lambda)\| \leq \frac{M_\varepsilon}{|\lambda| + 1}, \quad \lambda \in \Sigma_\omega^\varepsilon. \tag{2.20$'$}$$

Then we have the estimates

$$\|U_z\| \leq M_0'(\varepsilon) e^{-\delta \cdot \mathrm{Re}\, z}, \quad z \in \Delta_\omega^{2\varepsilon}, \tag{2.23$'$}$$

$$\|AU_z\| \leq \frac{M_1'(\varepsilon)}{|z|} e^{-\delta \cdot \mathrm{Re}\, z}, \quad z \in \Delta_\omega^{2\varepsilon}, \tag{2.24$'$}$$

with some constant $\delta > 0$.

Proof. Take a real number δ such that

$$0 < \delta < \frac{1}{M_\varepsilon} .$$

Then we have, by estimate (2.20′),

$$\delta \left\|(A - \lambda I)^{-1}\right\| \le \frac{\delta M_\varepsilon}{|\lambda| + 1} \le \delta M_\varepsilon < 1, \quad \lambda \in \Sigma_\omega^\varepsilon .$$

Hence it follows that the operator $(A + \delta I) - \lambda I$ has, as a Neumann series, the inverse

$$((A + \delta I) - \lambda I)^{-1} = (I + \delta(A - \lambda I)^{-1})^{-1}(A - \lambda I)^{-1} ,$$

and

$$\begin{aligned}\|((A + \delta I) - \lambda I)^{-1}\| &\le \|(I + \delta(A - \lambda I)^{-1})^{-1}\| \cdot \|(A - \lambda I)^{-1}\| \\ &\le \frac{M_\varepsilon}{|\lambda| + 1} \cdot \frac{1}{1 - \|\delta(A - \lambda I)^{-1}\|} \\ &\le \frac{M_\varepsilon}{|\lambda| + 1} \cdot \frac{1}{1 - \delta M_\varepsilon} \\ &\le \left(\frac{M_\varepsilon}{1 - \delta M_\varepsilon}\right) \frac{1}{|\lambda|} .\end{aligned}$$

This proves that the operator $A + \delta I$ satisfies condition (2.20), so that estimates (2.23) and (2.24) remain valid for the operator $A + \delta I$.

$$\|V_z\| \le M_0(\varepsilon), \quad z \in \Delta_\omega^{2\varepsilon} , \tag{2.33}$$

$$\|(A + \delta I)V_z\| \le \frac{M_1(\varepsilon)}{|z|}, \quad z \in \Delta_\omega^{2\varepsilon} , \tag{2.34}$$

where

$$V_z = -\frac{1}{2\pi i} \int_\Gamma e^{\lambda z}(A + \delta I - \lambda I)^{-1} \, d\lambda .$$

However, we have, by Cauchy's theorem,

$$\begin{aligned}V_z &= -\frac{1}{2\pi i} \int_\Gamma e^{\lambda z}(A + \delta I - \lambda I)^{-1} \, d\lambda \\ &= -\frac{1}{2\pi i} \int_{\Gamma + \delta} e^{\lambda z}(A + \delta I - \lambda I)^{-1} \, d\lambda \\ &= -\frac{1}{2\pi i} \int_\Gamma e^{\mu z} e^{\delta z}(A - \mu I)^{-1} \, d\mu = e^{\delta z} U_z .\end{aligned} \tag{2.35}$$

In view of formula (2.35), the desired estimates (2.23′) and (2.24′) follow from estimates (2.33) and (2.34). □

3

Markov Processes and Semigroups

In this chapter we introduce a class of semigroups associated with Markov processes, called Feller semigroups, and prove generation theorems for Feller semigroups (Theorems 3.3 and Theorem 3.5) which form a functional analytic background for the proof of Theorem 1.5 in Chap. 13. The results discussed here are adapted from Blumenthal-Getoor [BG], Dynkin [Dy], Lamperti [La] and Taira [Ta2] (see also Dynkin-Yushkevich [DY], Ethier-Kurtz [EK], Feller [Fe1], [Fe2], Ikeda-Watanabe [IW], Itô-McKean, Jr. [IM], Revuz-Yor [RY]).

3.1 Markov Processes

In 1828 the English botanist R. Brown observed that pollen grains suspended in water move chaotically, incessantly changing their direction of motion. The physical explanation of this phenomenon is that a single grain suffers innumerable collisions with the randomly moving molecules of the surrounding water. A mathematical theory for Brownian motion was put forward by A. Einstein in 1905. Let $p(t,x,y)$ be the probability density function that a one-dimensional Brownian particle starting at position x will be found at position y at time t. Einstein derived the following formula from statistical mechanical considerations:

$$p(t,x,y) = \frac{1}{\sqrt{2\pi Dt}} \exp\left[-\frac{(y-x)^2}{2Dt}\right] .$$

Here D is a positive constant determined by the radius of the particle, the interaction of the particle with surrounding molecules, temperature and the Boltzmann constant. This gives an accurate method of measuring Avogadro's number by observing particles. Einstein's theory was experimentally tested by J. Perrin between 1906 and 1909.

Brownian motion was put on a firm mathematical foundation for the first time by N. Wiener in 1923. Let Ω be the space of continuous functions $\omega\colon [0,\infty) \mapsto \mathbf{R}$ with coordinates $x_t(\omega) = \omega(t)$ and let $\mathcal{F}$ be the smallest σ-algebra in Ω which contains all sets of the form

$$\{\omega \in \Omega: a \le x_t(\omega) < b\}\,, \quad t \ge 0,\ a < b\,.$$

Wiener constructed probability measures P_x, $x \in \mathbf{R}$, on $\mathcal{F}$ for which the following formula holds:

$$\begin{aligned}
&P_x\{\omega \in \Omega: a_1 \le x_{t_1}(\omega) < b_1, a_2 \le x_{t_2}(\omega) < b_2, \ldots, a_n \le x_{t_n}(\omega) < b_n\} \\
&= \int_{a_1}^{b_1} \int_{a_2}^{b_2} \cdots \int_{a_n}^{b_n} p(t_1, x, y_1) p(t_2 - t_1, y_1, y_2) \ldots \\
&\qquad\qquad p(t_n - t_{n-1}, y_{n-1}, y_n)\, dy_1\, dy_2 \ldots dy_n\,, \\
&\quad 0 < t_1 < t_2 < \ldots < t_n < \infty\,.
\end{aligned}$$

This formula expresses the "starting afresh" property of Brownian motion that if a Brownian particle reaches a position, then it behaves subsequently as though that position had been its initial position. The measure P_x is called the *Wiener measure* starting at x.

Let $(\Omega, \mathcal{F})$ be a measurable space. A non-negative measure P on $\mathcal{F}$ is called a *probability measure* if $P(\Omega) = 1$. The triple $(\Omega, \mathcal{F}, P)$ is called a *probability space*. The elements of Ω are known as sample points, those of $\mathcal{F}$ as events and the values $P(A)$, $A \in \mathcal{F}$, are their probabilities. An extended real-valued, $\mathcal{F}$-measurable function X on Ω is called a *random variable*. The integral

$$\int_\Omega X\, dP$$

(if it exists) is called the *expectation* of X, and is denoted by $E(X)$.

Let $(\Omega, \mathcal{F}, P)$ be a probability space, $\mathcal{G}$ a σ-algebra contained in $\mathcal{F}$ and X an integrable random variable. The *conditional expectation* of X for given $\mathcal{G}$ is any random variable Y which satisfies the following two conditions:

(i) The function Y is $\mathcal{G}$-measurable.

(ii) $\displaystyle\int_A Y\, dP = \int_A X\, dP$, $A \in \mathcal{G}$.

We can verify that conditions (i) and (ii) determine Y up to a set in $\mathcal{G}$ of measure zero. We write $Y = E(X \mid \mathcal{G})$. When X is the characteristic function χ_B of a set $B \in \mathcal{F}$, we write $P(B \mid \mathcal{G})$ instead of $E(\chi_B \mid \mathcal{G})$. The function $P(B \mid \mathcal{G})$ is called the *conditional probability* of B for given $\mathcal{G}$. This function can also be characterized as a $\mathcal{G}$-measurable function which satisfies the condition

$$P(A \cap B) = \int_A P(B \mid \mathcal{G})\, dP\,, \quad A \in \mathcal{G}\,.$$

Now let K be a locally compact, separable metric space and $\mathcal{B}$ the σ-algebra of all Borel sets in K, that is, the smallest σ-algebra containing all open sets in K. Let $(\Omega, \mathcal{F}, P)$ be a probability space. A function X defined on Ω taking values in K is called a *random variable* if it satisfies the condition

$$\{X \in E\} = X^{-1}(E) \in \mathcal{F} \quad \text{for all } E \in \mathcal{B}\,.$$

We express this by saying that X is $\mathcal{F}/\mathcal{B}$-measurable. A family $\{x_t\}_{t\geq 0}$ of random variables is called a *stochastic process*, and it may be thought of as the motion in time of a physical particle. The space K is called the *state space* and Ω the *sample space*. For a fixed $\omega \in \Omega$, the function $x_t(\omega)$, $t \geq 0$, defines in the state space K a *trajectory* or *path* of the process corresponding to the sample point ω.

In this generality the notion of a stochastic process is of course not so interesting. The most important class of stochastic processes is the class of Markov processes which is characterized by the Markov property. Intuitively, this is the principle of the lack of any "memory" in the system. Markov processes are an abstraction of the idea of Brownian motion. More precisely, (temporally homogeneous) *Markov property* is that the prediction of subsequent motion of a physical particle, knowing its position at time t, depends neither on the value of t nor on what has been observed during the time interval $[0, t)$; that is, a physical particle "starts afresh".

Now we introduce a class of Markov processes which we will deal with in this book.

Assume that we are given the following:

(1) A locally compact, separable metric space K and the σ-algebra $\mathcal{B}$ of all Borel sets in K. A point ∂ is adjoined to K as the *point at infinity* if K is not compact, and as an isolated point if K is compact (see Fig. 3.1). We let

$$K_\partial = K \cup \{\partial\},$$
$$\mathcal{B}_\partial = \text{the } \sigma\text{-algebra in } K_\partial \text{ generated by } \mathcal{B}.$$

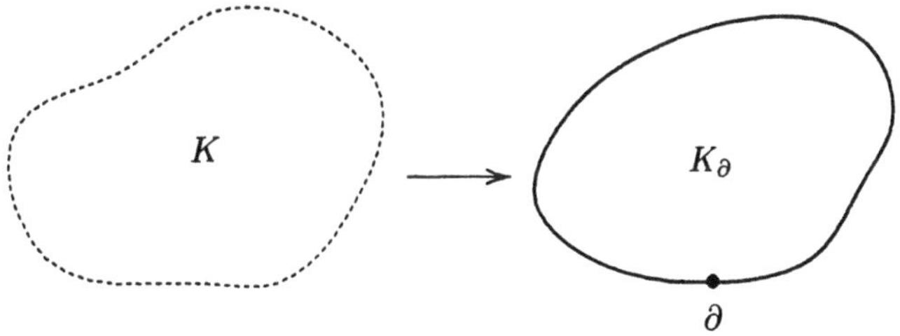

Fig. 3.1.

(2) The space Ω of all mappings $\omega\colon [0, \infty] \to K_\partial$ such that $\omega(\infty) = \partial$ and that if $\omega(t) = \partial$ then $\omega(s) = \partial$ for all $s \geq t$. Let ω_∂ be the constant map $\omega_\partial(t) = \partial$ for all $t \in [0, \infty]$.

(3) For each $t \in [0, \infty]$, the coordinate map x_t defined by $x_t(\omega) = \omega(t)$, $\omega \in \Omega$.

(4) For each $t \in [0,\infty]$, a pathwise shift mapping $\theta_t: \Omega \to \Omega$ defined by the formula $\theta_t\omega(s) = \omega(t+s)$, $\omega \in \Omega$. Note that $\theta_\infty\omega = \omega_\partial$ and $x_t \circ \theta_s = x_{t+s}$ for all t, $s \in [0,\infty]$.
(5) A σ-algebra $\mathcal{F}$ in Ω and an increasing family $\{\mathcal{F}_t\}_{0\le t\le\infty}$ of sub-σ-algebras of $\mathcal{F}$.
(6) For each $x \in K_\partial$, a probability measure P_x on $(\Omega, \mathcal{F})$.

We say that these elements define a (temporally homogeneous) *Markov process* $\mathcal{X} = (x_t, \mathcal{F}, \mathcal{F}_t, P_x)$ if the following conditions are satisfied:

(i) For each $0 \le t < \infty$, the function x_t is $\mathcal{F}_t/\mathcal{B}_\partial$-measurable, that is,

$$\{x_t \in E\} \in \mathcal{F}_t \quad \text{for all } E \in \mathcal{B}_\partial .$$

(ii) For each $0 \le t < \infty$ and $E \in \mathcal{B}$, the function

$$p_t(x,E) = P_x\{x_t \in E\} \tag{3.1}$$

is a Borel measurable function of $x \in K$.
(iii) $P_x\{\omega \in \Omega: x_0(\omega) = x\} = 1$ for each $x \in K_\partial$.
(iv) For all t, $h \in [0,\infty]$, $x \in K_\partial$ and $E \in \mathcal{B}_\partial$, we have

$$P_x\{x_{t+h} \in E \mid \mathcal{F}_t\} = p_h(x_t, E) \quad \text{a. e.},$$

or equivalently

$$P_x(A \cap \{x_{t+h} \in E\}) = \int_A p_h(x_t(\omega), E)\, dP_x(\omega) , \quad A \in \mathcal{F}_t .$$

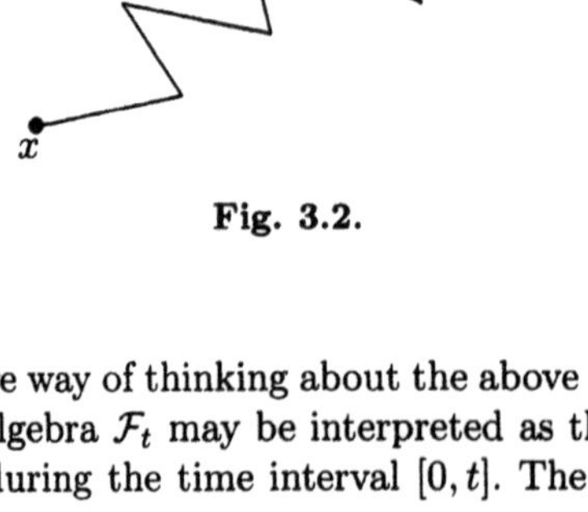

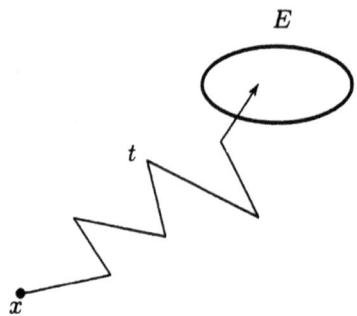

Fig. 3.2.

Here is an intuitive way of thinking about the above definition of a Markov process. The sub-σ-algebra $\mathcal{F}_t$ may be interpreted as the collection of events which are observed during the time interval $[0,t]$. The value $P_x(A)$, $A \in \mathcal{F}$, may be interpreted as the probability of the event A under the condition that a particle starts at position x; hence the value $p_t(x, E)$ expresses the transition probability that a particle starting at position x will be found in the set E at time t (see Fig. 3.2). The function $p_t(x, \cdot)$ is called the *transition function* of the process $\mathcal{X}$. The transition function $p_t(x, \cdot)$ specifies the probability structure of the process. The intuitive meaning of the crucial condition (iv) is that the future behavior of a particle, knowing its history up to time t, is the same as the behavior of a particle starting at $x_t(\omega)$, that is, a particle starts afresh.

A particle moves in the space K until it "dies" or "disappear" at which time it reaches the point ∂; hence the point ∂ is called the *terminal point* or . *cemetery*. With this interpretation in mind, we let

$$\zeta(\omega) = \inf\{t \in [0, \infty]: x_t(\omega) = \partial\} .$$

The random variable ζ is called the *lifetime* of the process $\mathcal{X}$. The process $\mathcal{X}$ is said to be *conservative* if $P_x\{\zeta = \infty\} = 1$ for all $x \in K$.

3.2 Transition Functions and Feller Semigroups

From the point of view of analysis, the transition function is something more convenient than the Markov process itself. In fact, it can be shown that the transition functions of Markov processes generate solutions of certain parabolic partial differential equations such as the classical diffusion equation; and, conversely, these differential equations can be used to construct and study the transition functions and the Markov processes themselves.

Let (K, ρ) be a locally compact, separable metric space and $\mathcal{B}$ the σ-algebra of all Borel sets in K.

A function $p_t(x, E)$, defined for all $t \geq 0$, $x \in K$ and $E \in \mathcal{B}$, is called a (temporally homogeneous) *Markov transition function* on K if it satisfies the following four conditions:

(a) $p_t(x, \cdot)$ is a non-negative measure on $\mathcal{B}$ and $p_t(x, K) \leq 1$ for each $t \geq 0$ and each $x \in K$.
(b) $p_t(\cdot, E)$ is a Borel measurable function for each $t \geq 0$ and each $E \in \mathcal{B}$.
(c) $p_0(x, \{x\}) = 1$ for each $x \in K$.
(d) (The Chapman-Kolmogorov equation) For any t, $s \geq 0$, any $x \in K$ and any $E \in \mathcal{B}$, we have

$$p_{t+s}(x, E) = \int_K p_t(x, dy) p_s(y, E) . \tag{3.2}$$

Here is an intuitive way of thinking about the above definition of a Markov transition function. The value $p_t(x, E)$ expresses the transition probability that a physical particle starting at position x will be found in the set E at

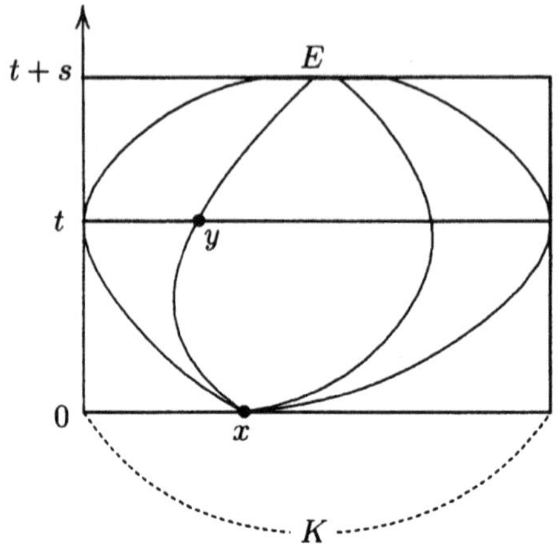

Fig. 3.3.

time t. Equation (3.2) expresses the idea that a transition from the position x to the set E in time $t+s$ is composed of a transition from x to some position y in time t, followed by a transition from y to the set E in the remaining time s; the latter transition has probability $p_s(y,E)$ which depends only on y (see Fig. 3.3). Thus a physical particle "starts afresh"; this property is called the *Markov property.*

The Chapman-Kolmogorov equation (3.2) tells us that $p_t(x,K)$ is monotonically increasing as $t \downarrow 0$, so that the limit $p_{+0}(x,K) = \lim_{t\downarrow 0} p_t(x,K)$ exists.

A Markov transition function $p_t(x,\cdot)$ is said to be *normal* if it satisfies the condition

$$p_{+0}(x,K) = \lim_{t\downarrow 0} p_t(x,K) = 1 \quad \text{for all } x \in K .$$

It is known (see [Dy, Chap. III, Sect. 2]) that, for every Markov process, the function $p_t(x,\cdot)$, defined by formula

$$p_t(x,E) = P_x\{x_t \in E\} , \quad x \in K, \;\; E \in \mathcal{B} , \;\; t \geq 0 ,$$

is a Markov transition function. Conversely, every normal Markov transition function corresponds to some Markov process.

Here are some important examples of normal transition functions on the line $\mathbf{R}$.

Example 3.1 (uniform motion). If $t \geq 0$, $x \in \mathbf{R}$ and $E \in \mathcal{B}$, we let

$$p_t(x,E) = \chi_E(x+vt) ,$$

where v is a constant, and $\chi_E(y) = 1$ if $y \in E$ and $= 0$ if $y \notin E$.

This process, starting at x, moves deterministically with constant velocity v.

Example 3.2 (Poisson process). If $t \geq 0$, $x \in \mathbf{R}$ and $E \in \mathcal{B}$, we let

$$p_t(x, E) = e^{-\lambda t} \sum_{n=0}^{\infty} \frac{(\lambda t)^n}{n!} \chi_E(x+n) ,$$

where λ is a positive constant.

This process, starting at x, advances one unit by jumps, and the probability of n jumps during the time 0 and t is equal to $e^{-\lambda t}(\lambda t)^n/n!$.

Example 3.3 (Brownian motion). If $t > 0$, $x \in \mathbf{R}$ and $E \in \mathcal{B}$, we let

$$p_t(x, E) = \frac{1}{\sqrt{2\pi t}} \int_E \exp\left[-\frac{(y-x)^2}{2t}\right] dy ,$$

and

$$p_0(x, E) = \chi_E(x) .$$

This is a mathematical model of one-dimensional Brownian motion. Its character is quite different from that of the Poisson process; the transition function $p_t(x, E)$ satisfies the condition

$$p_t(x, [x-\varepsilon, x+\varepsilon]) = 1 - o(t)$$

for every $\varepsilon > 0$ and every $x \in \mathbf{R}$. This means that the process never stands still, as does the Poisson process. In fact, this process changes state not by jumps but by *continuous* motion. A Markov process with this property is called a *diffusion process.*

Example 3.4 (Brownian motion with constant drift). If $t > 0$, $x \in \mathbf{R}$ and $E \in \mathcal{B}$, we let

$$p_t(x, E) = \frac{1}{\sqrt{2\pi t}} \int_E \exp\left[-\frac{(y-mt-x)^2}{2t}\right] dy ,$$

and

$$p_0(x, E) = \chi_E(x) ,$$

where m is a constant.

This represents Brownian motion with a constant drift of magnitude m superimposed; the process can be represented as $\{x_t + mt\}$, where $\{x_t\}$ is Brownian motion on $\mathbf{R}$.

Example 3.5 (Cauchy process). If $t > 0$, $x \in \mathbf{R}$ and $E \in \mathcal{B}$, we let

$$p_t(x, E) = \frac{1}{\pi} \int_E \frac{t}{t^2 + (y-x)^2} \, dy ,$$

and

$$p_0(x, E) = \chi_E(x) .$$

This process can be thought of as the "trace" on the real line of trajectories of two-dimensional Brownian motion, and it moves by jumps (see [Kn, Lemma 2.12]). More precisely, if $B_1(t)$ and $B_2(t)$ are two independent Brownian motions and if T is the first passage time of $B_1(t)$ to x, then $B_2(T)$ has the Cauchy density

$$\frac{1}{\pi} \frac{|x|}{x^2 + y^2} , \quad -\infty < y < \infty .$$

Here are two more examples of diffusion processes on the half line $\overline{\mathbf{R}^+} = [0, \infty)$ in which we must take account of the effect of the boundary point 0:

Example 3.6 (reflecting barrier Brownian motion). If $t > 0$, $x \in \overline{\mathbf{R}^+}$ and $E \in \mathcal{B}$, we let

$$p_t(x, E) = \frac{1}{\sqrt{2\pi t}} \left(\int_E \exp\left[-\frac{(y-x)^2}{2t} \right] dy + \int_E \exp\left[-\frac{(y+x)^2}{2t} \right] dy \right),$$

and

$$p_0(x, E) = \chi_E(x) .$$

This represents Brownian motion with a reflecting barrier at $x = 0$; the process may be represented as $\{|x_t|\}$, where $\{x_t\}$ is Brownian motion on $\mathbf{R}$. Indeed, since $\{|x_t|\}$ goes from x to y if $\{x_t\}$ goes from x to $\pm y$ due to the symmetry of the transition function in Example 3.3 about $x = 0$, it follows that

$$\begin{aligned} p_t(x, E) &= P_x\{|x_t| \in E\} \\ &= \frac{1}{\sqrt{2\pi t}} \left(\int_E \exp\left[-\frac{(y-x)^2}{2t} \right] dy + \int_E \exp\left[-\frac{(y+x)^2}{2t} \right] dy \right). \end{aligned}$$

Example 3.7 (sticking barrier Brownian motion). If $t > 0$, $x \in \overline{\mathbf{R}^+}$ and $E \in \mathcal{B}$, we let

$$\begin{aligned} p_t(x, E) = {} & \frac{1}{\sqrt{2\pi t}} \left(\int_E \exp\left[-\frac{(y-x)^2}{2t} \right] dy - \int_E \exp\left[-\frac{(y+x)^2}{2t} \right] dy \right) \\ & + \left(1 - \frac{1}{\sqrt{2\pi t}} \int_{-x}^{x} \exp\left[-\frac{z^2}{2t} \right] dz \right) \chi_E(0) , \end{aligned}$$

and

$$p_0(x, E) = \chi_E(x) .$$

This represents Brownian motion with a sticking barrier at $x = 0$. When a Brownian particle reaches the boundary point 0 for the first time, instead of reflecting it sticks there forever; in this case the state 0 is called a *trap*.

It was assumed so far that $p_t(x, K) \leq 1$ for each $t \geq 0$ and each $x \in K$. This implies that a Markovian particle may die or disappear in a finite time. A Markov transition function $p_t(x, \cdot)$ is said to be *conservative* if it satisfies the condition

$$p_t(x, K) = 1 \quad \text{for each } t \geq 0 \text{ and each } x \in K .$$

There is a simple trick which allows to turn the general case into the conservative case. We add a new point ∂ to the locally compact space K as the point at infinity if K is not compact, and as an isolated point if K is compact; so the space $K_\partial = K \cup \{\partial\}$ is compact. Then we can extend a Markov transition function $p_t(x, \cdot)$ on K to a Markov transition function $p'_t(x, \cdot)$ on K_∂ by the formulas

$$\begin{cases} p'_t(x, E) = p_t(x, E) , & x \in K , \ E \in \mathcal{B}; \\ p'_t(x, \{\partial\}) = 1 - p_t(x, K) , & x \in K; \\ p'_t(\partial, K) = 0 , \ p'_t(\partial, \{\partial\}) = 1 . \end{cases}$$

Intuitively, this means that a Markovian particle moves in the space K until it dies at which time it reaches the point ∂; hence the point ∂ is the terminal point or cemetery.

In the sequel, we will not distinguish in our notation between $p_t(x, \cdot)$ and $p'_t(x, \cdot)$; in the cases of interest for us the point ∂ will be absorbing.

Let $C(K)$ be the space of real-valued, bounded continuous functions f on K; the space $C(K)$ is a Banach space with the supremum norm

$$\|f\|_\infty = \sup_{x \in K} |f(x)| .$$

We say that a function $f \in C(K)$ converges to zero as $x \to \partial$ if, for each $\varepsilon > 0$, there exists a compact subset E of K such that

$$|f(x)| < \varepsilon , \quad x \in K \setminus E ,$$

and we then write $\lim_{x \to \partial} f(x) = 0$. We let

$$C_0(K) = \left\{ f \in C(K) : \lim_{x \to \partial} f(x) = 0 \right\} .$$

The space $C_0(K)$ is a closed subspace of $C(K)$; hence it is a Banach space. It should be emphasized that $C_0(K)$ may be identified with $C(K)$ if K is compact.

Moreover, we introduce a useful convention

Any real-valued function f on K is extended to the compact space $K_\partial = K \cup \{\partial\}$ by setting $f(\partial) = 0$.

From this point of view, the space $C_0(K)$ is identified with a subspace of the space $C(K_\partial)$ of real-valued, continuous functions on K_∂ as follows:

$$C_0(K) = \{f \in C(K_\partial): f(\partial) = 0\}.$$

Now we introduce some conditions on the measures $p_t(x, \cdot)$ related to continuity in $x \in K$, for fixed $t \geq 0$.

A Markov transition function $p_t(x, \cdot)$ is called a *Feller function* if the function

$$T_t f(x) = \int_K p_t(x, dy) f(y)$$

is a continuous function of $x \in K$ whenever f is in $C(K)$, that is, if we have

$$f \in C(K) \Longrightarrow T_t f \in C(K).$$

In other words, the Feller property is equivalent to saying that the measures $p_t(x, \cdot)$ depend continuously on $x \in K$ in the usual weak topology, for every fixed $t \geq 0$.

We say that $p_t(x, \cdot)$ is a C_0-*function* if the space $C_0(K)$ is an invariant subspace of $C(K)$ for the operators T_t:

$$f \in C_0(K) \Longrightarrow T_t f \in C_0(K).$$

The Feller or C_0-property deals with continuity of a Markov transition function $p_t(x, E)$ in x, and does not, by itself, have no concern with continuity in t. We give a necessary and sufficient condition on $p_t(x, E)$ in order that its associated operators $\{T_t\}_{t\geq 0}$ be strongly continuous in t on the space $C_0(K)$:

$$\lim_{s\downarrow 0} \|T_{t+s} f - T_t f\|_\infty = 0, \quad f \in C_0(K). \tag{3.3}$$

A Markov transition function $p_t(x, \cdot)$ on K is said to be *uniformly stochastically continuous* on K if the following condition is satisfied: For each $\varepsilon > 0$ and each compact $E \subset K$, we have

$$\lim_{t\downarrow 0} \sup_{x\in E} [1 - p_t(x, U_\varepsilon(x))] = 0, \tag{3.4}$$

where $U_\varepsilon(x) = \{y \in K: \rho(x, y) < \varepsilon\}$ is an ε-neighborhood of x.

Then we have the following:

Theorem 3.1. *Let $p_t(x, \cdot)$ be a C_0-transition function on K. Then the associated operators $\{T_t\}_{t\geq 0}$, defined by the formula*

$$T_t f(x) = \int_K p_t(x, dy) f(y), \quad f \in C_0(K), \tag{3.5}$$

are strongly continuous in t on $C_0(K)$ if and only if $p_t(x, \cdot)$ is uniformly stochastically continuous on K and satisfies the following condition:

(L) For each $s > 0$ and each compact $E \subset K$, we have

$$\lim_{x\to\partial} \sup_{0\leq t\leq s} p_t(x, E) = 0. \tag{3.6}$$

Proof. The proof is divided into two steps.

Step 1: First, we prove the "if" part of the theorem. Since continuous functions with compact support are dense in $C_0(K)$, it suffices to prove the strong continuity of $\{T_t\}$ at $t = 0$

$$\lim_{t \downarrow 0} \|T_t f - f\|_\infty = 0 \tag{3.3'}$$

for all such functions f.

For any compact subset E of K containing $\operatorname{supp} f$, we have

$$\begin{aligned}\|T_t f - f\|_\infty &\le \sup_{x \in E} |T_t f(x) - f(x)| + \sup_{x \in K \setminus E} |T_t f(x)| \\ &\le \sup_{x \in E} |T_t f(x) - f(x)| + \|f\|_\infty \sup_{x \in K \setminus E} p_t(x, \operatorname{supp} f)\,.\end{aligned} \tag{3.7}$$

However, condition (L) implies that, for each $\varepsilon > 0$, we can find a compact subset E of K such that, for all sufficiently small $t > 0$,

$$\sup_{x \in K \setminus E} p_t(x, \operatorname{supp} f) < \varepsilon\,. \tag{3.8}$$

On the other hand, we have, for each $\delta > 0$,

$$\begin{aligned}T_t f(x) - f(x) &= \int_{U_\delta(x)} p_t(x, dy)(f(y) - f(x)) \\ &\quad + \int_{K \setminus U_\delta(x)} p_t(x, dy)(f(y) - f(x)) - f(x)(1 - p_t(x, K))\,,\end{aligned}$$

and hence

$$\begin{aligned}&\sup_{x \in E} |T_t f(x) - f(x)| \\ &\le \sup_{\rho(x,y) < \delta} |f(y) - f(x)| + 3\|f\|_\infty \sup_{x \in E} [1 - p_t(x, U_\delta(x))]\,.\end{aligned}$$

Since f is uniformly continuous, we can choose a constant $\delta > 0$ such that

$$\sup_{\rho(x,y) < \delta} |f(y) - f(x)| < \varepsilon\,.$$

Furthermore, it follows from condition (3.4) with $\varepsilon := \delta$ that, for all sufficiently small $t > 0$,

$$\sup_{x \in E} [1 - p_t(x, U_\delta(x))] < \varepsilon\,.$$

Hence we have, for all sufficiently small $t > 0$,

$$\sup_{x \in E} |T_t f(x) - f(x)| < \varepsilon\,(1 + 3\|f\|_\infty)\,. \tag{3.9}$$

Therefore, carrying inequalities (3.8) and (3.9) into inequality (3.7) we obtain that, for all sufficiently small $t > 0$,

$$\|T_t f - f\|_\infty < \varepsilon\,(1 + 4\|f\|_\infty)\,.$$

This proves formula (3.3′), that is, the strong continuity of $\{T_t\}$.

Step 2: Next, we prove the "only if" part of the theorem.

(1) Let x be an arbitrary point of K. For any $\varepsilon > 0$, we define (see Fig. 3.4)

$$f_x(y) = \begin{cases} 1 - \dfrac{1}{\varepsilon}\rho(x,y) & \text{if } \rho(x,y) \leq \varepsilon\,, \\ 0 & \text{if } \rho(x,y) > \varepsilon\,. \end{cases} \tag{3.10}$$

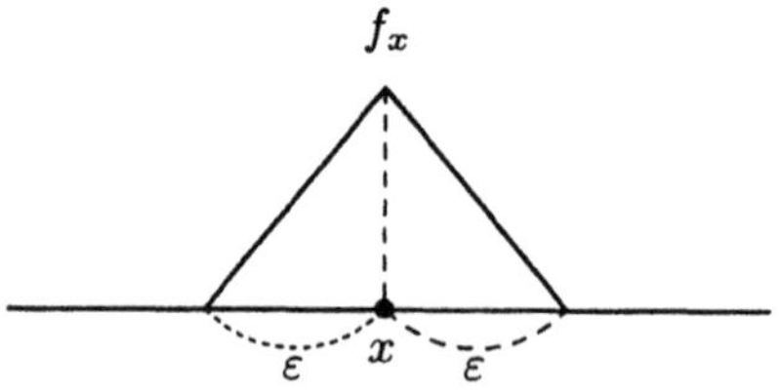

Fig. 3.4.

If E is a compact subset of K, then the functions f_x, $x \in E$, are in $C_0(K)$, for all sufficiently small $\varepsilon > 0$, and satisfy the condition

$$\|f_x - f_z\|_\infty \leq \frac{1}{\varepsilon}\rho(x,z)\,, \quad x, z \in E\,. \tag{3.11}$$

However, for any $\delta > 0$, by the compactness of E we can find a finite number of points $x_1, x_2, \ldots, x_n$ of E such that

$$E \subset \bigcup_{k=1}^{n} U_{\delta\varepsilon/4}(x_k)\,,$$

and hence

$$\min_{1 \leq k \leq n} \rho(x, x_k) \leq \frac{\delta\varepsilon}{4} \quad \text{for all } x \in E\,.$$

Thus, combining this with inequality (3.11) we obtain that

$$\min_{1 \leq k \leq n} \|f_x - f_{x_k}\|_\infty \leq \frac{\delta}{4} \quad \text{for all } x \in E\,. \tag{3.12}$$

Now we have, by formula (3.10),

$$\begin{aligned}
0 \leq 1 - p_t(x, U_\varepsilon(x)) &\leq 1 - \int_{K_\partial} p_t(x, dy) f_x(y) \\
&= f_x(x) - T_t f_x(x) \\
&\leq \|f_x - T_t f_x\|_\infty \\
&\leq \|f_x - f_{x_k}\|_\infty + \|f_{x_k} - T_t f_{x_k}\|_\infty \\
&\qquad + \|T_t f_{x_k} - T_t f_x\|_\infty \\
&\leq 2\|f_x - f_{x_k}\|_\infty + \|f_{x_k} - T_t f_{x_k}\|_\infty .
\end{aligned}$$

In view of inequality (3.12), the first term on the last inequality is bounded by $\delta/2$ for the right choice of k. Further it follows from the strong continuity (3.3′) of $\{T_t\}$ that the second term tends to zero as $t \downarrow 0$, for each $k = 1, \cdots, n$.

Consequently, we have, for all sufficiently small $t > 0$,

$$\sup_{x \in E} [1 - p_t(x, U_\varepsilon(x))] \leq \delta .$$

This proves condition (3.4), that is, the uniform stochastic continuity of $p_t(x, \cdot)$.

(2) It remains to verify condition (L). We assume, to the contrary, that:

For some $s > 0$ and some compact $E \subset K$, there exist a constant $\varepsilon_0 > 0$, a sequence $\{t_k\}$, $t_k \downarrow t$ $(0 \leq t \leq s)$ and a sequence $\{x_k\}$, $x_k \to \partial$, such that

$$p_{t_k}(x_k, E) \geq \varepsilon_0 . \tag{3.13}$$

Now we take a relatively compact subset U of K containing E, and let (see Fig. 3.5)

$$f(x) = \frac{\rho(x, K \setminus U)}{\rho(x, E) + \rho(x, K \setminus U)} .$$

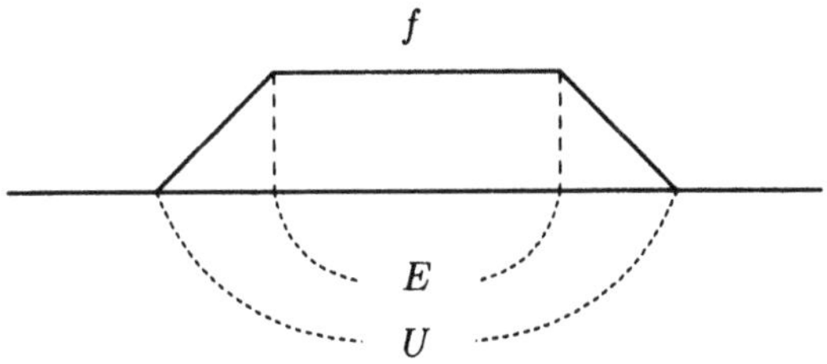

Fig. 3.5.

Then it follows that the function $f(x)$ is in $C_0(K)$ and satisfies the condition

$$T_t f(x) = \int_K p_t(x, dy) f(y) \geq p_t(x, E) \geq 0 .$$

Therefore, combining this with inequality (3.13) we obtain that

$$T_{t_k} f(x_k) \geq p_{t_k}(x_k, E) \geq \varepsilon_0 \,. \tag{3.14}$$

However, we have

$$T_{t_k} f(x_k) \leq \|T_{t_k} f - T_t f\|_\infty + T_t f(x_k) \,. \tag{3.15}$$

Since the semigroup $\{T_t\}$ is strongly continuous and $T_t f \in C_0(K)$, we can let $k \to \infty$ in inequality (3.15) to obtain that

$$\limsup_{k \to \infty} T_{t_k} f(x_k) = 0 \,.$$

This contradicts condition (3.14).

The proof of Theorem 3.1 is complete. □

A family $\{T_t\}_{t \geq 0}$ of bounded linear operators acting on the space $C_0(K)$ is called a *Feller semigroup* on K if it satisfies the following three conditions:

(i) $T_{t+s} = T_t \cdot T_s$, $t, s \geq 0$ (the semigroup property); $T_0 = I$.
(ii) The family $\{T_t\}$ is strongly continuous in t for $t \geq 0$:

$$\lim_{s \downarrow 0} \|T_{t+s} f - T_t f\|_\infty = 0 \,, \quad f \in C_0(K) \,.$$

(iii) The family $\{T_t\}$ is non-negative and contractive on $C_0(K)$:

$$f \in C_0(K) \,, \ 0 \leq f \leq 1 \quad \text{on } K \Longrightarrow 0 \leq T_t f \leq 1 \quad \text{on } K \,.$$

Rephrased, Theorem 3.1 gives a characterization of Feller semigroups in terms of Markov transition functions:

Theorem 3.2. *If $p_t(x, \cdot)$ is a uniformly stochastically continuous C_0-transition function on K and satisfies condition (L), then its associated operators $\{T_t\}_{t \geq 0}$ form a Feller semigroup on K.*

Conversely, if $\{T_t\}_{t \geq 0}$ is a Feller semigroup on K, then there exists a uniformly stochastically continuous C_0-transition $p_t(x, \cdot)$ on K, satisfying condition (L), such that formula (3.5) holds.

3.3 Generation Theorems for Feller Semigroups

If $\{T_t\}_{t \geq 0}$ is a Feller semigroup on K, we define its *infinitesimal generator* A by

$$Au = \lim_{t \downarrow 0} \frac{T_t u - u}{t} \,, \tag{3.16}$$

provided that the limit (3.16) exists in the space $C_0(K)$. More precisely, the generator A is a linear operator from $C_0(K)$ into itself defined as follows:

(1) The domain $\mathcal{D}(A)$ of A is the set

$$\mathcal{D}(A) = \{u \in C_0(K) : \text{the limit (3.16) exists in } C_0(K)\} .$$

(2) $Au = \lim_{t \downarrow 0} \frac{T_t u - u}{t}, \; u \in \mathcal{D}(A).$

The next theorem is a version of the Hille-Yosida theorem (Theorem 2.9) adapted to the present context:

Theorem 3.3. *(i) Let $\{T_t\}_{t \geq 0}$ be a Feller semigroup on K and A its infinitesimal generator. Then we have the following four assertions:*

(a) The domain $\mathcal{D}(A)$ is dense in the space $C_0(K)$.
(b) For each $\alpha > 0$, the equation $(\alpha I - A)u = f$ has a unique solution u in $\mathcal{D}(A)$ for any $f \in C_0(K)$. Hence, for each $\alpha > 0$ the Green operator $(\alpha I - A)^{-1}: C_0(K) \to C_0(K)$ can be defined by the formula

$$u = (\alpha I - A)^{-1} f , \quad f \in C_0(K) .$$

(c) For each $\alpha > 0$, the operator $(\alpha I - A)^{-1}$ is non-negative on $C_0(K)$:

$$f \in C_0(K) , \quad f \geq 0 \quad \text{on } K \Longrightarrow (\alpha I - A)^{-1} f \geq 0 \quad \text{on } K .$$

(d) For each $\alpha > 0$, the operator $(\alpha I - A)^{-1}$ is bounded on $C_0(K)$ with norm

$$\|(\alpha I - A)^{-1}\| \leq \frac{1}{\alpha} .$$

(ii) Conversely, if A is a linear operator from $C_0(K)$ into itself satisfying condition (a) and if there is a constant $\alpha_0 \geq 0$ such that, for all $\alpha > \alpha_0$, conditions (b) through (d) are satisfied, then A is the infinitesimal generator of some Feller semigroup $\{T_t\}_{t \geq 0}$ on K.

Proof. In view of Theorem 2.9, it suffices to show that the semigroup $\{T_t\}_{t \geq 0}$ is non-negative if and only if its resolvents $\{(\alpha I - A)^{-1}\}_{\alpha > \alpha_0}$ are non-negative.

The "only if" part is an immediate consequence of expression (1.12) of $(\alpha I - A)^{-1}$ in terms of the semigroup $\{T_t\}$:

$$(\alpha I - A)^{-1} = \int_0^\infty e^{-\alpha t} T_t \, dt , \quad \alpha > 0 .$$

On the other hand, the "if" part follows from expression (2.17) of the semigroup $T_t(\alpha)$ in terms of the Yosida approximation $J_\alpha = \alpha(\alpha I - A)^{-1}$:

$$T_t(\alpha) = e^{-\alpha t} \exp [\alpha t J_\alpha] = e^{-\alpha t} \sum_{n=0}^{\infty} \frac{(\alpha t)^n}{n!} J_\alpha^n ,$$

and definition (2.18) of the semigroup T_t:

$$T_t = \lim_{\alpha \to \infty} T_t(\alpha) . \quad \square$$

Corollary 3.4. *Let K be a* compact *metric space and let A be the infinitesimal generator of a Feller semigroup on K. Assume that the constant function 1 belongs to the domain $\mathcal{D}(A)$ of A and that we have, for some constant c,*

$$A1 \leq -c \quad \text{on } K . \tag{3.17}$$

Then the operator $A' = A + cI$ is the infinitesimal generator of some Feller semigroup on K.

Proof. It follows from an application of part (i) of Theorem 3.3 that, for all $\alpha > c$ the operators

$$(\alpha I - A')^{-1} = ((\alpha - c)I - A)^{-1}$$

are defined and non-negative on the whole space $C(K)$. However, in view of inequality (3.17) we obtain that

$$\alpha \leq \alpha - (A1 + c) = (\alpha I - A')1 \quad \text{on } K ,$$

so that

$$\alpha(\alpha I - A')^{-1}1 \leq (\alpha I - A')^{-1}(\alpha I - A')1 = 1 \quad \text{on } K .$$

Hence we have, for all $\alpha > c$,

$$\|(\alpha I - A')^{-1}\| = \|(\alpha I - A')^{-1}1\|_\infty \leq \frac{1}{\alpha} .$$

Therefore, applying part (ii) of Theorem 3.3 to the operator $A' = A + cI$ we find that A' is the infinitesimal generator of some Feller semigroup on K. □

Now we write down explicitly the infinitesimal generators of Feller semigroups associated with the transition functions in Examples 3.1 through 3.7.

Example 3.1 (uniform motion). $K = \mathbf{R}$ and

$$\begin{cases} \mathcal{D}(A) = \{f \in C_0(K) \cap C^1(K) : f' \in C_0(K)\} , \\ Af = vf' , \quad f \in \mathcal{D}(A) . \end{cases}$$

Example 3.2 (Poisson process). $K = \mathbf{R}$ and

$$\begin{cases} \mathcal{D}(A) = C_0(K) , \\ Af(x) = \lambda(f(x+1) - f(x)) , \quad f \in \mathcal{D}(A) . \end{cases}$$

The operator A is not "local"; the value $Af(x)$ depends on the values $f(x)$ and $f(x+1)$. This reflects the fact that the Poisson process changes state by jumps.

Example 3.3 (Brownian motion). $K = \mathbf{R}$ and

$$\begin{cases} \mathcal{D}(A) = \{f \in C_0(K) \cap C^2(K) \colon f' \in C_0(K), f'' \in C_0(K)\}, \\ Af = \dfrac{1}{2}f'', \quad f \in \mathcal{D}(A). \end{cases}$$

The operator A is "local", that is, the value $Af(x)$ is determined by the values of f in an arbitrary small neighborhood of x. This reflects the fact that Brownian motion changes state by continuous motion.

Example 3.4 (Brownian motion with constant drift). $K = \mathbf{R}$ and

$$\begin{cases} \mathcal{D}(A) = \{f \in C_0(K) \cap C^2(K) \colon f' \in C_0(K), f'' \in C_0(K)\}, \\ Af = \dfrac{1}{2}f'' + mf', \quad f \in \mathcal{D}(A). \end{cases}$$

Example 3.5 (Cauchy process). $K = \mathbf{R}$ and, the domain $\mathcal{D}(A)$ contains C^2 functions on K with compact support, and the infinitesimal generator A is of the form

$$Af(x) = \frac{1}{\pi} \int_{-\infty}^{+\infty} (f(x+y) - f(x)) \frac{dy}{y^2}.$$

The operator A is not "local", which reflects the fact that the Cauchy process changes state by jumps.

Example 3.6 (reflecting barrier Brownian motion). $K = [0, \infty)$ and

$$\begin{cases} \mathcal{D}(A) = \{f \in C_0(K) \cap C^2(K) \colon f' \in C_0(K), f'' \in C_0(K), f'(0) = 0\}, \\ Af = \dfrac{1}{2}f'', \quad f \in \mathcal{D}(A). \end{cases}$$

Example 3.7 (sticking barrier Brownian motion). $K = [0, \infty)$ and

$$\begin{cases} \mathcal{D}(A) = \{f \in C_0(K) \cap C^2(K) \colon f' \in C_0(K), f'' \in C_0(K), f''(0) = 0\}, \\ Af = \dfrac{1}{2}f'', \quad f \in \mathcal{D}(A). \end{cases}$$

Finally, here are two more examples where it is difficult to begin with a transition function and the infinitesimal generator is the basic tool of describing the process.

Example 3.8 (sticky barrier Brownian motion). $K = [0, \infty)$ and

$$\begin{cases} \mathcal{D}(A) = \{f \in C_0(K) \cap C^2(K) \colon f' \in C_0(K), \\ \qquad\qquad f'' \in C_0(K), f'(0) - \alpha f''(0) = 0\}, \\ Af = \dfrac{1}{2}f'', \quad f \in \mathcal{D}(A). \end{cases}$$

Here α is a positive constant.

This process $\{x_t\}$ may be thought of as a "combination" of the reflecting and sticking Brownian motions; the reflecting and sticking cases are formally obtained by letting $\alpha \to 0$ and $\alpha \to \infty$, respectively. Upon hitting $x = 0$, a Brownian particle leaves immediately, but it spends a positive duration of time there. It should be emphasized that the set $\{t > 0 : x_t = 0\}$ is somewhat analogous to Cantor-like sets of positive Lebesgue measure.

Example 3.9 (absorbing barrier Brownian motion). $K = [0, \infty)$ where the boundary point 0 is identified with the point at infinity ∂.

$$\begin{cases} \mathcal{D}(A) = \{f \in C_0(K) \cap C^2(K) : f' \in C_0(K)\,, f'' \in C_0(K)\,, f(0) = 0\}\,, \\ Af = \dfrac{1}{2} f''\,, \quad f \in \mathcal{D}(A)\,. \end{cases}$$

This represents Brownian motion with an absorbing barrier at $x = 0$; a Brownian particle dies at the first moment when it hits the boundary $x = 0$. Namely, the point 0 is the terminal point.

Although Theorem 3.3 tells us precisely when a linear operator A is the infinitesimal generator of some Feller semigroup, it is usually difficult to verify conditions (b) through (d). So we give useful criteria in terms of the *maximum principle* (see Sato-Ueno [SU, Theorem 1.2]):

Theorem 3.5. *Let K be a* compact *metric space. Then we have the following two assertions:*

(i) Let B be a linear operator from $C(K) = C_0(K)$ into itself, and assume that

   (α) The domain $\mathcal{D}(B)$ of B is dense in the space $C(K)$.

   (β) There exists an open and dense subset K_0 of K such that if $u \in \mathcal{D}(B)$ takes a positive *maximum at a point x_0 of K_0, then we have*

$$Bu(x_0) \leq 0\,.$$

   Then the operator B is closable in the space $C(K)$.

(ii) Let B be as in part (i), and further assume that

   (β') If $u \in \mathcal{D}(B)$ takes a positive *maximum at a point x' of K, then we have*

$$Bu(x') \leq 0\,.$$

   (γ) For some $\alpha_0 \geq 0$, the range $\mathcal{R}(\alpha_0 I - B)$ of $\alpha_0 I - B$ is dense in the space $C(K)$.

Then the minimal closed extension $\overline{B}$ of B is the infinitesimal generator of some Feller semigroup on K.

Proof. Assertion (i): It suffices to show that

$$\{u_n\} \subset \mathcal{D}(B), u_n \to 0 \text{ and } Bu_n \to v \quad \text{in } C(K) \Longrightarrow v = 0 .$$

Replacing v by $-v$ if necessary, we assume, to the contrary, that:

The function v takes a positive value at some point of K.

Then, since K_0 is open and dense in K, we can find a point x_0 of K_0, a neighborhood U of x_0 contained in K_0 and a constant $\varepsilon_0 > 0$ such that, for sufficiently large n,

$$Bu_n(x) > \varepsilon_0 , \quad x \in U . \tag{3.18}$$

On the other hand, by condition (α) there exists a function $h \in \mathcal{D}(B)$ such that

$$\begin{cases} h(x_0) > 1 , \\ h(x) < 0 , \quad x \in K \setminus U . \end{cases}$$

Therefore, since $u_n \to 0$ in $C(K)$, it follows that the function

$$u'_n(x) = u_n(x) + \frac{\varepsilon_0 h(x)}{1 + \|Bh\|_\infty}$$

satisfies the conditions

$$u'_n(x_0) = u_n(x_0) + \frac{\varepsilon_0 h(x_0)}{1 + \|Bh\|_\infty} > 0 ,$$
$$u'_n(x) = u_n(x) + \frac{\varepsilon_0 h(x)}{1 + \|Bh\|_\infty} < 0 , \quad x \in K \setminus U ,$$

if n is sufficiently large. This implies that the function $u'_n \in \mathcal{D}(B)$ takes its positive maximum at a point x'_n of $U \subset K_0$. Hence we have, by condition (β),

$$Bu'_n(x'_n) \leq 0 .$$

However it follows from inequality (3.18) that

$$Bu'_n(x'_n) = Bu_n(x'_n) + \varepsilon_0 \frac{Bh(x'_n)}{1 + \|Bh\|_\infty} > Bu_n(x'_n) - \varepsilon_0 > 0 .$$

This is a contradiction.

Assertion (ii): We apply part (ii) of Theorem 3.3 to the operator $\overline{B}$. The proof is divided into several steps.

Step 1: First, we show that

$$u \in \mathcal{D}(B) , \quad (\alpha_0 I - B)u \geq 0 \quad \text{on } K \Longrightarrow u \geq 0 \quad \text{on } K . \tag{3.19}$$

By condition (γ), we can find a function $v \in \mathcal{D}(B)$ such that

$$(\alpha_0 I - B)v \geq 1 \quad \text{on } K . \tag{3.20}$$

Then we have, for any $\varepsilon > 0$,

$$\begin{cases} u + \varepsilon v \in \mathcal{D}(B) , \\ (\alpha_0 I - B)(u + \varepsilon v) \geq \varepsilon \quad \text{on } K . \end{cases}$$

In view of condition (β'), this implies that the function $-(u + \varepsilon v)$ does not take any positive maximum on K, so that

$$u + \varepsilon v \geq 0 \quad \text{on } K .$$

Thus, letting $\varepsilon \downarrow 0$ we obtain that

$$u \geq 0 \quad \text{on } K .$$

This proves assertion (3.19).

Step 2: It follows from assertion (3.19) that the inverse $(\alpha_0 I - B)^{-1}$ of $\alpha_0 I - B$ is defined and non-negative on the range $\mathcal{R}(\alpha_0 I - B)$. Moreover, it is bounded with norm

$$\|(\alpha_0 I - B)^{-1}\| \leq \|v\|_\infty . \tag{3.21}$$

Here v is a function which satisfies condition (3.20).

Indeed, since $g = (\alpha_0 I - B)v \geq 1$ on K, it follows that, for all $f \in C(K)$

$$-\|f\|_\infty g \leq f \leq \|f\|_\infty g \quad \text{on } K .$$

Hence, by the non-negativity of $(\alpha_0 I - B)^{-1}$ we have, for all $f \in \mathcal{R}(\alpha_0 I - B)$,

$$-\|f\|_\infty v \leq (\alpha_0 I - B)^{-1} f \leq \|f\|_\infty v \quad \text{on } K .$$

This proves inequality (3.21).

Step 3: Next, we show that

$$\mathcal{R}(\alpha_0 I - \overline{B}) = C(K) . \tag{3.22}$$

Let f be an arbitrary element of $C(K)$. By condition (γ), we can find a sequence $\{u_n\}$ in $\mathcal{D}(B)$ such that $f_n = (\alpha_0 I - B)u_n \to f$ in $C(K)$. Since the inverse $(\alpha_0 I - B)^{-1}$ is bounded, it follows that, for some $u \in C(K)$,

$$u_n = (\alpha_0 I - B)^{-1} f_n \longrightarrow u \quad \text{in } C(K) ,$$

so that

$$B u_n = \alpha_0 u_n - f_n \longrightarrow \alpha_0 u - f \quad \text{in } C(K) .$$

Thus we have

$$\begin{cases} u \in \mathcal{D}(\overline{B}) , \\ \overline{B} u = \alpha_0 u - f , \end{cases}$$

and hence

$$(\alpha_0 I - \overline{B})u = f .$$

This proves assertion (3.22).

Step 4: Furthermore, we show that

$$u \in \mathcal{D}(\overline{B}) , \quad (\alpha_0 I - \overline{B})u \geq 0 \quad \text{on } K \Longrightarrow u \geq 0 \quad \text{on } K . \tag{3.19'}$$

Since $\mathcal{R}(\alpha_0 I - \overline{B}) = C(K)$, in view of the proof of assertion (3.19) it suffices to show the following:

If $u \in \mathcal{D}(\overline{B})$ takes a positive maximum at a point x' of K, then we have

$$\overline{B}u(x') \leq 0 . \tag{3.23}$$

Assume, to the contrary, that

$$\overline{B}u(x') > 0 .$$

Since there exists a sequence $\{u_n\}$ in $\mathcal{D}(B)$ such that $u_n \to u$ and $Bu_n \to \overline{B}u$ in $C(K)$, we can find a neighborhood U of x' and a constant $\varepsilon_0 > 0$ such that, for sufficiently large n,

$$Bu_n(x) > \varepsilon_0 , \quad x \in U . \tag{3.24}$$

Furthermore, by condition (α) we can find a function $h \in \mathcal{D}(B)$ such that

$$\begin{cases} h(x') > 1 , \\ h(x) < 0 , \quad x \in K \setminus U . \end{cases}$$

Then the function

$$u'_n(x) = u_n(x) + \frac{\varepsilon_0 h(x)}{1 + \|Bh\|_\infty}$$

satisfies the condition

$$\begin{cases} u'_n(x') > u(x') > 0 , \\ u'_n(x) < u(x') , \quad x \in K \setminus U , \end{cases}$$

if n is sufficiently large. This implies that the function $u'_n \in \mathcal{D}(B)$ takes its positive maximum at a point x'_n of U. Hence we have, by condition (β'),

$$Bu'_n(x'_n) \leq 0 .$$

However it follows from inequality (3.24) that

$$Bu'_n(x'_n) = Bu_n(x'_n) + \varepsilon_0 \frac{Bh(x'_n)}{1 + \|Bh\|_\infty} > Bu_n(x'_n) - \varepsilon_0 > 0 .$$

This is a contradiction.

Step 5: In view of Steps 3 and 4, we obtain that the inverse $(\alpha_0 I - \overline{B})^{-1}$ of $\alpha_0 I - \overline{B}$ is defined on the whole space $C(K)$, and is bounded with norm

$$\|(\alpha_0 I - \overline{B})^{-1}\| = \|(\alpha_0 I - \overline{B})^{-1} 1\|_\infty .$$

Step 6: Finally, we show that

For all $\alpha > \alpha_0$, the inverse $(\alpha I - \overline{B})^{-1}$ of $\alpha I - \overline{B}$ is defined on the whole space $C(K)$, and is non-negative and bounded with norm

$$\|(\alpha I - \overline{B})^{-1}\| \leq \frac{1}{\alpha} . \tag{3.25}$$

We let

$$G_{\alpha_0} = (\alpha_0 I - \overline{B})^{-1} .$$

First, we choose a constant $\alpha_1 > \alpha_0$ such that

$$(\alpha_1 - \alpha_0)\|G_{\alpha_0}\| < 1 ,$$

and let

$$\alpha_0 < \alpha \leq \alpha_1 .$$

Then, for any $f \in C(K)$, the Neumann series

$$u = \left(I + \sum_{n=1}^{\infty} (\alpha_0 - \alpha)^n G_{\alpha_0} \right) G_{\alpha_0} f$$

converges in $C(K)$, and is a solution of the equation

$$u - (\alpha_0 - \alpha) G_{\alpha_0} u = G_{\alpha_0} f .$$

Hence we have

$$\begin{cases} u \in \mathcal{D}(\overline{B}) , \\ (\alpha I - \overline{B})u = f . \end{cases}$$

This proves that

$$\mathcal{R}(\alpha I - \overline{B}) = C(K) , \quad \alpha_0 < \alpha \leq \alpha_1 . \tag{3.26}$$

Thus, arguing as in the proof of Step 1 we obtain that, for $\alpha_0 < \alpha \leq \alpha_1$,

$$u \in \mathcal{D}(\overline{B}) , \ (\alpha I - \overline{B})u \geq 0 \text{ on } K \Longrightarrow u \geq 0 \text{ on } K . \tag{3.27}$$

Combining assertions (3.26) and (3.27), we find that, for $\alpha_0 < \alpha \leq \alpha_1$, the inverse $(\alpha I - \overline{B})^{-1}$ is defined and non-negative on the whole space $C(K)$.

We let

$$G_\alpha = (\alpha I - \overline{B})^{-1} , \quad \alpha_0 < \alpha \leq \alpha_1 .$$

Then the operator G_α is bounded with norm

$$\|G_\alpha\| \leq \frac{1}{\alpha}. \tag{3.28}$$

Indeed, in view of assertion (3.23) it follows that if $u \in \mathcal{D}(\overline{B})$ takes a positive maximum at a point x' of K, then we have

$$\overline{B}u(x') \leq 0,$$

so that

$$\max_K u = u(x') \leq \frac{1}{\alpha}(\alpha I - \overline{B})u(x') \leq \frac{1}{\alpha}\|(\alpha I - \overline{B})u\|_\infty. \tag{3.29}$$

Similarly, if u takes a negative minimum at a point of K, then (replacing u by $-u$) we have

$$-\min_K u = \max_K(-u) \leq \frac{1}{\alpha}\|(\alpha I - \overline{B})u\|_\infty. \tag{3.30}$$

Inequality (3.28) follows from inequalities (3.29) and (3.30).

Summing up, we have proved assertion (3.25) for $\alpha_0 < \alpha \leq \alpha_1$.

Now assume that assertion (3.25) is proved for $\alpha_0 < \alpha \leq \alpha_{n-1}$, $n = 2, 3, \ldots$. Then, by taking

$$\alpha_n = 2\alpha_{n-1} - \frac{\alpha_1}{2},$$

or equivalently

$$\alpha_n = \left(2^{n-2} + \frac{1}{2}\right)\alpha_1,$$

we have, for $\alpha_{n-1} < \alpha \leq \alpha_n$,

$$\begin{aligned}(\alpha - \alpha_{n-1})\|G_{\alpha_{n-1}}\| &\leq \frac{\alpha - \alpha_{n-1}}{\alpha_{n-1}} \\ &\leq \frac{\alpha_n - \alpha_{n-1}}{\alpha_{n-1}} \\ &= \frac{1}{1 + 2^{2-n}} \\ &< 1.\end{aligned}$$

Hence assertion (3.25) for $\alpha_{n-1} < \alpha \leq \alpha_n$ is proved just as in the proof of assertion (3.25) for $\alpha_0 < \alpha \leq \alpha_1$. This proves assertion (3.25).

Consequently, applying part (ii) of Theorem 3.3 to the operator $\overline{B}$, we obtain that $\overline{B}$ is the infinitesimal generator of some Feller semigroup on K.

The proof of Theorem 3.5 is now complete. □

Corollary 3.6. *Let B be the infinitesimal generator of a Feller semigroup on a compact metric space K and C a bounded linear operator on $C(K)$ into itself. Assume that either C or $B + C$ satisfies condition (β'). Then the operator $A = B + C$ is the infinitesimal generator of some Feller semigroup on K.*

Proof. We apply part (ii) of Theorem 3.5 to the operator A.

First, note that $A = B + C$ is a densely defined, closed linear operator from $C(K)$ into itself. Since the operator B generates a non-negative and contractive semigroup $\{T_t\}_{t\geq 0}$, we find that if $u \in \mathcal{D}(B)$ takes a positive maximum at a point x' of K, then we have

$$Bu(x') = \lim_{t\downarrow 0} \frac{T_t u(x') - u(x')}{t} \leq 0 .$$

This implies that if C satisfies condition (β'), so does $A = B + C$.

We let

$$G_{\alpha_0} = (\alpha_0 I - B)^{-1} , \quad \alpha_0 > 0 .$$

If α_0 is so large that

$$\|G_{\alpha_0} C\| \leq \|G_{\alpha_0}\| \|C\| \leq \frac{\|C\|}{\alpha_0} < 1 ,$$

then the Neumann series

$$u = \left(I + \sum_{n=1}^{\infty} (G_{\alpha_0} C)^n \right) G_{\alpha_0} f$$

converges in $C(K)$ for any $f \in C(K)$, and is a solution of the equation

$$u - G_{\alpha_0} C u = G_{\alpha_0} f .$$

Hence we have

$$\begin{cases} u \in \mathcal{D}(B) = \mathcal{D}(A) , \\ (\alpha_0 I - A) u = f . \end{cases}$$

This proves that

$$\mathcal{R}(\alpha_0 I - A) = C(K) .$$

Therefore, applying part (ii) of Theorem 3.5 to the operator A we obtain that A is the infinitesimal generator of some Feller semigroup on K. □

3.4 Borel Kernels and the Maximum Principle

In this final subsection, following Bony-Courrège-Priouret [BCP] we state (without proof) useful characterizations of linear operators which satisfy the positive maximum principle related to condition (β') in Theorem 3.5.

Let Ω be an open subset of $\mathbf{R}^n$, and let

$$B_{\text{loc}}(\Omega) = \text{the space of Borel-measurable functions in } \Omega \text{ which are bounded on compact subsets of } \Omega.$$

Let $\mathcal{B}$ be the σ-algebra of all Borel sets in Ω. A *positive Borel kernel* on Ω is a mapping

$$x \longmapsto s(x, dy)$$

of Ω into the space of non-negative measures on $\mathcal{B}$ such that, for each $X \in \mathcal{B}$, the function $s(\cdot, X)$ is Borel-measurable in Ω.

Now we assume that a positive Borel kernel $s(x, dy)$ satisfies the following two conditions:

(NS1) $s(x, \{x\}) = 0$ for all $x \in \Omega$.
(NS2) For all non-negative functions $f(x)$ in $C_0(\Omega)$, the function

$$x \longmapsto \int_\Omega s(x, dy)|y - x|^2 f(y), \quad x \in \Omega,$$

belongs to the space $B_{\mathrm{loc}}(\Omega)$. Here $C_0(\Omega)$ is the space of functions in $C(\Omega)$ with compact support in Ω (see Sect. 4.2 in Chap. 4).

A *local unity function* on Ω is a smooth function $\sigma(x, y)$ in $\Omega \times \Omega$ which satisfies the following three conditions:

(a) $0 \leq \sigma(x, y) \leq 1$ in $\Omega \times \Omega$.
(b) $\sigma(x, y) = 1$ in a neighborhood of the diagonal $\Delta_\Omega = \{(x, x) : x \in \Omega\}$ in $\Omega \times \Omega$.
(c) For any compact subset K of Ω, there exists a compact subset K' of Ω such that the functions $\sigma_x(\cdot) = \sigma(x, \cdot)$, $x \in K$, have their support in K'.

Then, using Taylor's formula and condition (NS2) we can define a linear operator

$$S : C_0^2(\Omega) \longrightarrow B_{\mathrm{loc}}(\Omega)$$

by the formula

$$Su(x) = \int_\Omega s(x, dy)\left[u(y) - \sigma(x, y)\left(u(x) + \sum_{i=1}^n \frac{\partial u}{\partial x_i}(x)(y_i - x_i)\right)\right],$$
$$u \in C_0^2(\Omega). \tag{3.31}$$

Here $C_0^2(\Omega)$ is the space of functions in $C^2(\Omega)$ with compact support in Ω (see Sect. 4.2).

The next three theorems give useful characterizations of linear operators A from $C_0^2(\Omega)$ into $B_{\mathrm{loc}}(\Omega)$ which satisfy the positive maximum principle (cf. [BCP, Théorème I, Théorème II and Théorème III]):

Theorem 3.7. *Let A be a linear operator from $C_0^2(\Omega)$ into $B_{\mathrm{loc}}(\Omega)$. Then the following two assertions are equivalent:*

(i) $A : C_0^2(\Omega) \to B_{\mathrm{loc}}(\Omega)$ is continuous and satisfies the condition

$$x \in \Omega,\ u \in C_0^2(\Omega),\ u \geq 0 \text{ in } \Omega \text{ and } x \notin \operatorname{supp} u \Longrightarrow Au(x) \geq 0. \tag{3.32}$$

(ii) *There exist a second-order differential operator* $P: C^2(\Omega) \longrightarrow B_{\text{loc}}(\Omega)$ *and a positive Borel kernel* $s(x, dy)$, *having properties (NS1) and (NS2), such that the operator* A *is written in the form*

$$Au(x) = Pu(x) + Su(x)\,, \quad x \in \Omega\,, u \in C_0^2(\Omega)\,. \tag{3.33}$$

Theorem 3.8. *Let* $\mathcal{V}$ *be a linear subspace of* $C_0^2(\Omega)$ *which contains* $C_0^\infty(\Omega)$. *Assume that* A *is a linear operator from* $\mathcal{V}$ *into* $B_{\text{loc}}(\Omega)$ *and satisfies the condition*

$$x \in \Omega\,, u \in \mathcal{V}\,, u \geq 0 \text{ in } \Omega \text{ and } u(x) = 0 \Longrightarrow Au(x) \geq 0\,. \tag{3.34}$$

Then the operator A *can be extended uniquely to a continuous linear operator* $A: C_0^2(\Omega) \longrightarrow B_{\text{loc}}(\Omega)$ *which still satisfies condition (3.34) for all* $u \in C_0^2(\Omega)$:

$$x \in \Omega\,, u \in C_0^2(\Omega)\,, u \geq 0 \text{ in } \Omega \text{ and } u(x) = 0 \Longrightarrow Au(x) \geq 0\,. \tag{3.34$'$}$$

Theorem 3.9. *Let* A *be a linear operator from* $C_0^2(\Omega)$ *into* $B_{\text{loc}}(\Omega)$ *of the form (3.33), that is,* $A = P + S$ *where* $P: C^2(\Omega) \longrightarrow B_{\text{loc}}(\Omega)$ *is a second-order differential operator on* Ω *and the operator* S *is defined by formula (3.31), with a positive Borel kernel* $s(x, dy)$ *having properties (NS1) and (NS2).*

Then we have the following two assertions:

(i) *The operator* A *satisfies condition (3.32) if and only if the principal symbol of* P *is non-positive.*
(ii) *The operator* A *satisfies the* positive maximum principle

$$x_0 \in \Omega\,, v \in C_0^2(\Omega) \text{ and } v(x_0) = \sup_\Omega v \geq 0 \Longrightarrow Av(x_0) \leq 0 \tag{PM}$$

if and only if the principal symbol of P *is non-positive and the following two conditions hold true:*

$$\begin{aligned} &P1(x) \leq 0\,, \quad x \in \Omega\,, \\ &A1(x) = P1(x) + \int_\Omega s(x, dy)\,[1 - \sigma(x, y)] \leq 0\,, \quad x \in \Omega\,. \end{aligned}$$

In Sect. 5.3 in Chap. 5, we shall give a precise definition of the *principal symbol* of a differential operator.

4

Theory of Distributions

In this chapter we present a brief description of the basic concepts and results of the theory of distributions or generalized functions which will be used in subsequent chapters. Distribution theory has become a convenient tool in the study of partial differential equations. Many problems in partial differential equations can be formulated in terms of abstract operators acting between suitable spaces of distributions, and these operators are then analyzed by the methods of functional analysis.

For detailed studies of distributions, the reader might be referred to Gel'fand–Shilov [GS], Hörmander [Ho3] and Treves [Tv].

4.1 Notation

Let $\mathbf{R}^n$ be the n-dimensional Euclidean space. We use the conventional notation

$$x = (x_1, x_2, \ldots, x_n) .$$

If $x = (x_1, x_2, \ldots, x_n)$ and $y = (y_1, y_2, \ldots, y_n)$ are points in $\mathbf{R}^n$, we set

$$x \cdot y = \sum_{j=1}^{n} x_j y_j ,$$

$$|x| = \left(\sum_{j=1}^{n} x_j^2 \right)^{1/2} .$$

Let $\alpha = (\alpha_1, \alpha_2, \ldots, \alpha_n)$ be an n-tuple of non-negative integers. Such an n-tuple α is called a *multi-index*. We let

$$|\alpha| = \alpha_1 + \alpha_2 + \cdots + \alpha_n ,$$
$$\alpha! = \alpha_1! \alpha_2! \cdots \alpha_n! .$$

If $\alpha = (\alpha_1, \alpha_2, \ldots, \alpha_n)$ and $\beta = (\beta_1, \beta_2, \ldots, \beta_n)$ are multi-indices, we define

$$\alpha + \beta = (\alpha_1 + \beta_1, \alpha_2 + \beta_2, \ldots, \alpha_n + \beta_n) .$$

The notation $\alpha \leq \beta$ means that $\alpha_j \leq \beta_j$ for each $1 \leq j \leq n$. Then we let

$$\binom{\beta}{\alpha} = \frac{\beta!}{(\beta - \alpha)!} = \binom{\beta_1}{\alpha_1}\binom{\beta_2}{\alpha_2} \cdots \binom{\beta_n}{\alpha_n} .$$

We use the shorthand

$$\partial_j = \frac{\partial}{\partial x_j} ,$$
$$D_j = \frac{1}{i}\frac{\partial}{\partial x_j} \quad (i = \sqrt{-1})$$

for derivatives on $\mathbf{R}^n$. Higher-order derivatives are expressed by multi-indices as follows:

$$\partial^\alpha = \partial_1^{\alpha_1} \partial_2^{\alpha_2} \cdots \partial_n^{\alpha_n} ,$$
$$D^\alpha = D_1^{\alpha_1} D_2^{\alpha_2} \cdots D_n^{\alpha_n} .$$

Similarly, if $x = (x_1, x_2, \cdots, x_n) \in \mathbf{R}^n$, we write

$$x^\alpha = x_1^{\alpha_1} x_2^{\alpha_2} \cdots x_n^{\alpha_n} .$$

4.2 L^p Spaces

Let Ω be an open subset of $\mathbf{R}^n$. Two Lebesgue measurable functions f, g on Ω are said to be *equivalent* if they are equal almost everywhere in Ω with respect to the Lebesgue measure dx, that is, if $f(x) = g(x)$ for all x outside a set of Lebesgue measure zero. This is obviously an equivalence relation.

If $1 \leq p < \infty$, we let

$$L^p(\Omega) = \text{the space of equivalence classes of Lebesgue measurable functions } f \text{ on } \Omega \text{ such that } |f|^p \text{ is integrable on } \Omega .$$

The space $L^p(\Omega)$ is a Banach space with the norm

$$\|f\|_p = \left(\int_\Omega |f(x)|^p \, dx\right)^{1/p} .$$

Furthermore, the space $L^2(\Omega)$ is a Hilbert space with the inner product

$$(f, g) = \int_\Omega f(x)\overline{g(x)} \, dx .$$

A Lebesgue measurable function f on Ω is said to be *essentially bounded* if there exists a constant $C > 0$ such that $|f(x)| \leq C$ almost everywhere (a. e.) in Ω. We define

$$\operatorname{ess\,sup}_{x\in\Omega}|f(x)| = \inf\{C: |f(x)| \leq C \quad \text{a. e. in } \Omega\} .$$

For $p = \infty$, we let

$$L^\infty(\Omega) = \text{ the space of equivalence classes of essentially bounded, Lebesgue measurable functions on } \Omega .$$

The space $L^\infty(\Omega)$ is a Banach space with the norm

$$\|f\|_\infty = \operatorname{ess\,sup}_{x\in\Omega}|f(x)| .$$

If $1 < p < \infty$, we let $p' = p/(p-1)$, so that $1 < p' < \infty$ and

$$\frac{1}{p} + \frac{1}{p'} = 1 .$$

The number p' is called the *exponent conjugate* to p.

Recall that the most basic inequality for L^p-functions is the following:

Theorem 4.1 (Hölder's inequality). *If $1 < p < \infty$ and $f \in L^p(\Omega)$, $g \in L^{p'}(\Omega)$, then we have $fg \in L^1(\Omega)$ and the inequality*

$$\|fg\|_1 \leq \|f\|_p \|g\|_{p'} . \tag{4.1}$$

It should be noticed that inequality (4.1) holds true for the two cases $p = 1$, $p' = \infty$ and $p = \infty$, $p' = 1$. Inequality (4.1) in the case $p = p' = 2$ is referred to as *Schwarz's inequality.*

We give a general theorem about integral operators on a measure space:

Theorem 4.2 (generalized Young's inequality). *Let $(X, \mathcal{M}, \mu)$ be a measure space. Assume that $K(x, y)$ is a measurable function on the product space $X \times X$ such that*

$$\sup_{x\in X} \int_X |K(x,y)|\, d\mu(y) \leq C$$

and

$$\sup_{y\in X} \int_X |K(x,y)|\, d\mu(x) \leq C$$

where $C > 0$ is a constant. If $f \in L^p(X)$ with $1 \leq p \leq \infty$, then the function Tf, defined by the formula

$$Tf(x) = \int_X K(x,y) f(y)\, d\mu(y) ,$$

is well-defined for almost all $x \in X$, and is in $L^p(X)$.

Furthermore, we have the inequality

$$\|Tf\|_p \leq C\|f\|_p .$$

Theorem 4.2 is an immediate consequence of Fubini's theorem and Theorem 4.1.

Corollary 4.3 (Young's inequality). *If $f \in L^1(\mathbf{R}^n)$ and $g \in L^p(\mathbf{R}^n)$ with $1 \le p \le \infty$, then the function $f * g$, defined by the formula*

$$(f * g)(x) = \int_{\mathbf{R}^n} f(x-y)g(y)\,dy\,,$$

is well-defined for almost all $x \in \mathbf{R}^n$, and is in $L^p(\mathbf{R}^n)$.

Furthermore, we have the inequality

$$\|f * g\|_p \le \|f\|_1 \|g\|_p\,.$$

The function $f * g$ is called the *convolution* of f and g.

Let Ω be an open subset of $\mathbf{R}^n$. We let

$$C(\Omega) = \text{the space of continuous functions on } \Omega\,.$$

If K is a compact subset of Ω, we define a seminorm p_K on $C(\Omega)$ by the formula

$$C(\Omega) \ni \varphi \longmapsto p_K(\varphi) = \sup_{x \in K} |\varphi(x)|\,.$$

We equip the space $C(\Omega)$ with the topology defined by the family $\{p_K\}$ of seminorms where K ranges over all compact subsets of Ω.

If k is a positive integer, we let

$$C^k(\Omega) = \text{the space of } C^k \text{ functions on } \Omega\,.$$

We define a seminorm $p_{K,k}$ on $C^k(\Omega)$ by the formula

$$C^k(\Omega) \ni \varphi \longmapsto p_{K,k}(\varphi) = \sup_{\substack{x \in K \\ |\alpha| \le k}} |\partial^\alpha \varphi(x)|\,.$$

We equip the space $C^k(\Omega)$ with the topology defined by the family $\{p_{K,k}\}$ of seminorms where K ranges over all compact subsets of Ω. This is the topology of uniform convergence on compact subsets of Ω of the functions and their derivatives of order $\le k$.

We set

$$C^\infty(\Omega) = \bigcap_{k=1}^{\infty} C^k(\Omega)\,,$$

and

$$C^0(\Omega) = C(\Omega)\,.$$

Let m be a non-negative integer or $m = \infty$. Let $\{K_\ell\}$ be a sequence of compact subsets of Ω such that K_ℓ is contained in the interior of $K_{\ell+1}$ for each ℓ and $\Omega = \cup_{\ell=1}^{\infty} K_\ell$. For example, we may take

$$K_\ell = \left\{ x \in \Omega \colon |x| \leq \ell \, , \ \operatorname{dist}(x, \partial\Omega) \geq \frac{1}{\ell} \right\} .$$

Such a sequence $\{K_\ell\}$ is called an exhaustive sequence of compact subsets of Ω. It is easy to see that the countable family

$$\{p_{K_\ell, j}\}_{\substack{\ell=1,2,\ldots \\ 0 \leq j \leq m}}$$

of seminorms suffices to define the topology on $C^m(\Omega)$ and further that $C^m(\Omega)$ is complete. Hence the space $C^m(\Omega)$ is a Fréchet space.

Furthermore, we let

$$C(\overline{\Omega}) = \text{the space of functions in } C(\Omega) \text{ having continuous extensions to the closure } \overline{\Omega} \text{ of } \Omega \, .$$

If k is a positive integer, we let

$$C^k(\overline{\Omega}) = \text{the space of functions in } C^k(\Omega) \text{ all of whose derivatives of order } \leq k \text{ have continuous extensions to } \overline{\Omega} \, .$$

We set

$$C^\infty(\overline{\Omega}) = \bigcap_{k=1}^{\infty} C^\infty(\overline{\Omega}) \, ,$$

and

$$C^0(\overline{\Omega}) = C(\overline{\Omega}) \, .$$

Let m be a non-negative integer or $m = \infty$. We equip the space $C^m(\overline{\Omega})$ with the topology defined by the family $\{p_{K,j}\}$ of seminorms where K ranges over all compact subsets of $\overline{\Omega}$ and $0 \leq j \leq m$.

Let $\{F_\ell\}$ be an increasing sequence of compact subsets of $\overline{\Omega}$ such that

$$\bigcup_{\ell=1}^{\infty} F_\ell = \overline{\Omega} \, .$$

For example, we may take

$$F_\ell = \{ x \in \overline{\Omega} \colon |x| \leq \ell \} \, .$$

Such a sequence $\{F_\ell\}$ is called an exhaustive sequence of compact subsets of $\overline{\Omega}$. It is easy to see that the countable family

$$\{p_{F_\ell, j}\}_{\substack{\ell=1,2,\ldots \\ 0 \leq j \leq m}}$$

of seminorms suffices to define the topology on $C^m(\overline{\Omega})$ and further that $C^m(\overline{\Omega})$ is complete. Hence the space $C^m(\overline{\Omega})$ is a Fréchet space.

If Ω is bounded and $0 \leq m < \infty$, then the space $C^m(\overline{\Omega})$ is a Banach space with the norm

$$\|\varphi\|_{C^m(\overline{\Omega})} = \sup_{\substack{x \in \overline{\Omega} \\ |\alpha| \leq m}} |\partial^\alpha \varphi(x)| \, .$$

Let Ω be an open subset of $\mathbf{R}^n$ and let u be a continuous function on Ω. The *support* of u, denoted $\operatorname{supp} u$, is the closure in Ω of the set $\{x \in \Omega: u(x) \neq 0\}$. In other words, the support of u is the smallest closed subset of Ω outside of which u vanishes.

Let m be a non-negative integer or $m = \infty$. If K is a compact subset of Ω, we let

$$C_K^m(\Omega) = \text{the space of functions in } C^m(\Omega) \text{ with support in } K \, .$$

The space $C_K^m(\Omega)$ is a closed subspace of $C^m(\Omega)$. Further we let

$$C_0^m(\Omega) = \bigcup_{K \subset \Omega} C_K^m(\Omega) \, ,$$

where K ranges over all compact subsets of Ω, so that $C_0^m(\Omega)$ is the space of functions in $C^m(\Omega)$ with compact support in Ω. It should be emphasized that the space $C_0^m(\Omega)$ can be identified with the space of functions in $C_0^m(\mathbf{R}^n)$ with support in Ω. If $\{K_\ell\}$ is an exhaustive sequence of compact subsets of Ω, we equip the space $C_0^m(\Omega)$ with the *inductive limit topology* of the spaces $C_{K_\ell}^m(\Omega)$, that is, the strongest locally convex linear space topology such that each injection

$$C_{K_\ell}^m(\Omega) \longrightarrow C_0^m(\Omega)$$

is continuous. We can verify that this topology on $C_0^m(\Omega)$ is independent of the sequence $\{K_\ell\}$ used.

We list some basic properties of the topology on $C_0^m(\Omega)$:

(1) A sequence $\{\varphi_j\}$ in $C_0^m(\Omega)$ converges to an element φ in $C_0^m(\Omega)$ if and only if the functions φ_j and φ are supported in a *common* compact subset K of Ω and $\varphi_j \to \varphi$ in $C_K^m(\Omega)$.
(2) A subset of $C_0^m(\Omega)$ is bounded if and only if it is bounded in $C_K^m(\Omega)$ for some compact $K \subset \Omega$.
(3) A linear mapping of $C_0^m(\Omega)$ into a linear topological space is continuous if and only if its restriction to $C_K^m(\Omega)$ for every compact $K \subset \Omega$ is continuous.

The elements of $C_0^\infty(\Omega)$ are often called *test functions*.

Let D be a subset of $\mathbf{R}^n$ and let $0 < \theta < 1$. A function φ defined on D is said to be *Hölder continuous* with exponent θ if the quantity

$$[\varphi]_{\theta;D} = \sup_{\substack{x,y \in D \\ x \neq y}} \frac{|\varphi(x) - \varphi(y)|}{|x-y|^\theta}$$

is finite. We say that φ is *locally Hölder continuous* with exponent θ if it is Hölder continuous with exponent θ on compact subsets of D. Hölder continuity may be viewed as a fractional differentiability.

Let Ω be an open subset of $\mathbf{R}^n$. We let

$$C^\theta(\Omega) = \text{the space of functions in } C(\Omega) \text{ which are locally Hölder continuous with exponent } \theta \text{ on } \Omega \,.$$

If k is a positive integer, we let

$$C^{k+\theta}(\Omega) = \text{the space of functions in } C^k(\Omega) \text{ all of whose } k\text{-th order derivatives are locally Hölder continuous with exponent } \theta \text{ on } \Omega \,.$$

If K is a compact subset of Ω, we define a seminorm $q_{K,k}$ on $C^{k+\theta}(\Omega)$ by the formula

$$C^{k+\theta}(\Omega) \ni \varphi \longmapsto q_{K,k}(\varphi) = \sup_{\substack{x \in K \\ |\alpha| \leq k}} |\partial^\alpha \varphi(x)| + \sup_{|\alpha|=k} [\partial^\alpha \varphi]_{\theta;K} \,.$$

It is easy to see that the Hölder space $C^{k+\theta}(\Omega)$ is a Fréchet space.

Furthermore, we let

$$C^\theta(\overline{\Omega}) = \text{the space of functions in } C(\overline{\Omega}) \text{ which are Hölder continuous with exponent } \theta \text{ on } \overline{\Omega} \,.$$

If k is a positive integer, we let

$$C^{k+\theta}(\overline{\Omega}) = \text{the space of functions in } C^k(\overline{\Omega}) \text{ all of whose } k\text{-th order derivatives are Hölder continuous with exponent } \theta \text{ on } \overline{\Omega} \,.$$

Let m be a non-negative integer. We equip the space $C^{m+\theta}(\overline{\Omega})$ with the topology defined by the family $\{q_{K,k}\}$ of seminorms where K ranges over all compact subsets of $\overline{\Omega}$. It is easy to see that the Hölder space $C^{m+\theta}(\overline{\Omega})$ is a Fréchet space.

If Ω is bounded, then $C^{m+\theta}(\overline{\Omega})$ is a Banach space with the norm

$$\begin{aligned} \|\varphi\|_{C^{m+\theta}(\overline{\Omega})} &= \|\varphi\|_{C^m(\overline{\Omega})} + \sup_{|\alpha|=m} [\partial^\alpha \varphi]_{\theta;\overline{\Omega}} \\ &= \sup_{\substack{x \in \overline{\Omega} \\ |\alpha| \leq m}} |\partial^\alpha \varphi(x)| + \sup_{|\alpha|=m} [\partial^\alpha \varphi]_{\theta;\overline{\Omega}} \,. \end{aligned}$$

Let $\rho(x)$ be a non-negative, C^∞ function on $\mathbf{R}^n$ satisfying the following two conditions:

$$\operatorname{supp} \rho = \{x \in \mathbf{R}^n : |x| \leq 1\} \,. \tag{4.2a}$$

$$\int_{\mathbf{R}^n} \rho(x)\, dx = 1 \,. \tag{4.2b}$$

For example, we may take

$$\rho(x) = \begin{cases} k \exp[-1/(1-|x|^2)] & \text{if } |x| < 1, \\ 0 & \text{if } |x| \geq 1, \end{cases}$$

where the constant factor k is so chosen that condition (3.2b) is satisfied.

For each $\varepsilon > 0$, we define

$$\rho_\varepsilon(x) = \frac{1}{\varepsilon^n} \rho\left(\frac{x}{\varepsilon}\right) ,$$

then $\rho_\varepsilon(x)$ is a non-negative, C^∞ function on $\mathbf{R}^n$, and satisfies the conditions

$$\operatorname{supp} \rho_\varepsilon = \{x \in \mathbf{R}^n : |x| \leq \varepsilon\}; \tag{4.2c}$$

$$\int_{\mathbf{R}^n} \rho_\varepsilon(x)\, dx = 1 . \tag{4.2d}$$

The functions ρ_ε are called *mollifiers.*

The next theorem shows how mollifiers can be used to approximate a function by smooth functions:

Theorem 4.4. *Let Ω be an open subset of $\mathbf{R}^n$. Then we have the following two assertions:*

(i) If $u \in L^p(\Omega)$ with $1 \leq p < \infty$ and vanishes outside a compact subset K of Ω, then $\rho_\varepsilon * u \in C_0^\infty(\Omega)$ provided that $\varepsilon < \operatorname{dist}(K, \partial\Omega)$, and $\rho_\varepsilon * u \to u$ in $L^p(\Omega)$ as $\varepsilon \downarrow 0$.
(ii) If $u \in C_0^p(\Omega)$ with $0 \leq m < \infty$, then $\rho_\varepsilon * u \in C_0^\infty(\Omega)$ provided that $\varepsilon < \operatorname{dist}(\operatorname{supp} u, \partial\Omega)$, and $\rho_\varepsilon * u \to u$ in $C_0^\infty(\Omega)$ as $\varepsilon \downarrow 0$.

The functions $\rho_\varepsilon * u$ are called *regularizations* of u.

Corollary 4.5. *The space $C_0^\infty(\Omega)$ is dense in $L^p(\Omega)$ for each $1 \leq p < \infty$.*

Corollary 4.5 is an immediate consequence of part (i) of Theorem 4.4, since L^p-functions with compact support are dense in $L^p(\Omega)$.

4.3 Distributions

Let Ω be an open subset of $\mathbf{R}^n$. A *distribution* on Ω is a continuous linear functional on $C_0^\infty(\Omega)$. The space of distributions on Ω is denoted by $\mathcal{D}'(\Omega)$. In other words, the space $\mathcal{D}'(\Omega)$ is the dual space of $C_0^\infty(\Omega)$. If $u \in \mathcal{D}'(\Omega)$ and $\varphi \in C_0^\infty(\Omega)$, we denote the action of u on φ by $\langle u, \varphi \rangle$ or sometimes by $\langle \varphi, u \rangle$.

We state useful characterizations of distributions:

Theorem 4.6. *Let u be a linear functional on $C_0^\infty(\Omega)$. Then the following three conditions are equivalent:*

(i) *The functional u is a distribution.*
(ii) *For any compact subset K of Ω, there exist a constant $C > 0$ and a non-negative integer m such that*

$$|\langle u, \varphi \rangle| \leq C p_{K,m}(\varphi)\,, \quad \varphi \in C_K^\infty(\Omega)\,,$$

where

$$p_{K,m}(\varphi) = \sup_{\substack{x \in K \\ |\alpha| \leq m}} |\partial^\alpha \varphi(x)|\,.$$

(iii) $\langle u, \varphi_j \rangle \to 0$ *whenever* $\varphi_j \to 0$ *in* $C_0^\infty(\Omega)$.

Part (ii) of Theorem 4.4 tells us that the space $C_0^\infty(\Omega)$ is a dense subspace of $C_0^m(\Omega)$ for $0 \leq m < \infty$. Also it is clear that the injection of $C_0^\infty(\Omega)$ into $C_0^m(\Omega)$ is continuous. Hence the dual space $\mathcal{D}'^m(\Omega)$ of $C_0^m(\Omega)$ can be identified with a linear subspace of $\mathcal{D}'(\Omega)$, by the identification of a continuous linear functional on $C_0^m(\Omega)$ with its restriction to $C_0^\infty(\Omega)$. The elements of $\mathcal{D}'^m(\Omega)$ are called distributions of *order* $\leq m$ on Ω. In other words, the distributions of order $\leq m$ on Ω are precisely those distributions on Ω that have continuous extensions to $C_0^m(\Omega)$.

Now we give some important examples of distributions.

Example 4.1. We let

$L^1_{\text{loc}}(\Omega)$ = the space of equivalence classes of Lebesgue measurable functions on Ω which are integrable on every compact subset of Ω.

The elements of $L^1_{\text{loc}}(\Omega)$ are called *locally integrable functions* on Ω. For example ($n = 1$), we have

$$\log |x|\,, \quad Y(x) \in L^1_{\text{loc}}(\mathbf{R})\,,$$

where $Y(x)$ is the Heaviside step function defined by the formula

$$Y(x) = \begin{cases} 1 & \text{for } x > 0, \\ 0 & \text{for } x < 0. \end{cases}$$

Every element f of $L^1_{\text{loc}}(\Omega)$ defines a distribution T_f of order zero on Ω by the formula

$$\langle T_f, \varphi \rangle = \int_\Omega f(x) \varphi(x)\, dx\,, \quad \varphi \in C_0^\infty(\Omega)\,.$$

Indeed, we have, for all $\varphi \in C_K^\infty(\Omega)$,

$$|\langle T_f, \varphi \rangle| \leq \left(\int_K |f(x)|\, dx \right) p_{K,0}(\varphi)\,.$$

Since the mapping $f \mapsto T_f$ induces an injection of $L^1_{\mathrm{loc}}(\Omega)$ into $\mathcal{D}'(\Omega)$, we can regard locally integrable functions as distributions. We say that such distributions "are" functions. In particular, the functions in $C^m(\Omega)$ $(0 \leq m \leq \infty)$ and in $L^p(\Omega)$ $(1 \leq p \leq \infty)$ are distributions on Ω.

Example 4.2. More generally, every complex Borel measure μ on Ω defines a distribution of order zero on Ω by the formula

$$\langle \mu, \varphi \rangle = \int_\Omega \varphi(x)\, d\mu(x)\,, \quad \varphi \in C_0^\infty(\Omega)\,.$$

In particular, if we take μ to be the point mass at a point x_0 of Ω, we obtain the *Dirac measure* δ_{x_0} defined by the formula

$$\langle \delta_{x_0}, \varphi \rangle = \varphi(x_0)\,, \quad \varphi \in C_0^\infty(\Omega)\,.$$

In other words, the Dirac measure δ_{x_0} is the point evaluation functional for $x_0 \in \Omega$. We denote just by δ in the case $x = 0$.

Example 4.3. Let $f(x)$ be a continuous function on $\mathbf{R}^n \setminus \{0\}$ which is positively homogeneous of degree $-n$ and has mean zero on the unit sphere Σ:

$$f(\lambda x) = \lambda^{-n} f(x)\,, \quad \lambda > 0\,, \tag{4.3a}$$

$$\int_\Sigma f(\sigma)\, d\sigma = 0\,. \tag{4.3b}$$

Here σ is the surface measure on Σ.

Then the formula

$$\langle \mathrm{v.\,p.}\, f(x), \varphi \rangle = \lim_{\varepsilon \downarrow 0} \int_{|x| > \varepsilon} f(x)\varphi(x)\, dx\,, \quad \varphi \in C_0^\infty(\mathbf{R}^n)\,,$$

defines a distribution on $\mathbf{R}^n$. Here "v.p." stands for Cauchy's "valeur principale" in French.

For example $(n = 1)$, the distribution $\mathrm{v.\,p.}(1/x)$ is defined by the formula

$$\begin{aligned} \left\langle \mathrm{v.\,p.} \frac{1}{x}, \varphi \right\rangle &= \lim_{\varepsilon \downarrow 0} \int_{|x| > \varepsilon} \frac{\varphi(x)}{x}\, dx \\ &= \int_0^\infty \frac{\varphi(x) - \varphi(-x)}{x}\, dx\,, \quad \varphi \in C_0^\infty(\mathbf{R})\,. \end{aligned}$$

We define various operations on distributions.

(a) *Restriction*: If $u \in \mathcal{D}'(\Omega)$ and V is an open subset of Ω, we define the restriction $u|_V$ to V of u by the formula

$$\langle u|_V, \varphi \rangle = \langle u, \varphi \rangle\,, \quad \varphi \in C_0^\infty(V)\,.$$

Then it follows that $u|_V \in \mathcal{D}'(V)$.

(b) *Differentiation*: The derivative $\partial^\alpha u$ of a distribution $u \in \mathcal{D}'(\Omega)$ is the distribution on Ω defined by the formula

$$\langle \partial^\alpha u, \varphi \rangle = (-1)^{|\alpha|} \langle u, \partial^\alpha \varphi \rangle \ , \quad \varphi \in C_0^\infty(\Omega) \ .$$

For example ($n = 1$), we have

$$Y(x)' = \delta \ ,$$
$$(\log |x|)' = \text{v. p.} \frac{1}{x} \ .$$

(c) *Multiplication by functions*: The product au of a function $a \in C^\infty(\Omega)$ and a distribution $u \in \mathcal{D}'(\Omega)$ is the distribution on Ω defined by the formula

$$\langle au, \varphi \rangle = \langle u, a\varphi \rangle \ , \quad \varphi \in C_0^\infty(\Omega) \ .$$

For example ($n = 1$), we have

$$x\delta(x) = 0,$$
$$x \text{ v. p.} \frac{1}{x} = 1 \ .$$

The *Leibniz formula* for the differentiation of a product remains valid:

$$D^\beta(au) = \sum_{\alpha \leq \beta} \binom{\beta}{\alpha} D^{\beta-\alpha}a \cdot D^\alpha u \ .$$

(d) We can combine operations (b) and (c). We let

$$P(x, D) = \sum_{|\alpha| \leq m} a_\alpha(x) D^\alpha \ , \quad a_\alpha \in C^\infty(\Omega)$$

be a differential operator of order m on Ω. If $u \in \mathcal{D}'(\Omega)$, we define $P(x, D)u$ by the formula

$$\langle P(x, D)u, \varphi \rangle = \left\langle u, \sum_{|\alpha| \leq m} (-1)^{|\alpha|} D^\alpha (a_\alpha \varphi) \right\rangle \ , \quad \varphi \in C_0^\infty(\Omega) \ .$$

Then it follows that $P(x, D)u \in \mathcal{D}'(\Omega)$.

(e) *Conjugation*: The conjugate $\overline{u}$ of a distribution $u \in \mathcal{D}'(\Omega)$ is the distribution on Ω defined by the formula

$$\langle \overline{u}, \varphi \rangle = \overline{\langle u, \overline{\varphi} \rangle} \ , \quad \varphi \in C_0^\infty(\Omega)$$

where $\bar{\cdot}$ denotes complex conjugation.

There are two natural topologies on the space $\mathcal{D}'(\Omega)$:

(1) *Weak* topology* τ_s: This is the topology of convergence at each element of $C_0^\infty(\Omega)$. The space $\mathcal{D}'(\Omega)$ endowed with this topology is denoted by $\mathcal{D}'(\Omega)_s$. A sequence $\{u_j\}$ of distributions converges to a distribution u in $\mathcal{D}'(\Omega)_s$ if and only if the sequence $\{\langle u_j, \varphi\rangle\}$ converges to $\langle u, \varphi\rangle$ for every $\varphi \in C_0^\infty(\Omega)$.

(2) *Strong topology* τ_b: This is the topology of uniform convergence on all bounded subsets of $C_0^\infty(\Omega)$. The space $\mathcal{D}'(\Omega)$ endowed with this topology is denoted by $\mathcal{D}'(\Omega)_b$. A sequence $\{u_j\}$ of distributions converges to a distribution u in $\mathcal{D}'(\Omega)_b$ if and only if the sequence $\{\langle u_j, \varphi\rangle\}$ converges to $\langle u, \varphi\rangle$ uniformly in φ over all bounded subsets of $C_0^\infty(\Omega)$.

We list some basic topological properties of $\mathcal{D}'(\Omega)$:

(I) In the case of a sequence of distributions, the two notions of convergence coincide, that is, $u_j \to u$ in $\mathcal{D}'(\Omega)_s$ if and only if $u_j \to u$ in $\mathcal{D}'(\Omega)_b$.

Let Ω_1 and Ω_2 be open subsets of $\mathbf{R}^{n_1}$ and $\mathbf{R}^{n_2}$, respectively and let A be a linear operator on $C_0^\infty(\Omega_2)$ into $\mathcal{D}'(\Omega_1)$. Then the continuity of A does not depend on the topology (τ_s or τ_b) on $\mathcal{D}'(\Omega_1)$. Indeed, $A: C_0^\infty(\Omega_2) \mapsto \mathcal{D}'(\Omega_1)$ is continuous if and only if its restriction to $C_{K_2}^\infty(\Omega_2)$ for every compact $K_2 \subset \Omega_2$ is continuous; so it suffices to base our reasoning on sequences.

(II) If $\{u_j\}$ is a sequence in $\mathcal{D}'(\Omega)$ and

$$\langle u, \varphi\rangle = \lim_{j\to\infty} \langle u_j, \varphi\rangle$$

exists for every $\varphi \in C_0^\infty(\Omega)$, then it follows that $u \in \mathcal{D}'(\Omega)$. Thus we have $u_j \to u$ in $\mathcal{D}'(\Omega)_s$ and hence in $\mathcal{D}'(\Omega)_b$. This is one of the important consequences of the Banach-Steinhaus theorem (see [Tv, Chap. 33]).

(III) The strong dual space of $\mathcal{D}'(\Omega)_b$ can be identified with $C_0^\infty(\Omega)$. This fact is referred to as the *reflexivity* of $C_0^\infty(\Omega)$.

Two distributions u_1 and u_2 on Ω are said to be equal in an open subset V of Ω if the restrictions $u_1|_V$ and $u_2|_V$ are equal. In particular, we have $u = 0$ in V if and only if $\langle u, \varphi\rangle = 0$ for all $\varphi \in C_0^\infty(V)$.

The local behavior of a distribution determines it completely. More precisely, we have the following:

Theorem 4.7. *The space $\mathcal{D}'(\Omega)$ has the* sheaf property; *this means the following two properties:*

(S1) If $\{V_j\}$ is an open covering of Ω and a distribution $u \in \mathcal{D}'(\Omega)$ is zero in every V_j, then $u = 0$ in Ω.

(S2) Given an open covering $\{V_j\}$ of Ω and a family of distributions $u_j \in \mathcal{D}'(V_j)$ such that $u_j = u_k$ in $V_j \cap V_k$, there exists a distribution $u \in \mathcal{D}'(\Omega)$ such that $u = u_j$ in each V_j.

If $u \in \mathcal{D}'(\Omega)$, the *support* of u is the smallest closed subset of Ω outside of which u is zero. The support of u is denoted by $\operatorname{supp} u$. We remark that if $\varphi \in C_0^\infty(\Omega)$ such that $\operatorname{supp}\varphi \cap \operatorname{supp} u = \emptyset$, then we have $\langle u, \varphi \rangle = 0$. It should be emphasized that the present definition of support coincides with the previous one if u is a continuous function on Ω.

The injection of $C_0^\infty(\Omega)$ into $C^\infty(\Omega)$ is continuous and the space $C_0^\infty(\Omega)$ is a dense subspace of $C^\infty(\Omega)$. Hence the dual space $\mathcal{E}'(\Omega)$ of $C^\infty(\Omega)$ can be identified with a linear subspace of $\mathcal{D}'(\Omega)$, by the identification of a continuous linear functional on $C^\infty(\Omega)$ with its restriction to $C_0^\infty(\Omega)$. In other words, the elements of $\mathcal{E}'(\Omega)$ are precisely those distributions that have continuous extensions to $C^\infty(\Omega)$.

More precisely, we have the following:

Theorem 4.8. *(i) The dual space $\mathcal{E}'(\Omega)$ of $C^\infty(\Omega)$ consists of those elements of $\mathcal{D}'(\Omega)$ with compact support.*
(ii) The dual space $\mathcal{E}'^m(\Omega)$ of $C_0^m(\Omega)$ $(0 \leq m < \infty)$ consists of those elements of $\mathcal{D}'^m(\Omega)$ with compact support, and $\mathcal{E}'(\Omega) = \cup_{m=0}^\infty \mathcal{E}'^m(\Omega)$.

As in the case of $\mathcal{D}'(\Omega)$, we equip the space $\mathcal{E}'(\Omega)$ with two natural topologies τ_s and τ_b, and denote $(\mathcal{E}'(\Omega), \tau_s)$ and $(\mathcal{E}'(\Omega), \tau_b)$ by $\mathcal{E}'(\Omega)_s$ and $\mathcal{E}'(\Omega)_b$ respectively. We have the same topological properties of $\mathcal{E}'(\Omega)$ as those of $\mathcal{D}'(\Omega)$.

Let Ω_1 and Ω_2 be open subsets of $\mathbf{R}^{n_1}$ and $\mathbf{R}^{n_2}$, respectively. If $\varphi \in C_0^\infty(\Omega_1)$ and $\psi \in C_0^\infty(\Omega_2)$, we define the *tensor product* $\varphi \otimes \psi$ of φ and ψ by the formula

$$(\varphi \otimes \psi)(x_1, x_2) = \varphi(x_1)\psi(x_2) .$$

It is clear that $\varphi \otimes \psi \in C_0^\infty(\Omega_1 \times \Omega_2)$. We let

$$\begin{aligned} C_0^\infty(\Omega_1) \otimes C_0^\infty(\Omega_2) = \; & \text{the space of finite combinations of the form } \varphi \otimes \psi \\ & \text{where } \varphi \in C_0^\infty(\Omega_1) \text{ and } \psi \in C_0^\infty(\Omega_2) . \end{aligned}$$

The space $C_0^\infty(\Omega_1) \otimes C_0^\infty(\Omega_2)$ is a linear subspace of $C_0^\infty(\Omega_1 \times \Omega_2)$. Further, it is sequentially dense in $C_0^\infty(\Omega_1 \times \Omega_2)$; that is, for every $\Phi \in C_0^\infty(\Omega_1 \times \Omega_2)$, there exists a sequence $\{\Phi_j\}$ in $C_0^\infty(\Omega_1) \otimes C_0^\infty(\Omega_2)$ such that $\Phi_j \longrightarrow \Phi$ in $C_0^\infty(\Omega_1 \times \Omega_2)$.

The sequential density of $C_0^\infty(\Omega_1) \otimes C_0^\infty(\Omega_2)$ in $C_0^\infty(\Omega_1 \times \Omega_2)$ allows us to obtain the following:

Theorem 4.9. *If $u \in \mathcal{D}'(\Omega_1)$ and $v \in \mathcal{D}'(\Omega_2)$, there exists a unique distribution $u \otimes v \in \mathcal{D}'(\Omega_1 \times \Omega_2)$ such that*

$$\langle u \otimes v, \Phi \rangle = \langle u, \varphi \rangle = \langle v, \psi \rangle , \quad \Phi \in C_0^\infty(\Omega_1 \times \Omega_2) ,$$

where $\varphi(x_1) = \langle v, \Phi(x_1, \cdot) \rangle$ and $\psi(x_2) = \langle u, \Phi(\cdot, x_2) \rangle$.

The distribution $u \otimes v$ is called the *tensor product* of u and v.

We list some basic properties of the tensor product:

(1) $\langle u \otimes v, \varphi \otimes \psi \rangle = \langle u, \varphi \rangle \langle v, \psi \rangle$, $\varphi \in C_0^\infty(\Omega_1)$, $\psi \in C_0^\infty(\Omega_2)$.
(2) $\operatorname{supp}(u \otimes v) = \operatorname{supp} u \times \operatorname{supp} v$.
(3) $D_{x_1}^\alpha D_{x_2}^\beta (u \otimes v) = D_{x_1}^\alpha u \otimes D_{x_2}^\beta v$.

The Young inequality (Corollary 4.3) tells us that if $u \in L^1(\mathbf{R}^n)$ and $v \in L^p(\mathbf{R}^n)$ with $1 \leq p \leq \infty$, then the convolution

$$(u * v)(x) = \int_{\mathbf{R}^n} u(x-y)v(y)\, dy$$

is well-defined for almost all $x \in \mathbf{R}^n$, and is in $L^p(\mathbf{R}^n)$. Furthermore, it follows from Fubini's theorem that

$$\langle u * v, \varphi \rangle = \int\int_{\mathbf{R}^n \times \mathbf{R}^n} u(x)v(y)\varphi(x+y)\, dxdy\,, \quad \varphi \in C_0^\infty(\mathbf{R}^n)\,.$$

We use this formula to extend the definition of convolution to the case of distributions.

Let u, $v \in \mathcal{D}'(\mathbf{R}^n)$ and assume that one of them has compact support. If $\varphi \in C_0^\infty(\mathbf{R}^n)$, then the support of the function

$$\tilde{\varphi}\colon (x, y) \longmapsto \varphi(x+y)$$

is contained in the strip

$$\{(x, y) \in \mathbf{R}^n \times \mathbf{R}^n : x + y \in \operatorname{supp} \varphi\}\,.$$

Thus it is easy to see that the intersection

$$\operatorname{supp}(u \otimes v) \cap \operatorname{supp} \tilde{\varphi}$$

is a compact subset of $\mathbf{R}^n \times \mathbf{R}^n$. We choose a function θ in $C_0^\infty(\mathbf{R}^n \times \mathbf{R}^n)$ such that $\theta = 1$ in a neighborhood of $\operatorname{supp}(u \otimes v) \cap \operatorname{supp} \tilde{\varphi}$, and define

$$\langle u \otimes v, \tilde{\varphi} \rangle = \langle u \otimes v, \theta \tilde{\varphi} \rangle\,.$$

Observe that $\langle u \otimes v, \theta \tilde{\varphi} \rangle$ is independent of the function θ chosen, and the mapping

$$C_0^\infty(\mathbf{R}^n) \ni \varphi \longmapsto \langle u \otimes v, \tilde{\varphi} \rangle$$

is continuous. This discussion justifies the following definition:

Definition 4.1. The *convolution* $u * v$ of two distributions u and v in $\mathcal{D}'(\mathbf{R}^n)$, one of which has compact support, is a distribution on $\mathbf{R}^n$ defined by the formula

$$\langle u * v, \varphi \rangle = \langle u \otimes v, \tilde{\varphi} \rangle\,, \quad \varphi \in C_0^\infty(\mathbf{R}^n)\,.$$

We state some basic facts concerning the convolution product:

(1) $u * v = v * u$.

(2) $\operatorname{supp}(u * v) \subset \operatorname{supp} u + \operatorname{supp} v = \{x + y : x \in \operatorname{supp} u, y \in \operatorname{supp} v\}$.
(3) $D^\alpha(u * v) = (D^\alpha u) * v = u * (D^\alpha v)$.
(4) If either $u \in \mathcal{D}'(\mathbf{R}^n)$, $v \in C_0^\infty(\mathbf{R}^n)$ or $u \in \mathcal{E}'(\mathbf{R}^n)$, $v \in C^\infty(\mathbf{R}^n)$, then we have $u * v \in C^\infty(\mathbf{R}^n)$ and $(u * v)(x) = \langle u_y, v(x-y)\rangle$, where u_y means that the distribution u operates on $v(x-y)$ as a function of y with x fixed.
(5) Let ρ be a non-negative, C^∞ function on $\mathbf{R}^n$ such that

$$\operatorname{supp} \rho = \{x \in \mathbf{R}^n : |x| \leq 1\} ,$$
$$\int_{\mathbf{R}^n} \rho(x)\, dx = 1 ,$$

and let

$$\rho_\varepsilon(x) = \frac{1}{\varepsilon^n} \rho\left(\frac{x}{\varepsilon}\right), \quad \varepsilon > 0.$$

If $u \in \mathcal{D}'(\mathbf{R}^n)$ (resp. $u \in \mathcal{E}'(\mathbf{R}^n)$), then it follows that the convolutions $u * \rho_\varepsilon$ are in $C^\infty(\mathbf{R}^n)$ (resp. $C_0^\infty(\mathbf{R}^n)$) and that

$$u * \rho_\varepsilon \longrightarrow u \text{in } \mathcal{D}'(\mathbf{R}^n) \text{ (resp. } \mathcal{E}'(\mathbf{R}^n)) \text{ as } \varepsilon \downarrow 0 .$$

The functions $u * \rho_\varepsilon$ are called *regularizations* of u.

4.4 The Fourier Transform

If $f \in L^1(\mathbf{R}^n)$, we define its (direct) *Fourier transform* $\widehat{f}$ by the formula

$$\widehat{f}(\xi) = \int_{\mathbf{R}^n} e^{-ix\cdot\xi} f(x)\, dx , \quad \xi = (\xi_1, \xi_2, \ldots, \xi_n) , \tag{4.4}$$

where $x \cdot \xi = x_1\xi_1 + x_2\xi_2 + \cdots + x_n\xi_n$. It follows from an application of the dominated convergence theorem that the function $\widehat{f}$ is continuous on $\mathbf{R}^n$, and further we have

$$\|\widehat{f}\|_\infty = \sup_{\mathbf{R}^n} |\widehat{f}(\xi)| \leq \|f\|_1 .$$

We also denote $\widehat{f}$ by $\mathcal{F}f$.

Similarly, if $g \in L^1(\mathbf{R}^n)$, we define the function $\check{g}$ by the formula

$$\check{g}(x) = \frac{1}{(2\pi)^n} \int_{\mathbf{R}^n} e^{ix\cdot\xi} g(\xi)\, d\xi .$$

The function $\check{g}$ is called the *inverse Fourier transform* of g. We also denote $\check{g}$ by $\mathcal{F}^* g$.

Now we introduce a subspace of $L^1(\mathbf{R}^n)$ which is invariant under the Fourier transform. We let

$$\mathcal{S}(\mathbf{R}^n) = \text{ the space of } C^\infty \text{ functions } \varphi \text{ on } \mathbf{R}^n \text{ such that,}$$
$$\text{for any non-negative integer } j \text{, the quantity}$$
$$p_j(\varphi) = \sup_{\substack{x \in \mathbf{R}^n \\ |\alpha| \le j}} \left\{ (1+|x|^2)^{j/2} |\partial^\alpha \varphi(x)| \right\}$$
$$\text{is finite.}$$

The space $\mathcal{S}(\mathbf{R}^n)$ is called the *Schwartz space* or *space of C^∞ functions on $\mathbf{R}^n$ rapidly decreasing at infinity.* We equip the space $\mathcal{S}(\mathbf{R}^n)$ with the topology defined by the countable family $\{p_j\}$ of seminorms. It is easy to verify that $\mathcal{S}(\mathbf{R}^n)$ is complete; so it is a Fréchet space.

Now we give one of the most typical examples of functions in $\mathcal{S}(\mathbf{R}^n)$:

Example 4.4. For $a > 0$, we have

$$\varphi(x) = e^{-a|x|^2} \in \mathcal{S}(\mathbf{R}^n) .$$

Furthermore, it is easy to verify the following:

$$\hat{\varphi}(\xi) = \int_{\mathbf{R}^n} e^{-ix\cdot\xi} e^{-a|x|^2}\, dx = \left(\frac{\pi}{a}\right)^{n/2} e^{-\frac{|x|^2}{4a}} ,$$
$$\check{\hat{\varphi}}(x) = \frac{1}{(2\pi)^n} \int_{\mathbf{R}^n} e^{ix\cdot\xi} \hat{\varphi}(\xi)\, d\xi = \varphi(x) .$$

The next theorem summarizes the basic properties of the Fourier transform:

Theorem 4.10. *(i) The Fourier transforms $\mathcal{F}$ and $\mathcal{F}^*$ map $\mathcal{S}(\mathbf{R}^n)$ continuously into itself. Furthermore, we have, for all $\varphi \in \mathcal{S}(\mathbf{R}^n)$,*

$$\widehat{D^\alpha \varphi}(\xi) = \xi^\alpha \widehat{\varphi}(\xi) ,$$
$$D^\beta \widehat{\varphi}(\xi) = \widehat{(-x)^\beta \varphi}(\xi)$$

for all multi-indices α and β.
(ii) The Fourier transforms $\mathcal{F}$ and $\mathcal{F}^*$ are isomorphisms of $\mathcal{S}(\mathbf{R}^n)$ onto itself; more precisely, $\mathcal{F}\mathcal{F}^* = \mathcal{F}^*\mathcal{F} = I$ on $\mathcal{S}(\mathbf{R}^n)$. In particular, we have the formula

$$\varphi(x) = \frac{1}{(2\pi)^n} \int_{\mathbf{R}^n} e^{ix\cdot\xi} \widehat{\varphi}(\xi)\, d\xi , \quad \varphi \in \mathcal{S}(\mathbf{R}^n) . \tag{4.5}$$

(iii) If φ, $\psi \in \mathcal{S}(\mathbf{R}^n)$, we have the formulas

$$\int_{\mathbf{R}^n} \varphi(x) \widehat{\psi}(x)\, dx = \int_{\mathbf{R}^n} \widehat{\varphi}(\xi) \psi(\xi)\, d\xi , \tag{4.6}$$
$$\int_{\mathbf{R}^n} \varphi(x) \psi(x)\, dx = \frac{1}{(2\pi)^n} \int_{\mathbf{R}^n} \varphi(\xi) \psi(\xi)\, d\xi . \tag{4.7}$$

Formula (4.5) is called the *Fourier inversion formula* and formula (4.7) is called the *Parseval formula.*

For the spaces $C_0^\infty(\mathbf{R}^n)$, $\mathcal{S}(\mathbf{R}^n)$ and $C^\infty(\mathbf{R}^n)$, we have the following inclusions:

(i) The injection of $C_0^\infty(\mathbf{R}^n)$ into $\mathcal{S}(\mathbf{R}^n)$ is continuous and the space $C_0^\infty(\mathbf{R}^n)$ is dense in $\mathcal{S}(\mathbf{R}^n)$.

(ii) The injection of $\mathcal{S}(\mathbf{R}^n)$ into $C^\infty(\mathbf{R}^n)$ is continuous and the space $\mathcal{S}(\mathbf{R}^n)$ is dense in $C^\infty(\mathbf{R}^n)$.

Hence the dual space $\mathcal{S}'(\mathbf{R}^n)$ of $\mathcal{S}(\mathbf{R}^n)$ can be identified with a linear subspace of $\mathcal{D}'(\mathbf{R}^n)$ containing $\mathcal{E}'(\mathbf{R}^n)$, by the identification of a continuous linear functional on $\mathcal{S}(\mathbf{R}^n)$ with its restriction to $C_0^\infty(\mathbf{R}^n)$. Namely, we have

$$\mathcal{E}'(\mathbf{R}^n) \subset \mathcal{S}'(\mathbf{R}^n) \subset \mathcal{D}'(\mathbf{R}^n) .$$

The elements of $\mathcal{S}'(\mathbf{R}^n)$ are called *tempered distributions* on $\mathbf{R}^n$. In other words, the tempered distributions are precisely those distributions on $\mathbf{R}^n$ that have continuous extensions to $\mathcal{S}(\mathbf{R}^n)$.

Roughly speaking, the tempered distributions are those which grow at most polynomially at infinity, since the functions in $\mathcal{S}(\mathbf{R}^n)$ die out faster than any power of x at infinity. More precisely, we have the following examples of tempered distributions:

(1) The functions in $L^p(\mathbf{R}^n)$ $(1 \leq p \leq \infty)$ are tempered distributions.

(2) A locally integrable function on $\mathbf{R}^n$ is a tempered distribution if it grows at most polynomially at infinity.

(3) If $u \in \mathcal{S}'(\mathbf{R}^n)$ and f is a C^∞ function on $\mathbf{R}^n$ all of whose derivatives grow at most polynomially at infinity, then the product fu is a tempered distribution.

(4) Any derivative of a tempered distribution is also a tempered distribution.

Now we give some concrete and important examples of distributions in the space $\mathcal{S}'(\mathbf{R}^n)$:

Examples 4.5. (a) Dirac measures: $\delta(x)$.

(b) Riesz potentials:

$$R_\alpha(x) = \frac{\Gamma((n-\alpha)/2)}{2^\alpha \pi^{n/2}\Gamma(\alpha/2)} \frac{1}{|x|^{n-\alpha}}, \quad 0 < \alpha < n .$$

(c) Newtonian potentials:

$$N(x) = \frac{\Gamma((n-2)/2)}{4\pi^{n/2}} \frac{1}{|x|^{n-2}}, \quad n \geq 3 .$$

(d) Bessel potentials:

$$G_\alpha(x) = \frac{1}{\Gamma(\alpha/2)} \frac{1}{(4\pi)^{n/2}} \int_0^\infty e^{-t-\frac{|x|^2}{4t}} t^{\frac{\alpha-n}{2}} \frac{dt}{t}, \quad \alpha > 0 .$$

It is known (see Aronszajn–Smith [AS]) that the function $G_\alpha(x)$ is represented as follows:

$$G_\alpha(x) = \frac{1}{2^{(n+\alpha-2)/2}\pi^{n/2}\Gamma(\alpha/2)} K_{(n-\alpha)/2}(|x|)|x|^{\frac{\alpha-n}{2}} ,$$

where $K_{(n-\alpha)/2}(z)$ is the modified Bessel function of the third kind (cf. Watson [Wt]).

(e) Riesz kernels:

$$R_j(x) = i\frac{\Gamma((n+1)/2)}{\pi^{(n+1)/2}} \text{v. p.} \frac{x_j}{|x|^{n+1}} , \quad 1 \leq j \leq n .$$

The distribution v. p. $(x_j/|x|^{n+1})$ is an extension of v. p. $(1/x)$ to the n-dimensional case.

(f) In the case $n = 1$, we have

$$Y(x) , \quad \text{v. p.} \frac{1}{x} \in \mathcal{S}'(\mathbf{R}) .$$

The importance of tempered distributions lies in the fact that they have Fourier transforms.

If $u \in \mathcal{S}'(\mathbf{R}^n)$, we define its (direct) *Fourier transform* $\mathcal{F}u$ by the formula

$$\langle \mathcal{F}u, \varphi \rangle = \langle u, \mathcal{F}\varphi \rangle , \quad \varphi \in \mathcal{S}(\mathbf{R}^n) . \tag{4.8}$$

Then we have $\mathcal{F}u \in \mathcal{S}'(\mathbf{R}^n)$, since the Fourier transform

$$\mathcal{F} : \mathcal{S}(\mathbf{R}^n) \longrightarrow \mathcal{S}(\mathbf{R}^n)$$

is an isomorphism. Further, in view of formula (4.6) it follows that the above definition (4.8) agrees with definition (4.4) if $u \in \mathcal{S}(\mathbf{R}^n)$. We also denote $\mathcal{F}u$ by $\widehat{u}$.

Similarly, if $v \in \mathcal{S}'(\mathbf{R}^n)$, we define its *inverse Fourier transform* $\mathcal{F}^*v$ by the formula

$$\langle \mathcal{F}^*v, \psi \rangle = \langle v, \mathcal{F}^*\psi \rangle , \quad \psi \in \mathcal{S}(\mathbf{R}^n) .$$

The next theorem, which is a consequence of Theorem 4.10, summarizes the basic properties of Fourier transforms in the space $\mathcal{S}'(\mathbf{R}^n)$:

Theorem 4.11. *(i) The Fourier transforms $\mathcal{F}$ and $\mathcal{F}^*$ map $\mathcal{S}'(\mathbf{R}^n)$ continuously into itself. Furthermore, we have, for all $u \in \mathcal{S}'(\mathbf{R}^n)$,*

$$\mathcal{F}(D^\alpha u)(\xi) = \xi^\alpha \mathcal{F}u(\xi) ,$$
$$D_\xi^\beta(\mathcal{F}u(\xi)) = \mathcal{F}((-x)^\beta u)(\xi) .$$

(ii) The Fourier transforms $\mathcal{F}$ and $\mathcal{F}^*$ are isomorphisms of $\mathcal{S}'(\mathbf{R}^n)$ onto itself; more precisely, $\mathcal{F}\mathcal{F}^* = \mathcal{F}^*\mathcal{F} = I$ on $\mathcal{S}'(\mathbf{R}^n)$.

(iii) The transforms $\mathcal{F}$ and $\mathcal{F}^*$ are norm-preserving operators on $L^2(\mathbf{R}^n)$ and $\mathcal{F}\mathcal{F}^* = \mathcal{F}^*\mathcal{F} = I$ on $L^2(\mathbf{R}^n)$.

Assertion (iii) is referred to as the *Plancherel theorem.*

We can calculate explicitly the Fourier transform of the tempered distributions in Examples 4.5 as follows:

Examples 4.6. (a) Dirac measures: $\widehat{\delta}(\xi) = 1$.
(b) Riesz potentials:

$$\widehat{R_\alpha}(\xi) = \frac{1}{|\xi|^\alpha}, \quad 0 < \alpha < n.$$

(c) Newtonian potentials:

$$\widehat{N}(\xi) = \frac{1}{|\xi|^2}, \quad n \geq 3.$$

(d) Bessel potentials:

$$\widehat{G_\alpha}(\xi) = \frac{1}{(1+|\xi|^2)^{\alpha/2}}, \quad \alpha > 0\,.$$

(e) Riesz kernels:

$$\widehat{R_j}(\xi) = \frac{\xi_j}{|\xi|}, \quad 1 \leq j \leq n\,.$$

(f) In the case $n = 1$, we have

$$\widehat{Y}(\xi) = \frac{1}{i}\mathrm{v.\,p.}\,\frac{1}{\xi} + \pi\delta(\xi),$$

$$\mathcal{F}\left(\mathrm{v.\,p.}\,\frac{1}{x}\right) = -\pi\, i \operatorname{sgn}\xi = \begin{cases} -\pi i & \text{for } \xi > 0, \\ \pi i & \text{for } \xi < 0. \end{cases}$$

Finally, as for distributions with compact support, we have the following:

Theorem 4.12. *If $u \in \mathcal{E}'(\mathbf{R}^n)$, then its Fourier transform $\mathcal{F}u$ is a C^∞ function on $\mathbf{R}^n$ given by the formula*

$$\mathcal{F}u(\xi) = \left\langle u, e^{-ix\cdot\xi}\right\rangle, \quad \xi \in \mathbf{R}^n\,.$$

Moreover, the function $\mathcal{F}u(\xi)$ is slowly increasing, that is, there exist constants $C > 0$ and $\mu \in \mathbf{R}$ such that

$$|\mathcal{F}u(\xi)| \leq C(1+|\xi|)^\mu, \quad \xi \in \mathbf{R}^n\,.$$

For example, we have, for the Dirac measure δ_{x_0} at a point $x_0 \in \mathbf{R}^n$,

$$\widehat{\delta_{x_0}}(\xi) = \left\langle \delta_{x_0}, e^{-ix\cdot\xi}\right\rangle = e^{-ix_0\cdot\xi}, \quad \xi \in \mathbf{R}^n,$$

and so

$$\left|\widehat{\delta_{x_0}}(\xi)\right| = 1, \quad \xi \in \mathbf{R}^n\,.$$

4.5 Operators and Kernels

Let Ω_1 and Ω_2 be open subset of $\mathbf{R}^{n_1}$ and $\mathbf{R}^{n_2}$, respectively. If K is a distribution in $\mathcal{D}'(\Omega_1 \times \Omega_2)$, we can define a continuous linear operator

$$A \in L(C_0^\infty(\Omega_2), \mathcal{D}'(\Omega_1))$$

by the formula

$$\langle A\psi, \varphi \rangle = \langle K, \varphi \otimes \psi \rangle \ , \quad \varphi \in C_0^\infty(\Omega_1) \ , \ \psi \in C_0^\infty(\Omega_2) \ .$$

We then write $A = \mathrm{Op}(K)$. Since the space $C_0^\infty(\Omega_1) \otimes C_0^\infty(\Omega_2)$ is sequentially dense in $C_0^\infty(\Omega_1 \times \Omega_2)$, it follows that the mapping

$$\mathcal{D}'(\Omega_1 \times \Omega_2) \ni K \longmapsto \mathrm{Op}(K) \in L(C_0^\infty(\Omega_2), \mathcal{D}'(\Omega_1))$$

is injective. The next theorem asserts that it is also surjective:

Theorem 4.13 (the Schwartz kernel theorem). *If A is a continuous linear operator on $C_0^\infty(\Omega_2)$ into $\mathcal{D}'(\Omega_1)$, then there exists a unique distribution $K_A \in \mathcal{D}'(\Omega_1 \times \Omega_2)$ such that $A = \mathrm{Op}(K_A)$.*

The distribution K_A is called the *kernel* of A. Formally we have

$$A\psi(x_1) = \int_{\Omega_2} K_A(x_1, x_2)\psi(x_2)\, dx_2 \ , \quad \psi \in C_0^\infty(\Omega_2) \ .$$

Now we give some important examples of distributions kernels:

Examples 4.7. (a) Riesz potentials: $\Omega_1 = \Omega_2 = \mathbf{R}^n$, $0 < \alpha < n$.

$$\begin{aligned}(-\Delta)^{-\alpha/2}u(x) &= R_\alpha * u(x) \\ &= \frac{\Gamma((n-\alpha)/2)}{2^\alpha \pi^{n/2} \Gamma(\alpha/2)} \int_{\mathbf{R}^n} \frac{1}{|x-y|^{n-\alpha}}\, u(y)\, dy \ , \quad u \in C_0^\infty(\mathbf{R}^n) \ .\end{aligned}$$

(b) Newtonian potentials: $\Omega_1 = \Omega_2 = \mathbf{R}^n$, $n \geq 3$.

$$\begin{aligned}(-\Delta)^{-1}u(x) &= N * u(x) \\ &= \frac{\Gamma((n-2)/2)}{4\pi^{n/2}} \int_{\mathbf{R}^n} \frac{1}{|x-y|^{n-2}}\, u(y)\, dy \ , \quad u \in C_0^\infty(\mathbf{R}^n) \ .\end{aligned}$$

(c) Bessel potentials: $\Omega_1 = \Omega_2 = \mathbf{R}^n$, $\alpha > 0$.

$$(I - \Delta)^{-\alpha/2}u(x) = G_\alpha * u(x) = \int_{\mathbf{R}^n} G_\alpha(x-y)\, u(y)\, dy \ , \quad u \in C_0^\infty(\mathbf{R}^n) \ .$$

(d) Riesz operators: $\Omega_1 = \Omega_2 = \mathbf{R}^n$, $1 \leq j \leq n$.

$$Y_j u(x) = R_j * u(x)$$
$$= i\,\frac{\Gamma((n+1)/2)}{\pi^{(n+1)/2}}\,\text{v. p.} \int_{\mathbf{R}^n} \frac{x_j - y_j}{|x-y|^{n+1}}\, u(y)\, dy\,, \quad u \in C_0^\infty(\mathbf{R}^n)\,.$$

(e) The Calderón–Zygmund integro-differential operator: $\Omega_1 = \Omega_2 = \mathbf{R}^n$.

$$(-\Delta)^{1/2} u(x) = \frac{1}{i} \sum_{j=1}^{n} Y_j \left(\frac{\partial u}{\partial x_j} \right)(x)$$
$$= \frac{\Gamma((n+1)/2)}{\pi^{(n+1)/2}} \sum_{j=1}^{n} \text{v. p.} \int_{\mathbf{R}^n} \frac{x_j - y_j}{|x-y|^{n+1}} \frac{\partial u}{\partial y_j}(y)\, dy\,,$$
$$u \in C_0^\infty(\mathbf{R}^n)\,.$$

4.6 Layer Potentials

The purpose of this section is to describe the classical layer potentials arising in the Dirichlet and Neumann problems for the Laplacian in the case of the half space. More detailed and concise accounts are given by the books of Folland [Fo] and McLean [Mc].

4.6.1 The Jump Formula

If $x = (x_1, x_2, \ldots, x_n)$ is a point of $\mathbf{R}^n$, we write

$$x = (x', x_n)\,, \quad x' = (x_1, x_2, \ldots, x_{n-1})\,.$$

If $u \in C^\infty(\overline{\mathbf{R}^n_+})$, we define its extension u^0 to the whole space $\mathbf{R}^n$ by the formula

$$u^0(x', x_n) = \begin{cases} u(x', x_n) & \text{for } x_n \geq 0, \\ 0 & \text{for } x_n < 0. \end{cases}$$

Then u^0 is a distribution on $\mathbf{R}^n$ and its j-th derivative $\partial_n^j(u^0)$ with respect to the normal variable x_n is expressed as follows:

$$\frac{\partial^j (u^0)}{\partial x_n^j} = \left(\frac{\partial^j u}{\partial x_n^j} \right)^0 + \sum_{k=0}^{j-1} \gamma_{j-k-1} u \otimes \delta^{(k)}(x_n)\,,$$

where $\gamma_k u$ is a C^∞ function on $\mathbf{R}^{n-1}_{x'}$ defined by the formula

$$(\gamma_k u)(x') = \frac{\partial^k u}{\partial x_n^k}(x', 0)\,, \quad x' \in \mathbf{R}^{n-1}\,,$$

and $\delta(x_n)$ is the Dirac measure at 0 on $\mathbf{R}_{x_n}$.

Furthermore, if Δ is the usual Laplacian

$$\Delta = \frac{\partial^2}{\partial x_1^2} + \frac{\partial^2}{\partial x_2^2} + \cdots + \frac{\partial^2}{\partial x_n^2} ,$$

then we have the following formula:

$$\begin{aligned}\Delta(u^0) &= (\Delta u)^0 + \gamma_1 u \otimes \delta(x_n) + \gamma_0 u \otimes \delta'(x_n) \\ &= (\Delta u)^0 + \frac{\partial u}{\partial x_n}(x', 0) \otimes \delta(x_n) + u(x', 0) \otimes \delta'(x_n) . \end{aligned} \tag{4.9}$$

This formula will be referred to as the *jump formula.*

4.6.2 Single and Double Layer Potentials

Recall that the Newtonian potential is defined by the formula

$$\begin{aligned}(-\Delta)^{-1} f(x) &= N * f(x) \\ &= \frac{\Gamma((n-2)/2)}{4\pi^{n/2}} \int_{\mathbf{R}^n} \frac{1}{|x-y|^{n-2}} f(y)\, dy \\ &= \frac{1}{(n-2)\omega_n} \int_{\mathbf{R}^n} \frac{1}{|x-y|^{n-2}} f(y)\, dy , \\ & \qquad f \in C_0^\infty(\mathbf{R}^n) . \end{aligned} \tag{4.10}$$

Here

$$\omega_n = \frac{2\pi^{n/2}}{\Gamma(n/2)} , \quad n \geq 3 ,$$

is the surface area of the unit sphere. In the case $n = 3$, we have the classical Newtonian potential

$$u(x) = \frac{1}{4\pi} \int_{\mathbf{R}^3} \frac{f(y)}{|x-y|}\, dy .$$

Up to an appropriate constant of proportionality, the Newtonian potential

$$\frac{1}{4\pi|x-y|}$$

is the gravitational potential at position x due to a unit point mass at position y, and so the function $u(x)$ is the gravitational potential due to a continuous mass distribution with density $f(x)$. In terms of electrostatics, the function $u(x)$ describes the electrostatic potential due to a charge distribution with density $f(x)$.

We define a *single layer potential* with density φ by the formula

$$\begin{aligned} N * (\varphi(x') \otimes \delta(x_n)) &= \frac{1}{(n-2)\omega_n} \int_{\mathbf{R}^{n-1}} \frac{\varphi(y')}{(|x'-y'|^2 + x_n^2)^{(n/2)/2}}\, dy' , \\ & \qquad \varphi \in C_0^\infty(\mathbf{R}^{n-1}) . \end{aligned} \tag{4.11}$$

In the case $n = 3$, the function $N * (\varphi \otimes \delta)$ is related to the distribution of electric charge on a conductor Ω. In equilibrium, mutual repulsion causes all the charge to reside on the surface $\partial\Omega$ of the conducting body with density φ, and $\partial\Omega$ is an equipotential surface.

We define a *double layer potential* with density ψ by the formula

$$N * (\psi(x') \otimes \delta'(x_n)) = \frac{1}{\omega_n} \int_{\mathbf{R}^{n-1}} \frac{x_n \psi(y')}{(|x' - y'|^2 + x_n^2)^{n/2}} \, dy' , \quad \psi \in C_0^\infty(\mathbf{R}^{n-1}) . \tag{4.12}$$

In the case $n = 3$, the function $N * (\psi \otimes \delta')$ is the potential induced by a distribution of dipoles on $\mathbf{R}^2$ with density $\psi(y')$, the axes of the dipoles being normal to $\mathbf{R}^2$.

4.6.3 The Green Representation Formula

Applying the Newtonian potential to both sides of the jump formula (4.9), we obtain that

$$\begin{aligned} u^0 &= ((N * (-\Delta))(u^0) \\ &= N * ((-\Delta u)^0) - N * (\gamma_1 u \otimes \delta(x_n)) - N * (\gamma_0 u \otimes \delta'(x_n)) \\ &= - \int_{\mathbf{R}^n} N(x - y) \Delta u(y) \, dy - \int_{\mathbf{R}^{n-1}} N(x' - y', x_n) \frac{\partial u}{\partial y_n}(y', 0) \, dy' \\ &\quad + \int_{\mathbf{R}^{n-1}} \frac{\partial N}{\partial y_n}(x' - y', x_n) u(y', 0) \, dy . \end{aligned}$$

Hence we arrive at *Green's representation formula*:

$$\begin{aligned} u(x) &= \frac{1}{(2-n)\omega_n} \int_{\mathbf{R}^n_+} \frac{1}{|x - y|^{n-2}} \Delta u(y) \, dy \\ &\quad + \frac{1}{(2-n)\omega_n} \int_{\mathbf{R}^{n-1}} \frac{1}{(|x' - y'|^2 + x_n^2)^{(n-2)/2}} \frac{\partial u}{\partial y_n}(y', 0) \, dy' \\ &\quad + \frac{1}{\omega_n} \int_{\mathbf{R}^{n-1}} \frac{x_n}{(|x' - y'|^2 + x_n^2)^{n/2}} u(y', 0) \, dy' , \quad x \in \mathbf{R}^n_+ . \end{aligned} \tag{4.13}$$

By formulas (4.10), (4.11) and (4.12), we find that the first term is the Newtonian potential and the second and third terms are the single and double layer potentials, respectively.

Furthermore, if φ is bounded and continuous on $\mathbf{R}^{n-1}$, then the function

$$u(x', x_n) = \frac{2}{\omega_n} \int_{\mathbf{R}^{n-1}} \frac{x_n}{(|x' - y'|^2 + x_n^2)^{n/2}} \varphi(y') \, dy' \tag{4.14}$$

is well-defined for $(x', x_n) \in \mathbf{R}^n_+$, and is a (unique) solution of the Dirichlet problem

$$\begin{cases} \Delta u = 0 & \text{in } \mathbf{R}^n_+, \\ u = \varphi & \text{on } \mathbf{R}^{n-1}. \end{cases}$$

Formula (4.14) is called the *Poisson integral formula* for the solution of the Dirichlet problem.

5

Theory of Pseudo-Differential Operators

In this chapter we present a brief description of the basic concepts and results of the theory of pseudo-differential operators — a modern theory of potentials — which will be used in the subsequent chapters. In recent years there has been a trend in the theory of partial differential equations towards constructive methods. The development of the theory of pseudo-differential operators has made possible such an approach to the study of elliptic differential operators.

For detailed studies of pseudo-differential operators, the reader might be referred to Chazarain–Piriou [CP], Èskin [Es], Hörmander [Ho3], Kumano-go [Ku], Rempel–Schulze [RS], Schulze [Su] and Taylor [Ty].

5.1 Function Spaces

Let Ω be a bounded domain of Euclidean space $\mathbf{R}^n$ with smooth boundary $\partial\Omega$. Its closure $\overline{\Omega} = \Omega \cup \partial\Omega$ is an n-dimensional, compact smooth manifold with boundary. We may assume that:

(a) The domain Ω is a relatively compact, open subset of an n-dimensional compact smooth manifold M without boundary (see Fig. 5.1):
(b) In a neighborhood W of $\partial\Omega$ in M a normal coordinate t is chosen so that the points of W are represented as (x', t), $x' \in \partial\Omega$, $-1 < t < 1$; $t > 0$ in Ω, $t < 0$ in $M \setminus \overline{\Omega}$ and $t = 0$ only on $\partial\Omega$ (see Fig. 5.2).
(c) The manifold M is equipped with a strictly positive density μ which, on W, is the product of a strictly positive density ω on $\partial\Omega$ and the Lebesgue measure dt on $(-1, 1)$. This manifold M is called the *double* of Ω.

The function spaces we shall treat are the following (see Bergh–Löfström [BL], Calderón [Ca], Stein [Sn2], Taibleson [Tb], Triebel [Tr]):

(i) The generalized Sobolev spaces $H^{s,p}(\Omega)$ and $H^{s,p}(M)$, consisting of all potentials of order s of L^p functions. When s is integral, these spaces coincide with the usual Sobolev spaces $W^{s,p}(\Omega)$ and $W^{s,p}(M)$, respectively.

(ii) The Besov spaces $B^{s,p}(\partial\Omega)$. These are functions spaces defined in terms of the L^p modulus of continuity, and enter naturally in connection with boundary value problems.

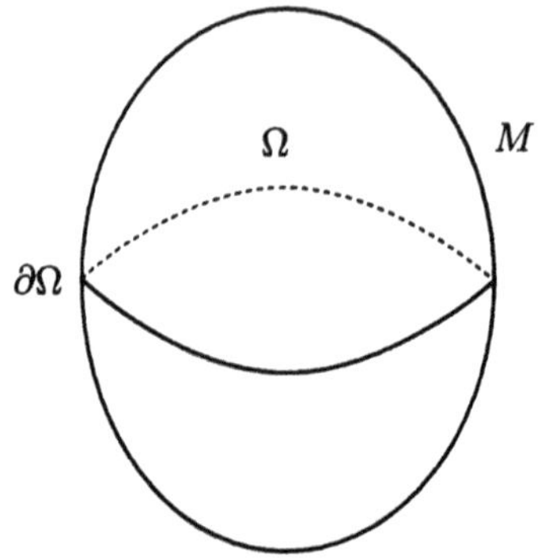

Fig. 5.1.

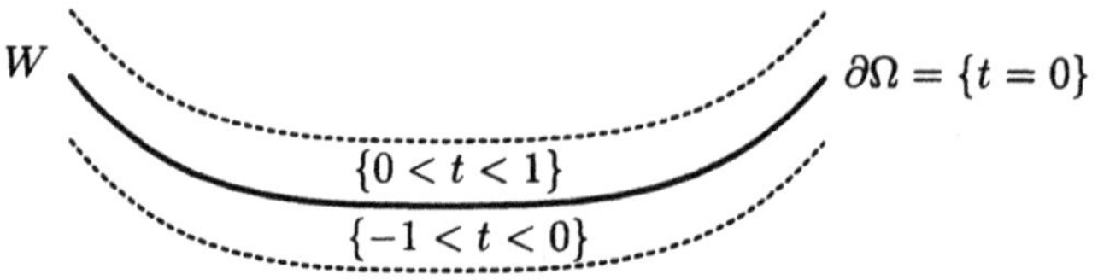

Fig. 5.2.

If $s \in \mathbf{R}$, we define a linear map

$$J^s = (I - \Delta)^{-s/2} : \mathcal{S}'(\mathbf{R}^n) \longrightarrow \mathcal{S}'(\mathbf{R}^n)$$

by the formula

$$J^s u = \mathcal{F}^* \left((1 + |\xi|^2)^{-s/2} \mathcal{F} u \right) , \quad u \in \mathcal{S}'(\mathbf{R}^n) .$$

Then the map J^s is an isomorphism of $\mathcal{S}'(\mathbf{R}^n)$ onto itself, and its inverse is the map J^{-s}. The function $J^s u$ is called the *Bessel potential* of order s of u (see Examples 4.5 and 4.6 in Chap. 4).

(I) Now, if $s \in \mathbf{R}$ and $1 < p < \infty$, we let

$$H^{s,p}(\mathbf{R}^n) = \text{the image of } L^p(\mathbf{R}^n) \text{ under the mapping } J^s .$$

We equip $H^{s,p}(\mathbf{R}^n)$ with the norm

$$\|u\|_{s,p} = \|J^{-s}u\|_p\,, \quad u \in H^{s,p}(\mathbf{R}^n)\,.$$

The space $H^{s,p}(\mathbf{R}^n)$ is called the Bessel-potential space of order s or the generalized *Sobolev space* of order s.

We list some basic topological properties of $H^{s,p}(\mathbf{R}^n)$:

(1) The Schwartz space $\mathcal{S}(\mathbf{R}^n)$ is dense in $H^{s,p}(\mathbf{R}^n)$.
(2) The space $H^{-s,p'}(\mathbf{R}^n)$ is the dual space of $H^{s,p}(\mathbf{R}^n)$, where $p' = p/(p-1)$ is the exponent conjugate to p.
(3) If $s > t$, then we have the inclusions

$$\mathcal{S}(\mathbf{R}^n) \subset H^{s,p}(\mathbf{R}^n) \subset H^{t,p}(\mathbf{R}^n) \subset \mathcal{S}'(\mathbf{R}^n)\,,$$

with continuous injections.
(4) If s is a non-negative integer, then the space $H^{s,p}(\mathbf{R}^n)$ is isomorphic to the usual Sobolev space $W^{s,p}(\mathbf{R}^n)$, that is, the space $H^{s,p}(\mathbf{R}^n)$ coincides with the space of functions $u \in L^p(\mathbf{R}^n)$ such that $D^\alpha u \in L^p(\mathbf{R}^n)$ for $|\alpha| \leq s$, and the norm $\|\cdot\|_{s,p}$ is equivalent to the norm

$$\left(\sum_{|\alpha|\leq s} \int_{\mathbf{R}^n} |D^\alpha u(x)|^p dx\right)^{1/p}.$$

(II) Next, if $1 \leq p < \infty$, we let

$$B^{1,p}(\mathbf{R}^{n-1}) = \text{ the space of (equivalence classes of) functions } \varphi \in L^p(\mathbf{R}^{n-1}) \text{ for which the integral}$$
$$\iint_{\mathbf{R}^{n-1}\times\mathbf{R}^{n-1}} \frac{|\varphi(x'+h') - 2\varphi(x') + \varphi(x'-h')|^p}{|h'|^{(n-1)+p}}\, dh'\, dx' \text{ is finite}\,.$$

The space $B^{1,p}(\mathbf{R}^{n-1})$ is a Banach space with respect to the norm

$$|\varphi|_{1,p} = \left(\int_{\mathbf{R}^{n-1}} |\varphi(x')|^p\, dx' + \iint_{\mathbf{R}^{n-1}\times\mathbf{R}^{n-1}} \frac{|\varphi(x'+h') - 2\varphi(x') + \varphi(x'-h')|^p}{|h'|^{(n-1)+p}}\, dh'\, dx'\right)^{1/p}.$$

If $p = \infty$, we let

$$B^{1,\infty}(\mathbf{R}^{n-1}) = \text{ the space of (equivalence classes of) functions } \varphi \in L^\infty(\mathbf{R}^{n-1}) \text{ for which the quantity}$$
$$\sup_{\substack{h'\in\mathbf{R}^{n-1}\\ h'\neq 0}} \frac{\|\varphi(\cdot + h') - 2\varphi(\cdot) + \varphi(\cdot - h')\|_\infty}{|h'|} \text{ is finite}\,.$$

The space $B^{1,\infty}(\mathbf{R}^{n-1})$ is a Banach space with respect to the norm

$$|\varphi|_{1,\infty} = \|\varphi\|_\infty + \sup_{\substack{h' \in \mathbf{R}^{n-1} \\ h' \neq 0}} \frac{\|\varphi(\cdot + h') - 2\varphi(\cdot) + \varphi(\cdot - h')\|_\infty}{|h'|}.$$

If $s \in \mathbf{R}$ and $1 \leq p \leq \infty$, we let

$B^{s,p}(\mathbf{R}^{n-1}) =$ the image of $B^{1,p}(\mathbf{R}^{n-1})$ under the mapping J'^{s-1}, where J'^{s-1} is the Bessel potential of order $s-1$ on $\mathbf{R}^{n-1}$.

We equip the space $B^{s,p}(\mathbf{R}^{n-1})$ with the norm

$$|\varphi|_{s,p} = \left|J'^{-s+1}\varphi\right|_{1,p}, \quad \varphi \in B^{s,p}(\mathbf{R}^{n-1}).$$

The space $B^{s,p}(\mathbf{R}^{n-1})$ is called the *Besov space* of order s.

We list some basic topological properties of $B^{s,p}(\mathbf{R}^{n-1})$:

(1) The Schwartz space $\mathcal{S}(\mathbf{R}^{n-1})$ is dense in $B^{s,p}(\mathbf{R}^{n-1})$.

(2) The space $B^{-s,p'}(\mathbf{R}^{n-1})$ is the dual space of $B^{s,p}(\mathbf{R}^{n-1})$, where $p' = p/(p-1)$.

(3) If $s > t$, then we have the inclusions

$$\mathcal{S}(\mathbf{R}^{n-1}) \subset B^{s,p}(\mathbf{R}^{n-1}) \subset B^{t,p}(\mathbf{R}^{n-1}) \subset \mathcal{S}'(\mathbf{R}^{n-1}),$$

with continuous injections.

(4) If $1 \leq p < \infty$ and if $s = m + \sigma$ with a non-negative integer m and $0 < \sigma < 1$, then the Besov space $B^{s,p}(\mathbf{R}^{n-1})$ coincides with the space of functions $\varphi \in H^{m,p}(\mathbf{R}^{n-1})$ such that, for $|\alpha| = m$, the integral

$$\iint_{\mathbf{R}^{n-1} \times \mathbf{R}^{n-1}} \frac{|D^\alpha \varphi(x') - D^\alpha \varphi(y')|^p}{|x' - y'|^{(n-1)+p\sigma}} \, dx' \, dy'$$

is finite. Furthermore, the norm $|\varphi|_{s,p}$ is equivalent to the norm

$$\left(\sum_{|\alpha| \leq m} \int_{\mathbf{R}^{n-1}} |D^\alpha \varphi(x')|^p \, dx' + \sum_{|\alpha| = m} \iint_{\mathbf{R}^{n-1} \times \mathbf{R}^{n-1}} \frac{|D^\alpha \varphi(x') - D^\alpha \varphi(y')|^p}{|x' - y'|^{(n-1)+p\sigma}} \, dx' \, dy' \right)^{1/p}.$$

(5) If $p = \infty$ and if $s = m + \sigma$ with a non-negative integer m and $0 < \sigma < 1$, then the Besov space $B^{s,\infty}(\mathbf{R}^{n-1})$ coincides with the Hölder space $C^s(\mathbf{R}^{n-1})$.

(III) Now we define the generalized Sobolev spaces $H^{s,p}(\Omega)$, $H^{s,p}(M)$ and the Besov spaces $B^{s,p}(\partial\Omega)$ for arbitrary values of s.

For each $s \in \mathbf{R}$, we define

$$H^{s,p}(\Omega) = \text{the space of restrictions to } \Omega \text{ of functions in } H^{s,p}(\mathbf{R}^n) .$$

We equip the space $H^{s,p}(\Omega)$ with the norm

$$\|u\|_{s,p} = \inf \|U\|_{s,p} ,$$

where the infimum is taken over all $U \in H^{s,p}(\mathbf{R}^n)$ which equal u in Ω. The space $H^{s,p}(\Omega)$ is a Banach space with respect to the norm $\|\cdot\|_{s,p}$. It should be noticed that

$$H^{0,p}(\Omega) = L^p(\Omega); \quad \|\cdot\|_{0,p} = \|\cdot\|_p .$$

The Sobolev spaces $H^{s,p}(M)$ are defined to be locally the Sobolev spaces $H^{s,p}(\mathbf{R}^n)$, upon using local coordinate systems flattening out M, together with a partition of unity. The Besov spaces $B^{s,p}(\partial\Omega)$ are defined similarly, with $H^{s,p}(\mathbf{R}^n)$ replaced by $B^{s,p}(\mathbf{R}^{n-1})$. The norms of $H^{s,p}(M)$ and $B^{s,p}(\partial\Omega)$ will be denoted by $\|\cdot\|_{s,p}$ and $|\cdot|_{s,p}$, respectively.

We state two important theorems which will be used in the study of boundary value problems:

(i) **(The trace theorem)** Let $1 < p < \infty$. Then the trace map

$$\begin{aligned} \gamma_0 \colon H^{s,p}(\Omega) &\longrightarrow B^{s-1/p,p}(\partial\Omega) \\ u &\longmapsto u|_{\partial\Omega} \end{aligned}$$

is continuous for all $s > 1/p$, and is surjective (cf. [BL], [Sn1], [Tb], [Tr]).

(ii) **(The Rellich theorem)** If $s > t$, then the injections

$$\begin{aligned} H^{s,p}(M) &\longrightarrow H^{t,p}(M), \\ B^{s,p}(\partial\Omega) &\longrightarrow B^{t,p}(\partial\Omega) \end{aligned}$$

are both compact (or completely continuous).

Finally, we introduce a space of distributions on Ω which behave locally just like the distributions in $H^{s,p}(\mathbf{R}^n)$:

$$\begin{aligned} H^{s,p}_{\text{loc}}(\Omega) = \; &\text{the space of distributions } u \in \mathcal{D}'(\Omega) \text{ such that} \\ &\varphi u \in H^{s,p}(\mathbf{R}^n) \text{ for all } \varphi \in C_0^\infty(\Omega) . \end{aligned}$$

We equip the space $H^{s,p}_{\text{loc}}(\Omega)$ with the topology defined by the seminorms $u \mapsto \|\varphi u\|_{s,p}$ as φ ranges over $C_0^\infty(\Omega)$. It is easy to verify that the *localized Sobolev space* $H^{s,p}_{\text{loc}}(\Omega)$ is a Fréchet space. The *localized Besov space* $B^{s,p}_{\text{loc}}(\partial\Omega)$ is defined similarly, with $H^{s,p}(\mathbf{R}^n)$ replaced by $B^{s,p}(\mathbf{R}^{n-1})$.

5.2 Fourier Integral Operators

5.2.1 Symbol Classes

Let Ω be an open subset of $\mathbf{R}^n$. If $m \in \mathbf{R}$ and $0 \leq \delta < \rho \leq 1$, we let

$S^m_{\rho,\delta}(\Omega \times \mathbf{R}^N) =$ the set of all functions $a(x,\theta) \in C^\infty(\Omega \times \mathbf{R}^N)$ with the property that, for any compact $K \subset \Omega$ and any multi-indices α, β, there exists a constant $C_{K,\alpha,\beta} > 0$ such that, for all $x \in K$ and $\theta \in \mathbf{R}^N$,

$$\left|\partial_\theta^\alpha \partial_x^\beta a(x,\theta)\right| \leq C_{K,\alpha,\beta}(1+|\theta|)^{m-\rho|\alpha|+\delta|\beta|} .$$

The elements of $S^m_{\rho,\delta}(\Omega \times \mathbf{R}^N)$ are called *symbols* of order m. We drop the $\Omega \times \mathbf{R}^N$ and use $S^m_{\rho,\delta}$ when the context is clear.

If K is a compact subset of Ω and j is a non-negative integer, we define a seminorm $p_{K,j,m}$ on $S^m_{\rho,\delta}(\Omega \times \mathbf{R}^N)$ by the formula

$$S^m_{\rho,\delta}(\Omega \times \mathbf{R}^N) \ni a \longmapsto p_{K,j,m}(a) = \sup_{\substack{x \in K \\ \theta \in \mathbf{R}^N \\ |\alpha|+|\beta| \leq j}} \frac{\left|\partial_\theta^\alpha \partial_x^\beta a(x,\theta)\right|}{(1+|\theta|)^{m-\rho|\alpha|+\delta|\beta|}} .$$

We equip the space $S^m_{\rho,\delta}(\Omega \times \mathbf{R}^N)$ with the topology defined by the family $\{p_{K,j,m}\}$ of seminorms where K ranges over all compact subsets of Ω and $j = 0, 1, \ldots$. The space $S^m_{\rho,\delta}(\Omega \times \mathbf{R}^N)$ is a Fréchet space.

Examples 5.1. (1) A polynomial $p(x,\xi) = \sum_{|\alpha| \leq m} a_\alpha(x)\xi^\alpha$ of order m with coefficients in $C^\infty(\Omega)$ is in the class $S^m_{1,0}(\Omega \times \mathbf{R}^n)$.
(2) If $m \in \mathbf{R}$, the function

$$\Omega \times \mathbf{R}^n \ni (x,\xi) \longmapsto \left(1+|\xi|^2\right)^{m/2}$$

is in the class $S^m_{1,0}(\Omega \times \mathbf{R}^n)$.
(3) A function $a(x,\theta) \in C^\infty(\Omega \times (\mathbf{R}^N \setminus \{0\}))$ is said to be positively homogeneous of degree m in θ if it satisfies the condition

$$a(x,t\theta) = t^m a(x,\theta) , \quad t > 0 .$$

If $a(x,\theta)$ is positively homogeneous of degree m in θ and if $\varphi(\theta)$ is a smooth function such that $\varphi(\theta) = 0$ for $|\theta| \leq 1/2$ and $\varphi(\theta) = 1$ for $|\theta| \geq 1$, then the function $\varphi(\theta)a(x,\theta)$ is in the class $S^m_{1,0}(\Omega \times \mathbf{R}^N)$.

We set

$$S^{-\infty}(\Omega \times \mathbf{R}^N) = \bigcap_{m \in \mathbf{R}} S^m_{\rho,\delta}(\Omega \times \mathbf{R}^N) .$$

For example, if $\varphi(\xi) \in \mathcal{S}(\mathbf{R}^N)$, then it follows that $\varphi(\xi) \in S^{-\infty}(\Omega \times \mathbf{R}^N)$. More precisely, we have the formula

$$S^{-\infty}(\Omega \times \mathbf{R}^N) = C^\infty(\Omega)\widehat{\otimes}_\pi \mathcal{S}(\mathbf{R}^N),$$

where the space $C^\infty(\Omega)\widehat{\otimes}_\pi \mathcal{S}(\mathbf{R}^N)$ is the completed π-topology (or projective topology) tensor product of the Fréchet spaces $C^\infty(\Omega)$ and $\mathcal{S}(\mathbf{R}^N)$ (see [Sc, Chap. III, Sect. 6], [Tv, Chap. 45]).

The next theorem gives a meaning to a formal sum of symbols of decreasing order:

Theorem 5.1. *Let $a_j(x,\theta) \in S^{m_j}_{\rho,\delta}(\Omega \times \mathbf{R}^N)$, $m_j \downarrow -\infty$, $j = 0, 1, \ldots$. Then there exists a symbol $a(x,\theta) \in S^{m_0}_{\rho,\delta}(\Omega \times \mathbf{R}^N)$, unique modulo $S^{-\infty}(\Omega \times \mathbf{R}^N)$, such that we have, for all $k > 0$,*

$$a(x,\theta) - \sum_{j=0}^{k-1} a_j(x,\theta) \in S^{m_k}_{\rho,\delta}(\Omega \times \mathbf{R}^N). \tag{5.1}$$

If formula (5.1) holds, we write

$$a(x,\theta) \sim \sum_{j=0}^{\infty} a_j(x,\theta).$$

The formal sum $\sum_j a_j(x,\theta)$ is called an *asymptotic expansion* of $a(x,\theta)$.

A symbol $a(x,\theta) \in S^m_{1,0}(\Omega \times \mathbf{R}^N)$ is said to be *classical* if there exist smooth functions $a_j(x,\theta)$, positively homogeneous of degree $m - j$ in θ for $|\theta| \geq 1$, such that

$$a(x,\theta) \sim \sum_{j=0}^{\infty} a_j(x,\theta).$$

The homogeneous function $a_0(x,\theta)$ of degree m is called the *principal part* of $a(x,\theta)$.

We let

$$S^m_{\mathrm{cl}}(\Omega \times \mathbf{R}^N) = \text{the set of all classical symbols of order } m.$$

It should be emphasized that the subspace $S^m_{\mathrm{cl}}(\mathbf{R}^N)$, defined as the set of all x-independent elements of $S^m_{\mathrm{cl}}(\Omega \times \mathbf{R}^N)$, is closed in the induced topology, and we have the formula

$$S^m_{\mathrm{cl}}(\Omega \times \mathbf{R}^N) = C^\infty(\Omega)\widehat{\otimes}_\pi S^m_{\mathrm{cl}}(\mathbf{R}^N).$$

Examples 5.2. The symbols in Examples 5.1 are all classical, and they have respectively as principal part the following functions:

(1) $p_m(x,\xi) = \sum_{|\alpha|=m} a_\alpha(x)\xi^\alpha$.
(2) $|\xi|^m$.
(3) $a(x,\theta)$.

A symbol $a(x,\theta)$ in $S^m_{\rho,\delta}(\Omega \times \mathbf{R}^N)$ is said to be *elliptic* of order m if, for any compact $K \subset \Omega$, there exists a constant $C_K > 0$ such that

$$|a(x,\theta)| \geq C_K(1+|\theta|)^m \,, \quad x \in K \,, |\theta| \geq \frac{1}{C_K} \,.$$

There is a simple criterion in the case of classical symbols:

Theorem 5.2. *Let $a(x,\theta)$ be in $S^m_{\mathrm{cl}}(\Omega \times \mathbf{R}^N)$ with principal part $a_0(x,\theta)$. Then $a(x,\theta)$ is elliptic if and only if we have the condition*

$$a_0(x,\theta) \neq 0 \,, \quad x \in \Omega \,, \quad |\theta| = 1 \,.$$

5.2.2 Phase Functions

Let Ω be an open subset of $\mathbf{R}^n$. A function $\varphi(x,\theta)$ in $C^\infty(\Omega \times (\mathbf{R}^N \setminus \{0\}))$ is called a *phase function* on $\Omega \times (\mathbf{R}^N \setminus \{0\})$ if it satisfies the following three conditions:

(a) $\varphi(x,\theta)$ is real-valued.
(b) $\varphi(x,\theta)$ is positively homogeneous of degree one in the variable θ.
(c) The differential $d\varphi$ does not vanish on the space $\Omega \times (\mathbf{R}^N \setminus \{0\})$.

Example 5.3. Let U be an open subset of $\mathbf{R}^p$ and $\Omega = U \times U$. The function

$$\varphi(x,y,\xi) = (x-y) \cdot \xi$$

is a phase function on the space $\Omega \times (\mathbf{R}^p \setminus \{0\})$ ($n = 2p$, $N = p$).

The next lemma will play a fundamental role in defining oscillatory integrals.

Lemma 5.3. *If $\varphi(x,\theta)$ is a phase function on $\Omega \times (\mathbf{R}^N \setminus \{0\})$, then there exists a first-order differential operator*

$$L = \sum_{j=1}^{N} a_j(x,\theta)\frac{\partial}{\partial\theta_j} + \sum_{k=1}^{n} b_k(x,\theta)\frac{\partial}{\partial x_k} + c(x,\theta)$$

such that

$$L(e^{i\varphi}) = e^{i\varphi} \,,$$

and its coefficients $a_j(x,\theta)$, $b_k(x,\theta)$, $c(x,\theta)$ enjoy the following properties:

$$a_j(x,\theta) \in S^0_{1,0}; b_k(x,\theta) \,, c(x,\theta) \in S^{-1}_{1,0} \,.$$

Furthermore, the transpose L' of L has coefficients $a'_j(x,\theta)$, $b'_k(x,\theta)$, $c'(x,\theta)$ in the same symbol classes as $a_j(x,\theta)$, $b_k(x,\theta)$, $c(x,\theta)$, respectively.

For example, if $\varphi(x,y,\xi)$ is a phase function as in Example 5.3

$$\varphi(x,y,\xi) = (x-y)\cdot\xi\,, \quad (x,y) \in U \times U\,, \xi \in (\mathbf{R}^p \setminus \{0\})\,,$$

then the operator L is given by the formula

$$L = \frac{1}{i}\frac{1-\rho(\xi)}{2+|x-y|^2}\left\{\sum_{j=1}^{p}(x_j-y_j)\frac{\partial}{\partial\xi_j} + \sum_{k=1}^{p}\frac{\xi_j}{|\xi|^2}\frac{\partial}{\partial x_k} + \sum_{k=1}^{p}\frac{-\xi_j}{|\xi|^2}\frac{\partial}{\partial y_k}\right\}$$
$$+\,\rho(\xi)\,,$$

where $\rho(\xi)$ is a function in $C_0^\infty(\mathbf{R}^p)$ such that $\rho(\xi)=1$ for $|\xi|\le 1$.

5.2.3 Oscillatory Integrals

We let

$$S^{\infty}_{\rho,\delta}\left(\Omega\times\mathbf{R}^N\right) = \bigcup_{m\in\mathbf{R}} S^{m}_{\rho,\delta}\left(\Omega\times\mathbf{R}^N\right).$$

If $\varphi(x,\theta)$ is a phase function on $\Omega\times(\mathbf{R}^N\setminus\{0\})$, we wish to give a meaning to the integral

$$I_\varphi(au) = \iint_{\Omega\times\mathbf{R}^N} e^{i\varphi(x,\theta)}a(x,\theta)u(x)\,dx\,d\theta\,, \quad u\in C_0^\infty(\Omega)\,, \tag{5.2}$$

for each symbol $a(x,\theta)\in S^{\infty}_{\rho,\delta}(\Omega\times\mathbf{R}^N)$.

By Lemma 5.3, we can replace $e^{i\varphi}$ in formula (5.2) by $L(e^{i\varphi})$. Then a *formal* integration by parts gives us that

$$I_\varphi(au) = \iint_{\Omega\times\mathbf{R}^N} e^{i\varphi(x,\theta)}L'(a(x,\theta)u(x))\,dx\,d\theta\,.$$

However, the properties of the coefficients of L' imply that L' maps $S^{r}_{\rho,\delta}$ continuously into $S^{r-\eta}_{\rho,\delta}$ for all $r\in\mathbf{R}$, where $\eta=\min(\rho,1-\delta)$. Continuing this process, we can reduce the growth of the integrand at infinity until it becomes integrable, and give a meaning to the integral (5.2) for each symbol $a(x,\theta)\in S^{\infty}_{\rho,\delta}(\Omega\times\mathbf{R}^N)$.

More precisely, we have the following:

Theorem 5.4. *(i) The linear functional*

$$S^{-\infty}\left(\Omega\times\mathbf{R}^N\right)\ni a\longmapsto I_\varphi(au)\in\mathbf{C}$$

extends uniquely to a linear functional ℓ on $S^{\infty}_{\rho,\delta}(\Omega\times\mathbf{R}^N)$ whose restriction to each $S^{m}_{\rho,\delta}(\Omega\times\mathbf{R}^N)$ is continuous. Furthermore, the restriction to $S^{m}_{\rho,\delta}(\Omega\times\mathbf{R}^N)$ of ℓ is expressed in the form

$$\ell(a) = \iint_{\Omega\times\mathbf{R}^N} e^{i\varphi(x,\theta)}(L')^k(a(x,\theta)u(x))\,dx\,d\theta\,,$$

where $k > (m+N)/\eta$, $\eta = \min(\rho, 1-\delta)$.
(ii) For any fixed $a(x,\theta) \in S^m_{\rho,\delta}(\Omega \times \mathbf{R}^N)$, *the mapping*

$$C_0^\infty(\Omega) \ni u \longmapsto I_\varphi(au) = \ell(a) \in \mathbf{C} \tag{5.3}$$

is a distribution of order $\leq k$ *for* $k > (m+N)/\eta$.

We call the linear functional ℓ on $S^\infty_{\rho,\delta}$ an *oscillatory integral*, but use the standard notation as in formula (5.2). The distribution (5.3) is called the *Fourier integral distribution* associated with the phase function $\varphi(x,\theta)$ and the amplitude $a(x,\theta)$, and is denoted by the formula

$$\int_{\mathbf{R}^N} e^{i\varphi(x,\theta)} a(x,\theta)\, d\theta\, .$$

Examples 5.4. (a) The Dirac measure $\delta(x)$ may be expressed as an oscillatory integral in the form

$$\delta(x) = \frac{1}{(2\pi)^n} \int_{\mathbf{R}^n} e^{ix\xi}\, d\xi\, .$$

This formula is called the *plane-wave expansion* of the delta function (cf. Gel'fand–Shilov [GS]).
(b) The distributions v.p. $(x_j/|x|^{n+1})$, $1 \leq j \leq n$, are expressed as

$$\text{v.p.}\, \frac{x_j}{|x|^{n+1}} = -\frac{i}{2^n \Gamma((n+1)/2)\pi^{(n-1)/2}} \int_{\mathbf{R}^n} e^{ix\xi} \frac{\xi_j}{|\xi|}\, d\xi\, , \quad 1 \leq j \leq n\, .$$

Indeed, it suffices to note that $\widehat{R_j}(\xi) = \xi_j/|\xi|$, $1 \leq j \leq n$, as stated in Examples 4.6.

If u is a distribution on Ω, the *singular support* of u is the smallest closed subset of Ω outside of which u is smooth. The singular support of u is denoted by $\operatorname{sing\,supp} u$.

The next theorem estimates the singular support of a Fourier integral distribution.

Theorem 5.5. *If* $\varphi(x,\theta)$ *is a phase function on the space* $\Omega \times (\mathbf{R}^N \setminus \{0\})$ *and if* $a(x,\theta)$ *is in* $S^\infty_{\rho,\delta}(\Omega \times \mathbf{R}^N)$, *then the distribution*

$$A = \int_{\mathbf{R}^N} e^{i\varphi(x,\theta)} a(x,\theta)\, d\theta \in \mathcal{D}'(\Omega)$$

satisfies the condition

$$\operatorname{sing\,supp} A \subset \left\{x \in \Omega \colon d_\theta\varphi(x,\theta) = 0 \text{ for some } \theta \in \mathbf{R}^N \setminus \{0\}\right\}\, .$$

5.2.4 Fourier Integral Operators

Let U and V be open subsets of $\mathbf{R}^p$ and $\mathbf{R}^q$, respectively. If $\varphi(x,y,\theta)$ is a phase function on $U \times V \times (\mathbf{R}^N \setminus \{0\})$ and if $a(x,y,\theta) \in S^{\infty}_{\rho,\delta}(U \times V \times \mathbf{R}^N)$, then there is associated a distribution $K \in \mathcal{D}'(U \times V)$ defined by the formula

$$K(x,y) = \int_{\mathbf{R}^N} e^{i\varphi(x,y,\theta)} a(x,y,\theta)\, d\theta \,.$$

Applying Theorem 5.5 to our situation, we obtain that

$$\operatorname{sing\,supp} K \subset \left\{(x,y) \in U \times V \colon d_\theta \varphi(x,y,\theta) = 0 \text{ for some } \theta \in \mathbf{R}^N \setminus \{0\}\right\} \,.$$

The distribution K defines a continuous linear operator

$$A \colon C_0^\infty(V) \longrightarrow \mathcal{D}'(U)$$

by the formula

$$\langle Av, u\rangle = \langle K, u \otimes v\rangle \,, \quad u \in C_0^\infty(U) \,, \ v \in C_0^\infty(V) \,.$$

The operator A is called the *Fourier integral operator* associated with the phase function $\varphi(x,y,\theta)$ and the amplitude $a(x,y,\theta)$, and is denoted by the formula

$$Av(x) = \iint_{V \times \mathbf{R}^N} e^{i\varphi(x,y,\theta)} a(x,y,\theta) v(y)\, dy\, d\theta \,, \quad v \in C_0^\infty(V) \,.$$

For example, the distribution

$$K(x,y) = \frac{1}{(2\pi)^n} \int_{\mathbf{R}^n} e^{i(x-y)\xi} \frac{1}{(1+|\xi|^2)^{s/2}}\, d\xi \in \mathcal{D}'(\mathbf{R}^n \times \mathbf{R}^n)$$

defines the Bessel potential $J^s = (I - \Delta)^{-s/2}$ for any $s > 0$. Indeed, it suffices to note that

$$\begin{aligned}
&(I - \Delta)^{-s/2} v(x) \\
&= G_s * v(x) \\
&= \frac{1}{(2\pi)^n} \int_{\mathbf{R}^n} e^{i(x-y)\xi} \widehat{G_s}(\xi) \widehat{v}(\xi)\, d\xi \\
&= \int_{\mathbf{R}^n} \left(\frac{1}{(2\pi)^n} \int_{\mathbf{R}^n} e^{i(x-y)\xi} \frac{1}{(1+|\xi|^2)^{s/2}}\, d\xi \right) v(y)\, dy \,, \\
&\qquad v \in C_0^\infty(\mathbf{R}^n) \,.
\end{aligned}$$

The next theorem summarizes some basic properties of the operator A.

Theorem 5.6. *(i) If $d_{y,\theta}\varphi(x,y,\theta) \neq 0$ on $U \times V \times (\mathbf{R}^N \setminus \{0\})$, then the operator A maps $C_0^\infty(V)$ continuously into $C^\infty(U)$.*
(ii) If $d_{x,\theta}\varphi(x,y,\theta) \neq 0$ on $U \times V \times (\mathbf{R}^N \setminus \{0\})$, then the operator A extends to a continuous linear operator on $\mathcal{E}'(V)$ into $\mathcal{D}'(U)$.
(iii) If $d_{y,\theta}\varphi(x,y,\theta) \neq 0$ and $d_{x,\theta}\varphi(x,y,\theta) \neq 0$ on $U \times V \times (\mathbf{R}^N \setminus \{0\})$, then we have, for all $v \in \mathcal{E}'(V)$,

$$\begin{aligned}&\operatorname{sing\,supp} Av \\ &\subset \left\{x \in U : d_\theta\varphi(x,y,\theta) = 0 \textit{for some } y \in \operatorname{sing\,supp} v \textit{ and } \theta \in \mathbf{R}^N \setminus \{0\}\right\}.\end{aligned}$$

5.3 Pseudo-Differential Operators

In this section we define pseudo-differential operators and study their basic properties such as the behavior of transposes, adjoints and compositions of such operators, and the effect of a change of coordinates on such operators. Furthermore, we formulate classical surface and volume potentials in terms of pseudo-differential operators. This calculus of pseudo-differential operators will be applied to elliptic boundary value problems in Chaps. 5 through 9.

Let Ω be an open subset of $\mathbf{R}^n$ and $m \in \mathbf{R}$. A *pseudo-differential operator* of order m on Ω is a Fourier integral operator of the form

$$Au(x) = \iint_{\Omega\times\mathbf{R}^n} e^{i(x-y)\cdot\xi} a(x,y,\xi) u(y)\, dy\, d\xi\,, \quad u \in C_0^\infty(\Omega)\,, \tag{5.4}$$

with some $a(x,y,\xi) \in S^m_{\rho,\delta}(\Omega \times \Omega \times \mathbf{R}^n)$. In other words, a pseudo-differential operator of order m is a Fourier integral operator associated with the phase function $\varphi(x,y,\xi) = (x-y)\cdot\xi$ and some amplitude $a(x,y,\xi) \in S^m_{\rho,\delta}(\Omega \times \Omega \times \mathbf{R}^n)$.

We let

$$L^m_{\rho,\delta}(\Omega) = \text{the set of all pseudo-differential operators of order } m \text{ on } \Omega\,.$$

Applying Theorems 5.5 and 5.6 to our situation, we obtain the following three assertions:

(1) A pseudo-differential operator A maps the space $C_0^\infty(\Omega)$ continuously into the space $C^\infty(\Omega)$ and extends to a continuous linear operator $A : \mathcal{E}'(\Omega) \to \mathcal{D}'(\Omega)$.
(2) The distribution kernel K_A of a pseudo-differential operator A satisfies the condition
$$\operatorname{sing\,supp} K_A \subset \{(x,x) : x \in \Omega\}\,,$$
that is, the kernel K_A is smooth off the diagonal $\Delta_\Omega = \{(x,x) : x \in \Omega\}$ in $\Omega \times \Omega$.
(3) $\operatorname{sing\,supp} Au \subset \operatorname{sing\,supp} u$, $u \in \mathcal{E}'(\Omega)$.
In other words, Au is smooth whenever u is. This property is referred to as the *pseudo-local property*.

We set

$$L^{-\infty}(\Omega) = \bigcap_{m \in \mathbf{R}} L^m_{\rho,\delta}(\Omega) .$$

The next theorem characterizes the class $L^{-\infty}(\Omega)$.

Theorem 5.7. *The following three conditions are equivalent:*

(i) $A \in L^{-\infty}(\Omega)$.
(ii) A is written in the form (5.4) with some $a(x, y, \xi) \in S^{-\infty}(\Omega \times \Omega \times \mathbf{R}^n)$.
(iii) A is a regularizer, *or equivalently, its distribution kernel* K_A *is in the space* $C^\infty(\Omega \times \Omega)$.

We recall that a continuous linear operator $A : C_0^\infty(\Omega) \to \mathcal{D}'(\Omega)$ is said to be *properly supported* if the following two conditions are satisfied:

(a) For any compact subset K of Ω, there exists a compact subset K' of Ω such that
$$\operatorname{supp} v \subset K \Longrightarrow \operatorname{supp} Av \subset K' .$$
(b) For any compact subset K' of Ω, there exists a compact subset $K \supset K'$ of Ω such that
$$\operatorname{supp} v \cap K = \emptyset \Longrightarrow \operatorname{supp} Av \cap K' = \emptyset .$$

If A is properly supported, then it maps $C_0^\infty(\Omega)$ continuously into $\mathcal{E}'(\Omega)$, and further it extends to a continuous linear operator on $C^\infty(\Omega)$ into $\mathcal{D}'(\Omega)$.

The next theorem states that every pseudo-differential operator can be written as the sum of a properly supported operator and a regularizer.

Theorem 5.8. *If* $A \in L^m_{\rho,\delta}(\Omega)$, *then we have*

$$A = A_0 + R ,$$

where $A_0 \in L^m_{\rho,\delta}(\Omega)$ *is properly supported and* $R \in L^{-\infty}(\Omega)$.

If $p(x, \xi) \in S^m_{\rho,\delta}(\Omega \times \mathbf{R}^n)$, then the operator $p(x, D)$, defined by

$$p(x, D)u(x) = \frac{1}{(2\pi)^n} \int_{\mathbf{R}^n} e^{ix\cdot\xi} p(x, \xi)\hat{u}(\xi)\, d\xi , \quad u \in C_0^\infty(\Omega) , \tag{5.5}$$

is a pseudo-differential operator of order m on Ω, that is, $p(x, D) \in L^m_{\rho,\delta}(\Omega)$.

The next theorem asserts that every properly supported pseudo-differential operator can be reduced to the form (5.5).

Theorem 5.9. *If* $A \in L^m_{\rho,\delta}(\Omega)$ *is properly supported, then we have*

$$p(x, \xi) = e^{-ix\cdot\xi} A(e^{ix\cdot\xi}) \in S^m_{\rho,\delta}(\Omega \times \mathbf{R}^n) ,$$

and

$$A = p(x, D) .$$

Furthermore, if $a(x,y,\xi) \in S^m_{\rho,\delta}(\Omega \times \Omega \times \mathbf{R}^n)$ is an amplitude for A, then we have the following asymptotic expansion:

$$p(x,\xi) \sim \sum_{\alpha \geq 0} \frac{1}{\alpha!} \partial_\xi^\alpha D_y^\alpha \left(a(x,y,\xi)\right)\Big|_{y=x} .$$

The function $p(x,\xi)$ is called the complete symbol *of A.*

We extend the notion of a complete symbol to the whole space $L^m_{\rho,\delta}(\Omega)$. If $A \in L^m_{\rho,\delta}(\Omega)$, we choose a properly supported operator $A_0 \in L^m_{\rho,\delta}(\Omega)$ such that $A - A_0 \in L^{-\infty}(\Omega)$, and define

$$\begin{aligned}\sigma(A) = {} & \text{the equivalence class of the complete symbol of } A_0 \text{ in} \\ & S^m_{\rho,\delta}(\Omega \times \mathbf{R}^n)/S^{-\infty}(\Omega \times \mathbf{R}^n).\end{aligned}$$

In view of Theorems 5.7 and 5.8, it follows that $\sigma(A)$ does not depend on the operator A_0 chosen. The equivalence class $\sigma(A)$ is called the *complete symbol* of A. It is easy to see that the mapping

$$L^m_{\rho,\delta}(\Omega) \ni A \longmapsto \sigma(A) \in S^m_{\rho,\delta}(\Omega \times \mathbf{R}^n)/S^{-\infty}(\Omega \times \mathbf{R}^n)$$

induces an isomorphism

$$L^m_{\rho,\delta}/L^{-\infty} \longrightarrow S^m_{\rho,\delta}/S^{-\infty} .$$

We shall often identify the complete symbol $\sigma(A)$ with a representative in the class $S^m_{\rho,\delta}(\Omega \times \mathbf{R}^n)$ for notational convenience, and call any member of $\sigma(A)$ a complete symbol of A.

A pseudo-differential operator $A \in L^m_{1,0}(\Omega)$ is said to be *classical* if its complete symbol $\sigma(A)$ has a representative in the class $S^m_{\mathrm{cl}}(\Omega \times \mathbf{R}^n)$.

We let

$$\begin{aligned}L^m_{\mathrm{cl}}(\Omega) = {} & \text{the set of all classical pseudo-differential operators of order } m \\ & \text{on } \Omega .\end{aligned}$$

Then the mapping

$$L^m_{\mathrm{cl}}(\Omega) \ni A \longmapsto \sigma(A) \in S^m_{\mathrm{cl}}(\Omega \times \mathbf{R}^n)/S^{-\infty}(\Omega \times \mathbf{R}^n)$$

induces an isomorphism

$$L^m_{\mathrm{cl}}/L^{-\infty} \longrightarrow S^m_{\mathrm{cl}}/S^{-\infty} .$$

Also we have

$$L^{-\infty}(\Omega) = \bigcap_{m \in \mathbf{R}} L^m_{\mathrm{cl}}(\Omega) .$$

If $A \in L^m_{\mathrm{cl}}(\Omega)$, then the principal part of $\sigma(A)$ has a canonical representative $\sigma_A(x,\xi) \in C^\infty(\Omega \times (\mathbf{R}^n \setminus \{0\}))$ which is positively homogeneous of degree

m in the variable ξ. The function $\sigma_A(x,\xi)$ is called the *homogeneous principal symbol* of A. For example, the Bessel potential $J^s = (I - \Delta)^{-s/2}$, $s \in \mathbf{R}^n$, is a classical pseudo-differential operator on $\mathbf{R}^n$ with homogeneous principal symbol $|\xi|^{-s}$.

The next two theorems assert that the class of pseudo-differential operators forms an algebra closed under the operations of composition of operators and taking the transpose or adjoint of an operator.

Theorem 5.10. *If $A \in L^m_{\rho,\delta}(\Omega)$, then its transpose A' and its adjoint A^* are both in $L^m_{\rho,\delta}(\Omega)$, and the complete symbols $\sigma(A')$ and $\sigma(A^*)$ have respectively the following asymptotic expansions:*

$$\sigma(A')(x,\xi) \sim \sum_{\alpha \geq 0} \frac{1}{\alpha!} \partial_\xi^\alpha D_x^\alpha \left(\sigma(A)(x,-\xi)\right),$$

$$\sigma(A^*)(x,\xi) \sim \sum_{\alpha \geq 0} \frac{1}{\alpha!} \partial_\xi^\alpha D_x^\alpha \left(\overline{\sigma(A)(x,\xi)}\right).$$

Theorem 5.11. *If $A \in L^{m'}_{\rho',\delta'}(\Omega)$ and $B \in L^{m''}_{\rho'',\delta''}(\Omega)$ where $0 \leq \delta' < \rho'' \leq 1$ and if one of them is properly supported, then the composition AB is in $L^{m'+m''}_{\rho,\delta}(\Omega)$ with $\rho = \min(\rho', \rho'')$, $\delta = \max(\delta', \delta'')$, and we have the following asymptotic expansion:*

$$\sigma(AB)(x,\xi) \sim \sum_{\alpha \geq 0} \frac{1}{\alpha!} \partial_\xi^\alpha \left(\sigma(A)(x,\xi)\right) \cdot D_x^\alpha \left(\sigma(B)(x,\xi)\right) .$$

A pseudo-differential operator $A \in L^m_{\rho,\delta}(\Omega)$ is said to be *elliptic* of order m if its complete symbol $\sigma(A)$ is elliptic of order m. In view of Theorem 5.2, it follows that a classical pseudo-differential operator $A \in L^m_{\mathrm{cl}}(\Omega)$ is elliptic if and only if its homogeneous principal symbol $\sigma_A(x,\xi)$ does not vanish on the space $\Omega \times (\mathbf{R}^n \setminus \{0\})$.

The next theorem states that elliptic operators are the "invertible" elements in the algebra of pseudo-differential operators.

Theorem 5.12. *An operator $A \in L^m_{\rho,\delta}(\Omega)$ is elliptic if and only if there exists a properly supported operator $B \in L^{-m}_{\rho,\delta}(\Omega)$ such that*

$$\begin{cases} AB \equiv I \bmod L^{-\infty}(\Omega), \\ BA \equiv I \bmod L^{-\infty}(\Omega). \end{cases}$$

Such an operator B is called a *parametrix* for A. In other words, a parametrix for A is a two-sided inverse of A modulo $L^{-\infty}(\Omega)$. We observe that a parametrix is unique modulo $L^{-\infty}(\Omega)$.

The next theorem proves the invariance of pseudo-differential operators under change of coordinates.

Theorem 5.13. *Let Ω_1, Ω_2 be two open subsets of $\mathbf{R}^n$ and $\chi: \Omega_1 \to \Omega_2$ a C^∞ diffeomorphism. If $A \in L^m_{\rho,\delta}(\Omega_1)$, where $1-\rho \le \delta < \rho \le 1$, then the mapping*

$$\begin{aligned} A_\chi: C_0^\infty(\Omega_2) &\longrightarrow C^\infty(\Omega_2) \\ v &\longmapsto A(v \circ \chi) \circ \chi^{-1} \end{aligned}$$

is in $L^m_{\rho,\delta}(\Omega_2)$, and we have the asymptotic expansion

$$\sigma(A_\chi)(y,\eta) \sim \sum_{\alpha \ge 0} \frac{1}{\alpha!} \left(\partial_\xi^\alpha \sigma(A)\right)(x, {}^t\chi'(x) \cdot \eta) \cdot D_z^\alpha \left(e^{ir(x,z,\eta)}\right)\Big|_{z=x} \tag{5.6}$$

with

$$r(x,z,\eta) = \langle \chi(z) - \chi(x) - \chi'(x)\cdot(z-x), \eta \rangle .$$

Here $x = \chi^{-1}(y)$, $\chi'(x)$ is the derivative of χ at x and ${}^t\chi'(x)$ its transpose.

The situation may be represented by the following diagram:

$$\begin{array}{ccc} C_0^\infty(\Omega_1) & \xrightarrow{\ A\ } & C^\infty(\Omega_1) \\ \chi^* \uparrow & & \downarrow \chi_* \\ C_0^\infty(\Omega_2) & \xrightarrow[A_\chi]{} & C^\infty(\Omega_2) \end{array}$$

Here $\chi^* v = v \circ \chi$ is the pull-back of v by χ and $\chi_* u = u \circ \chi^{-1}$ is the push-forward of u by χ, respectively.

Remark 5.1. Formula (5.6) shows that

$$\sigma(A_\chi)(y,\eta) \equiv \sigma(A)\left(x, {}^t\chi'(x)\cdot\eta\right) \quad \mathrm{mod}\, S^{m-(\rho-\delta)}_{\rho,\delta} .$$

Note that the mapping

$$\Omega_2 \times \mathbf{R}^n \ni (y,\eta) \longmapsto \left(x, {}^t\chi'(x)\cdot\eta\right) \in \Omega_1 \times \mathbf{R}^n$$

is just a transition map of the cotangent bundle $T^*(\mathbf{R}^n)$. This implies that the principal symbol $\sigma_m(A)$ of $A \in L^m_{\rho,\delta}(\mathbf{R}^n)$ can be invariantly defined on $T^*(\mathbf{R}^n)$ when $1-\rho \le \delta < \rho \le 1$.

A differential operator of order m with smooth coefficients on Ω is continuous on $H^{s,p}_{\mathrm{loc}}(\Omega)$ (resp. $B^{s,p}_{\mathrm{loc}}(\Omega)$) into $H^{s-m,p}_{\mathrm{loc}}(\Omega)$ (resp. $B^{s-m,p}_{\mathrm{loc}}(\Omega)$) for all $s \in \mathbf{R}$. This result extends to pseudo-differential operators (see Bourdaud [Bo, Theorem 1]):

Theorem 5.14. *Every properly supported operator $A \in L^m_{1,\delta}(\Omega)$, $0 \le \delta < 1$, extends to a continuous linear operator $A: H^{s,p}_{\mathrm{loc}}(\Omega) \to H^{s-m,p}_{\mathrm{loc}}(\Omega)$ for all $s \in \mathbf{R}$ and $1 < p < \infty$, and also it extends to a continuous linear operator $A: B^{s,p}_{\mathrm{loc}}(\Omega) \to B^{s-m,p}_{\mathrm{loc}}(\Omega)$ for all $s \in \mathbf{R}$ and $1 \le p \le \infty$.*

We prove a *global version* of Theorem 5.14 in Appendix A (Theorem A.6), due to its length.

Now we define the concept of a pseudo-differential operator on a manifold, and transfer all the machinery of pseudo-differential operators to manifolds. Let M be an n-dimensional, *compact* smooth manifold without boundary. Theorem 5.13 leads us to the following:

Definition 5.1. Let $1 - \rho \leq \delta < \rho \leq 1$. A continuous linear operator $A: C^\infty(M) \to C^\infty(M)$ is called a *pseudo-differential operator* of order $m \in \mathbf{R}$ if it satisfies the following two conditions:

(i) The distribution kernel of A is smooth off the diagonal $\Delta_M = \{(x, x): x \in M\}$ in $M \times M$.
(ii) For any chart (U, χ) on M, the mapping

$$\begin{aligned} A_\chi: C_0^\infty(\chi(U)) &\longrightarrow C^\infty(\chi(U)) \\ u &\longmapsto A(u \circ \chi) \circ \chi^{-1} \end{aligned}$$

belongs to the class $L^m_{\rho,\delta}(\chi(U))$.

We let

$$L^m_{\rho,\delta}(M) = \text{the set of all pseudo-differential operators of order } m \text{ on } M,$$

and set

$$L^{-\infty}(M) = \bigcap_{m \in \mathbf{R}} L^m_{\rho,\delta}(M).$$

Some results about pseudo-differential operators on $\mathbf{R}^n$ stated above are also true for pseudo-differential operators on M. In fact, pseudo-differential operators on M are defined to be locally pseudo-differential operators on $\mathbf{R}^n$.

For example, we have the following five results:

(1) A pseudo-differential operator A extends to a continuous linear operator $A: \mathcal{D}'(M) \to \mathcal{D}'(M)$.
(2) $\operatorname{sing\,supp} Au \subset \operatorname{sing\,supp} u$, $u \in \mathcal{D}'(M)$.
(3) A continuous linear operator $A: C^\infty(M) \to \mathcal{D}'(M)$ is a regularizer if and only if it is in $L^{-\infty}(M)$.
(4) The class $L^m_{\rho,\delta}(M)$ is stable under the operations of composition of operators and taking the transpose or adjoint of an operator.
(5) A pseudo-differential operator $A \in L^m_{1,\delta}(M)$, $0 \leq \delta < 1$, extends to a continuous linear operator $A: H^{s,p}(M) \to H^{s-m,p}(M)$ for all $s \in \mathbf{R}$ and $1 < p < \infty$ and also a continuous linear operator $A: B^{s,p}(M) \to B^{s-m,p}(M)$ for all $s \in \mathbf{R}$ and $1 \leq p \leq \infty$.

A pseudo-differential operator $A \in L^m_{1,0}(M)$ is said to be *classical* if, for any chart (U, χ) on M, the mapping $A_\chi: C_0^\infty(\chi(U)) \to C^\infty(\chi(U))$ belongs to the class $L^m_{\mathrm{cl}}(\chi(U))$.

We let

$L^m_{\text{cl}}(M)$ = the set of all classical pseudo-differential operators of order m on M .

We observe that

$$L^{-\infty}(M) = \bigcap_{m \in \mathbf{R}} L^m_{\text{cl}}(M) .$$

Let $A \in L^m_{\text{cl}}(M)$. If (U, χ) is a chart on M, there is associated a homogeneous principal symbol $\sigma_{A_\chi} \in C^\infty(\chi(U) \times (\mathbf{R}^n \setminus \{0\}))$. In view of Remark 5.1, by smoothly patching together the functions σ_{A_χ} we can obtain a smooth function $\sigma_A(x, \xi)$ on $T^*(M) \setminus \{0\} = \{(x, \xi) \in T^*(M) : \xi \neq 0\}$, which is positively homogeneous of degree m in the variable ξ. The function $\sigma_A(x, \xi)$ is called the *homogeneous principal symbol* of A.

A classical pseudo-differential operator $A \in L^m_{\text{cl}}(M)$ is said to be *elliptic* of order m if its homogeneous principal symbol $\sigma_A(x, \xi)$ does not vanish on the bundle $T^*(M) \setminus \{0\}$ of non-zero cotangent vectors.

Then we have the following result:

(6) An operator $A \in L^m_{\text{cl}}(M)$ is elliptic if and only if there exists a parametrix $B \in L^{-m}_{\text{cl}}(M)$ for A:

$$\begin{cases} AB \equiv I \mod L^{-\infty}(M), \\ BA \equiv I \mod L^{-\infty}(M). \end{cases}$$

Let Ω be an open subset of $\mathbf{R}^n$. A properly supported pseudo-differential operator A on Ω is said to be *hypoelliptic* if it satisfies the condition

$$\text{sing supp}\, u = \text{sing supp}\, Au , \quad u \in \mathcal{D}'(\Omega) .$$

For example, Theorem 5.12 tells us that elliptic operators are hypoelliptic. It should be noticed that this notion may be transferred to manifolds.

The following criterion for hypoellipticity is due to Hörmander (see [Ho2, Theorem 4.2]):

Theorem 5.15. *Let $1 - \rho \leq \delta < \rho \leq 1$, and let $A = p(x, D) \in L^m_{\rho,\delta}(\Omega)$ be properly supported. Assume that, for any compact $K \subset \Omega$ and any multi-indices α, β, there exist constants $C_{K,\alpha,\beta} > 0$, $C_K > 0$ and $\mu \in \mathbf{R}$ such that we have, for all $x \in K$ and $|\xi| \geq C_K$,*

$$\left| D_\xi^\alpha D_x^\beta p(x, \xi) \right| \leq C_{K,\alpha,\beta} |p(x, \xi)| (1 + |\xi|)^{-\rho|\alpha| + \delta|\beta|} , \tag{5.7a}$$

$$|p(x, \xi)|^{-1} \leq C_K (1 + |\xi|)^\mu . \tag{5.7b}$$

Then there exists a parametrix $B \in L^\mu_{\rho,\delta}(\Omega)$ for A.

5.4 Potentials and Pseudo-Differential Operators

The purpose of this section is to describe, in terms of pseudo-differential operators, the surface and volume potentials arising in boundary value problems for elliptic differential operators.

We give a formal description of a background. Let Ω be a bounded domain in Euclidean space $\mathbf{R}^n$ with smooth boundary. Its closure $\overline{\Omega}$ is an n-dimensional compact smooth manifold with boundary. We may assume that $\overline{\Omega}$ is the closure of a relatively compact, open subset Ω of an n-dimensional compact smooth manifold M without boundary in which Ω has a smooth boundary $\partial\Omega$ (see Fig. 5.1).

Let P be a differential operator of order m with smooth coefficients on M. Then we have the jump formula

$$P(u^0) = (Pu)^0 + \widetilde{P}\gamma u\,, \quad u \in C^\infty(\overline{\Omega})\,, \tag{5.8}$$

where u^0 is the extension of u to M by zero outside $\overline{\Omega}$, and $\widetilde{P}\gamma u$ is a distribution on M with support in $\partial\Omega$. If P admits an "inverse" Q, then the function u may be expressed as follows:

$$u = Q((Pu)^0)|_\Omega + Q(\widetilde{P}\gamma u)|_\Omega\,. \tag{5.9}$$

The first term on the right-hand side is a volume potential and the second term is a surface potential with m "layers". For example, if P is the usual Laplacian Δ and if $\Omega = \mathbf{R}^n_+$, then formulas (5.8) and (5.9) coincide with formulas (4.9) and (4.13), respectively.

First, we state a theorem which covers surface potentials (cf. [CP], [Es], [Ho2], [RS], [Se1], [Ty]):

Theorem 5.16. *Let $A \in L^m_{\mathrm{cl}}(M)$ be properly supported. Assume that*

$$\text{Every term in the complete symbol } \textstyle\sum_{j=0}^{\infty} a_j(x,\xi) \text{ of } A \text{ is a rational function of } \xi \tag{5.10}$$

Then we have the following two assertions:

(i) The operator

$$H\colon v \longmapsto A(v \otimes \delta)|_\Omega$$

is continuous on $C^\infty(\partial\Omega)$ into $C^\infty(\overline{\Omega})$. If $v \in \mathcal{D}'(\partial\Omega)$, the distribution Hv has sectional traces on $\partial\Omega$ of any order.

(ii) The operator

$$\begin{aligned} S\colon C^\infty(\partial\Omega) &\longrightarrow C^\infty(\partial\Omega) \\ v &\longmapsto Hv|_{\partial\Omega} \end{aligned}$$

belongs to the class $L^{m+1}_{\mathrm{cl}}(\partial\Omega)$. Furthermore, its homogeneous principal symbol is given by the formula

$$(x', \xi') \longmapsto \frac{1}{2\pi} \int_\Gamma a_0(x', 0, \xi', \xi_n)\, d\xi_n$$

where $a_0(x', x_n, \xi', \xi_n) \in C^\infty(T^*(M) \setminus \{0\})$ *is the homogeneous principal symbol of* A*, and* Γ *is a circle in the plane* $\{\xi_n \in \mathbf{C}: \operatorname{Im} \xi_n > 0\}$ *which encloses the poles* ξ_n *of* $a_0(x', 0, \xi', \xi_n)$ *there.*

(iii) If $1 < p < \infty$*, then the operator* H *extends to a continuous linear operator*

$$H: B^{s,p}(\partial\Omega) \longrightarrow H^{s-m-1/p,p}(\Omega)$$

for all $s \in \mathbf{R}$.

It should be emphasized that condition (5.10) is invariant under change of coordinates. Furthermore, it is easy to see that every parametrix for an elliptic differential operator satisfies condition (5.10).

The next theorem covers volume potentials (cf. [CP], [Ho2], [RS], [Se1], [Ty]):

Theorem 5.17. *Let* $A \in L^m_{\mathrm{cl}}(M)$ *be as in Theorem 5.16. Then we have the following two assertions:*

(i) The operator

$$G: f \longmapsto A(f^0)|_\Omega$$

is continuous on $C^\infty(\overline{\Omega})$ *into itself.*

(ii) If $1 < p < \infty$*, then the operator* G *extends to a continuous linear operator*

$$G: H^{s,p}(\Omega) \longrightarrow H^{s-m,p}(\Omega)$$

for all $s > -1/p$.

5.5 The Transmission Property

One of the important questions in the theory of pseudo-differential operators in a domain is that of the smoothness of a solution near the boundary. In this section we introduce a subclass of pseudo-differential operators A whose solutions u are smooth up to the boundary $\partial\Omega$ of a domain Ω if Au are smooth on the closure $\overline{\Omega} = \Omega \cup \partial\Omega$ (cf. [Es, Chap. III], [Ho3, Sect. 18.2]).

Following Boutet de Monvel [Bt], we impose a condition about symbols in the normal direction at the boundary in order to ensure the stated regularity property (see [RS]).

We let

$$S^m_{1,0}(\overline{\mathbf{R}^n_+} \times \mathbf{R}^n) = \text{the space of symbols } a(x,\xi) \text{ in } S^m_{1,0}(\mathbf{R}^n_+ \times \mathbf{R}^n) \text{ which have an extension in } S^m_{1,0}(\mathbf{R}^n \times \mathbf{R}^n)\,.$$

A symbol $a(x, \xi) \in S^m_{1,0}(\overline{\mathbf{R}^n_+} \times \mathbf{R}^n)$ is said to have the *transmission property* with respect to the boundary $\mathbf{R}^{n-1}$ if all its derivatives $(\partial^{\alpha_n}_{x_n} a)(x', 0, \xi', \nu)$, $\alpha_n \geq 0$, admit an expansion of the form

$$\left(\frac{\partial}{\partial x_n}\right)^{\alpha_n} a(x',0,\xi',\nu)$$
$$= \sum_{j=0}^{m} b_j(x',\xi')\nu^j + \sum_{k=-\infty}^{\infty} a_k(x',\xi') \frac{\left(\langle\xi'\rangle - \sqrt{-1}\nu\right)^k}{\left(\langle\xi'\rangle + \sqrt{-1}\nu\right)^{k+1}}, \quad \nu \in \mathbf{R},$$

where $b_j(x',\xi') \in S^{m-j}_{1,0}(\mathbf{R}^{n-1} \times \mathbf{R}^{n-1})$ and the $a_k(x',\xi')$ form a rapidly decreasing sequence in $S^{m+1}_{1,0}(\mathbf{R}^{n-1} \times \mathbf{R}^{n-1})$ with respect to k, and $\langle\xi'\rangle = (1+|\xi'|^2)^{1/2}$ (cf. [RS, p. 119, Proposition 3]).

For example, if $a(x,\xi) \in S^m_{1,0}(\mathbf{R}^n \times \mathbf{R}^n)$ is a classical symbol of order $m \in \mathbf{Z}$ with an asymptotic expansion

$$a(x,\xi) \sim \sum_{j=0}^{\infty} a_{m-j}(x,\xi),$$

where $a_{m-j}(x,\xi) \in S^{m-j}_{1,0}$ is positively homogeneous of degree $m-j$ for $|\xi| \geq 1$, then it is easy to verify (see [RS, p. 123, Proposition 1], [Ho3, Lemma 18.2.14]) that $a(x,\xi)$ has the transmission property if and only if we have, for all multi-indices $\alpha = (\alpha', \alpha_n)$,

$$\left(\frac{\partial}{\partial x_n}\right)^{\alpha_n} \left(\frac{\partial}{\partial \xi'}\right)^{\alpha'} a_{m-j}(x',0,0,+1)$$
$$= (-1)^{m-j-|\alpha'|} \left(\frac{\partial}{\partial x_n}\right)^{\alpha_n} \left(\frac{\partial}{\partial \xi'}\right)^{\alpha'} a_{m-j}(x',0,0,-1).$$

We let

$L^m_{1,0}(\overline{\mathbf{R}^n_+})$ = the space of pseudo-differential operators in $L^m_{1,0}(\mathbf{R}^n_+)$ which can be extended to a pseudo-differential operator in $L^m_{1,0}(\mathbf{R}^n)$.

A pseudo-differential operator $A \in L^m_{1,0}(\overline{\mathbf{R}^n_+})$ is said to have the *transmission property* with respect to the boundary $\mathbf{R}^{n-1}$ if any complete symbol of A has the transmission property with respect to the boundary $\mathbf{R}^{n-1}$.

Now we illustrate how the transmission property of the symbol ensures that the associated operator preserves smoothness up to the boundary. If A is a pseudo-differential operator in $L^m_{1,0}(\overline{\mathbf{R}^n_+})$, then we define a new operator

$$\begin{aligned} A_{\mathbf{R}^n_+} : C^\infty_0\left(\overline{\mathbf{R}^n_+}\right) &\longrightarrow C^\infty\left(\mathbf{R}^n_+\right) \\ u &\longmapsto (Au^0)|_{\mathbf{R}^n_+}, \end{aligned}$$

where u^0 is the extension of u to $\mathbf{R}^n$ by zero outside $\overline{\mathbf{R}^n_+}$. The transmission property implies that if u is smooth up to the boundary, then so is $A_{\mathbf{R}^n_+} u$. More precisely, we have the following two assertions (see [RS, p. 137, Corollary 3; p 168, Theorem 8]):

(I) If a pseudo-differential operator $A \in L^m_{1,0}(\overline{\mathbf{R}^n_+})$ has the transmission property with respect to the boundary $\mathbf{R}^{n-1}$, then $A_{\mathbf{R}^n_+}$ maps $C_0^\infty(\overline{\mathbf{R}^n_+})$ continuously into $C^\infty(\overline{\mathbf{R}^n_+})$.

(II) If a pseudo-differential operator $A \in L^m_{1,0}(\overline{\mathbf{R}^n_+})$ has the transmission property, then $A_{\mathbf{R}^n_+}$ maps $C^{k+\theta}_{\mathrm{com}}(\overline{\mathbf{R}^n_+})$ continuously into $C^{k-m+\theta}_{\mathrm{loc}}(\overline{\mathbf{R}^n_+})$ for any integer $k \geq m$. Here $C^{k+\theta}_{\mathrm{com}}(\overline{\mathbf{R}^n_+})$ is the space of functions in $C^k(\overline{\mathbf{R}^n_+})$ with compact support in $\overline{\mathbf{R}^n_+}$ and all of whose k-th order derivatives are Hölder continuous with exponent θ, and $C^{k-m+\theta}_{\mathrm{loc}}(\overline{\mathbf{R}^n_+})$ is the space of functions in $C^k(\overline{\mathbf{R}^n_+})$ all of whose k-th order derivatives are locally Hölder continuous with exponent θ, respectively.

Moreover, it should be noticed that the notion of transmission property is invariant under a change of coordinates which preserves the boundary. Hence this notion can be transferred to manifolds with boundary as follows. Indeed, if Ω is a relatively compact, open subset of an n-dimensional paracompact smooth manifold M without boundary (see Fig. 5.1), then the notion of transmission property can be extended to the class $L^m_{1,0}(M)$, upon using local coordinate systems flattening out the boundary $\partial\Omega$.

Then we have the following two assertions (see [RS, p. 139, Theorem 4; p. 176, Theorem 1]):

(III) If a pseudo-differential operator $A \in L^m_{1,0}(M)$ has the transmission property with respect to the boundary $\partial\Omega$, then the operator

$$\begin{aligned} A_\Omega : C^\infty(\overline{\Omega}) &\longrightarrow C^\infty(\Omega) \\ u &\longmapsto A(u^0)|_\Omega \end{aligned}$$

maps $C^\infty(\overline{\Omega})$ continuously into itself, where u^0 is the extension of u to M by zero outside $\overline{\Omega}$.

(IV) If a pseudo-differential operator $A \in L^m_{1,0}(M)$ has the transmission property, then the operator A_Ω maps $C^{k+\theta}(\overline{\Omega})$ continuously into $C^{k-m+\theta}(\overline{\Omega})$ for any integer $k \geq m$ and $0 < \theta < 1$.

5.6 The Boutet de Monvel Calculus

Elliptic boundary value problems can not be treated directly by pseudo-differential operator methods. It was Boutet de Monvel [Bt] who brought in the operator-algebraic aspect with his calculus in 1971. He constructed a relatively small "algebra" which contains the boundary value problems for elliptic differential operators as well as their parametrices. More detailed and concise accounts are given in the books of Rempel–Schulze [RS] and Schrohe [Sr].

Let Ω be a relatively compact, open subset of an n-dimensional compact smooth manifold M without boundary. Boutet de Monvel [Bt] introduced matrices of operators

$$\mathcal{A} = \begin{pmatrix} P_\Omega + G & K \\ T & S \end{pmatrix} : \begin{matrix} C^\infty(\overline{\Omega}) \\ \bigoplus \\ C^\infty(\partial\Omega) \end{matrix} \longrightarrow \begin{matrix} C^\infty(\overline{\Omega}) \\ \bigoplus \\ C^\infty(\partial\Omega) \end{matrix} .$$

Here:

(1) P is a pseudo-differential operator on the full manifold M and

$$P_\Omega u = P(u^0)|_\Omega , \quad u \in C^\infty(\overline{\Omega}) ,$$

where u^0 is the extension of u by zero to M. The crucial requirement is that the symbol of P has the *transmission property* in order that P_Ω map $C^\infty(\overline{\Omega})$ into itself.

(2) S is a pseudo-differential operator on $\partial\Omega$.

(3) K and T are generalizations of the potentials and trace operators known from the theory of elliptic boundary value problems.

(4) G, a so-called *singular Green operator*, is an operator which is smoothing in the interior Ω while it acts like a pseudo-differential operator in directions tangential to the boundary $\partial\Omega$.

Boutet de Monvel [Bt] proved that these operator matrices form an algebra in the following sense (see [RS, p. 175, Theorem 1; p. 195, Theorem 2]): Given another element of the calculus, say,

$$\mathcal{A}' = \begin{pmatrix} P'_\Omega + G' & K' \\ T' & S' \end{pmatrix} : \begin{matrix} C^\infty(\overline{\Omega}) \\ \bigoplus \\ C^\infty(\partial\Omega) \end{matrix} \longrightarrow \begin{matrix} C^\infty(\overline{\Omega}) \\ \bigoplus \\ C^\infty(\partial\Omega) \end{matrix} ,$$

the composition $\mathcal{A}'\mathcal{A}$ is again an operator matrix of the type described above. It is worth pointing out here that the product $P'_\Omega P_\Omega$ does not coincide with $(P'P)_\Omega$; in fact, the difference $P'_\Omega P_\Omega - (P'P)_\Omega$ turns out to be a singular Green operator (see [RS, p. 100, Lemma 5]).

For example, we consider the Dirichlet problem

$$\begin{cases} \Delta u = f & \text{in } \Omega, \\ u = \varphi & \text{on } \partial\Omega. \end{cases} \tag{D}$$

Then problem (D) corresponds to the map

$$\mathcal{D} = \begin{pmatrix} \Delta_\Omega \\ \gamma_0 \end{pmatrix} : C^\infty(\overline{\Omega}) \longrightarrow \begin{matrix} C^\infty(\overline{\Omega}) \\ \bigoplus \\ C^\infty(\partial\Omega) \end{matrix} ,$$

where $\gamma_0 u = u|_{\partial\Omega}$ is the trace operator on $\partial\Omega$. The map $\mathcal{D}$ defines an isomorphism and its inverse $\mathcal{D}^{-1}$ has the form

$$\mathcal{D}^{-1} = \begin{pmatrix} P_\Omega + G & L \end{pmatrix} : \begin{matrix} C^\infty(\overline{\Omega}) \\ \bigoplus \\ C^\infty(\partial\Omega) \end{matrix} \longrightarrow C^\infty(\overline{\Omega}) ,$$

with a pseudo-differential operator P, a singular Green operator G and a potential operator L. More precisely, if $g(x,z)$ and $\ell(x,y)$ are the classical Green and Poisson kernels of problem (D) respectively, then it follows that

$$u(x) = \int_{\Omega} g(x,z) f(z)\, dz + \int_{\partial\Omega} \ell(x,y)\varphi(y)\, d\omega(y) ,$$

so that

$$(P_{\Omega} + G) f = \int_{\Omega} g(x,z) f(z)\, dz ,$$
$$L\varphi = \int_{\partial\Omega} \ell(x,y)\varphi(y)\, d\omega(y) .$$

Furthermore, it should be noticed that

$$\begin{pmatrix} \Delta_{\Omega} \\ \gamma_0 \end{pmatrix} \begin{pmatrix} P_{\Omega} + G & L \end{pmatrix} = \begin{pmatrix} \Delta_{\Omega}(P_{\Omega} + G) & \Delta_{\Omega} L \\ \gamma_0(P_{\Omega} + G) & \gamma_0 L \end{pmatrix} = \begin{pmatrix} I & 0 \\ 0 & I \end{pmatrix} , \tag{5.11}$$

since $\mathcal{D}^{-1}$ is the two-sided inverse of $\mathcal{D}$.

Let

$$\mathcal{A} = \begin{pmatrix} \Delta_{\Omega} \\ B \end{pmatrix} : C^{\infty}(\overline{\Omega}) \longrightarrow \begin{matrix} C^{\infty}(\overline{\Omega}) \\ \bigoplus \\ C^{\infty}(\partial\Omega) \end{matrix}$$

be another elliptic boundary value problem such as the Neumann problem. Then we have, by formula (5.11),

$$\mathcal{A}\mathcal{D}^{-1} = \begin{pmatrix} \Delta_{\Omega} \\ B \end{pmatrix} \begin{pmatrix} P_{\Omega} + G & L \end{pmatrix} = \begin{pmatrix} I & 0 \\ B(P_{\Omega} + G) & BL \end{pmatrix} ,$$

and the right lower corner

$$Q = BL \colon C^{\infty}(\partial\Omega) \longrightarrow C^{\infty}(\partial\Omega)$$

is a pseudo-differential operator on $\partial\Omega$. The ellipticity of $\mathcal{A}$ and $\mathcal{D}$ implies that Q is elliptic. If R is a *parametrix* for Q and if we let

$$\mathcal{C} = \begin{pmatrix} I & 0 \\ -RB(P_{\Omega} + G) & R \end{pmatrix} ,$$

then it follows that

$$\mathcal{A}\left(\mathcal{D}^{-1}\mathcal{C}\right) = \begin{pmatrix} I & 0 \\ B(P_{\Omega} + G) & BL \end{pmatrix} \begin{pmatrix} I & 0 \\ -RB(P_{\Omega} + G) & R \end{pmatrix} \equiv \begin{pmatrix} I & 0 \\ 0 & I \end{pmatrix} .$$

This proves that $\mathcal{D}^{-1}\mathcal{C}$ is a parametrix for $\mathcal{A}$.

Therefore, starting at the map $\mathcal{D}$ we can find the parametrix of another elliptic map $\mathcal{A}$ by calculating the parametrix of an elliptic pseudo-differential operator Q on the boundary. In Sect. 6.3 we will return this reduction to the boundary in a more general setting (see Theorems 6.9, 6.10 and 6.11).

6

Elliptic Boundary Value Problems

This chapter is devoted to general boundary value problems for second-order elliptic differential operators. We begin in Sect. 6.1 with a summary of the basic facts about existence, uniqueness and regularity of solutions of the Dirichlet problem in the framework of Hölder spaces. In Sect 6.2, using the calculus of pseudo-differential operators, we prove existence, uniqueness and regularity theorems for the Dirichlet problem in the framework of Sobolev spaces. In Sect. 6.3 we formulate general boundary value problems, and show that these problems can be reduced to the study of pseudo-differential operators on the boundary. The virtue of this reduction is that there is no difficulty in taking adjoints after restricting the attention to the boundary, whereas boundary value problems in general do not have adjoints. This allows us to discuss the existence theory more easily.

For more thorough treatments of this subject, the reader might refer to Chazarain–Piriou [CP, Chap. 5], Èskin [Es, Chap. VI], Kumano-go [Ku, Chap. 6], Hörmander [Ho3, Chap. XX], Rempel–Schulze [RS, Chap. 3], Schulze [Sc, Chap. 4] and Taylor [Ty, Chap. XI].

6.1 The Dirichlet Problem

In this section we shall consider the Dirichlet problem in the framework of Sobolev spaces of L^p style. This is a generalization of the classical potential approach to the Dirichlet problem.

Let Ω be a bounded domain of Euclidean space $\mathbf{R}^n$ with smooth boundary $\partial\Omega$. Its closure $\overline{\Omega} = \Omega \cup \partial\Omega$ is an n-dimensional compact smooth manifold with boundary. We may assume (see Figs. 5.1 and 5.2):

(a) The domain Ω is a relatively compact, open subset of an n-dimensional compact smooth manifold M without boundary.

(b) In a neighborhood W of $\partial\Omega$ in M a normal coordinate t is chosen so that the points of W are represented as (x', t), $x' \in \partial\Omega$, $-1 < t < 1$; $t > 0$ in Ω, $t < 0$ in $M \setminus \overline{\Omega}$ and $t = 0$ only on $\partial\Omega$. We remark that $\partial/\partial\mathbf{n} = \partial/\partial t$.

(c) The manifold M is equipped with a strictly positive density μ which, on W, is the product of a strictly positive density ω on $\partial\Omega$ and the Lebesgue measure dt on $(-1, 1)$.
We let

$$Au(x) = \sum_{i,j=1}^{n} a^{ij}(x)\frac{\partial^2 u}{\partial x_i \partial x_j}(x) + \sum_{i=1}^{n} b^i(x)\frac{\partial u}{\partial x_i}(x) + c(x)u(x) \tag{6.1}$$

be a second-order, *elliptic* differential operator with real coefficients such that

(1) $a^{ij}(x) \in C^\infty(M)$, $a^{ij}(x) = a^{ji}(x)$ and there exists a constant $a_0 > 0$ such that

$$\sum_{i,j=1}^{n} a^{ij}(x)\xi_i\xi_j \geq a_0|\xi|^2 \quad \text{on } T^*(M).$$

Here $T^*(M)$ is the cotangent bundle of M.
(2) $b^i(x) \in C^\infty(M)$.
(3) $c(x) \in C^\infty(M)$ and $c(x) \leq 0$ in M.

Furthermore, for simplicity, assume that

$$\text{The function } c(x) \text{ does not vanish } \textit{identically} \text{ on } M\,. \tag{6.2}$$

First, we consider the non-homogeneous Dirichlet problem

$$\begin{cases} Au = f & \text{in } \Omega, \\ u = \varphi & \text{on } \partial\Omega. \end{cases} \tag{D}$$

To do this, recall the following definition:

Definition 6.1. Let X, Y be Banach spaces and let $T: X \to Y$ be a densely defined, closed linear operator with domain $\mathcal{D}(T)$. We say that T is a *Fredholm operator* if it satisfies the following three conditions:

(a) The null space $\mathcal{N}(T)$ of T has finite dimension, that is, $\dim \mathcal{N}(T)$ is finite.
(b) The range $\mathcal{R}(T)$ of T is closed in Y.
(c) The range $\mathcal{R}(T)$ has finite codimension in Y, that is, $\operatorname{codim} \mathcal{R}(T) = \dim Y/\mathcal{R}(T)$ is finite.

In this case, the *index* of T is defined by the formula

$$\operatorname{ind} T = \dim \mathcal{N}(T) - \operatorname{codim} \mathcal{R}(T)\,.$$

By using Theorem 5.17, we can construct a volume potential for A, which plays the same role for A as the Newtonian potential plays for the Laplacian (cf. Seeley [Se1, Theorem 1]; [Ta2, Theorem 8.2.1]):

Theorem 6.1. *(i) The operator $A: C^\infty(M) \to C^\infty(M)$ is bijective, and its inverse Q is a classical, elliptic pseudo-differential operator of order -2 on M.*
(ii) The operators A and Q extend respectively to isomorphisms

$$A: H^{s,p}(M) \longrightarrow H^{s-2,p}(M) ,$$
$$Q: H^{s-2,p}(M) \longrightarrow H^{s,p}(M)$$

for all $s \in \mathbf{R}$ and $1 < p < \infty$, which are still inverses of each other.

Proof. Since the principal symbol of A is real, we find (see [Ta2, Corollary 6.7.12]) that

$$\operatorname{ind} A = 0 .$$

Therefore, it suffices to show that

$$\mathcal{N}(A) := \{u \in C^\infty(M): Au = 0 \text{ in } M\} = \{0\} .$$

However, applying the strong maximum principle (Theorem C.3 in Appendix C) we obtain that

$$\mathcal{N}(A) \subset \{\text{constants functions}\} .$$

By virtue of condition (6.2), this implies that

$$\mathcal{N}(A) = \{0\} . \qquad \square$$

Next we construct a surface potential for A, which is a generalization of the classical Poisson kernel for the Laplacian.

We let

$$K\psi = Q(\psi \otimes \delta)|_{\partial\Omega} , \quad \psi \in C^\infty(\partial\Omega) ,$$

where $\psi \otimes \delta$ is a distribution on M defined by the formula

$$\langle \psi \otimes \delta, \varphi \cdot \mu \rangle = \langle \psi, \varphi(\cdot, 0) \cdot \omega \rangle , \quad \varphi \in C^\infty(M) .$$

Then, by using Theorem 5.16 we can prove the following (see [Ta2, Theorem 8.2.2]):

Theorem 6.2. *(i) The operator K is a classical, elliptic pseudo-differential operator of order -1 on $\partial\Omega$.*
(ii) The operator $K: C^\infty(\partial\Omega) \to C^\infty(\partial\Omega)$ is bijective, and its inverse L is a first-order, classical elliptic pseudo-differential operator on $\partial\Omega$. Furthermore, the operators K and L extend respectively to isomorphisms

$$K: B^{\sigma,p}(\partial\Omega) \longrightarrow B^{\sigma+1,p}(\partial\Omega),$$
$$L: B^{\sigma+1,p}(\partial\Omega) \longrightarrow B^{\sigma,p}(\partial\Omega)$$

for all $\sigma \in \mathbf{R}$ and $1 < p < \infty$, which are still inverses of each other.

Proof. Assertion (i): We calculate the homogeneous principal symbol of $K \in L^{-1}_{cl}(\partial\Omega)$. In a neighborhood W of $\partial\Omega$ in M, we can write the operator $A = A(x, D)$ uniquely in the form

$$A(x, D) = A_2(x)D_t^2 + A_1(x, D_{x'})D_t + A_0(x, D_{x'}) , \quad x = (x', t) , \tag{6.3}$$

where $A_j(x, D_{x'})$ $(j = 0, 1, 2)$ is a differential operator of order $2 - j$ acting along the surfaces parallel to $\partial\Omega$. We denote by $a_1(x, \xi')$ and $a_0(x, \xi')$ the principal symbols of $A_1(x, D_{x'})$ and $A_0(x, D_{x'})$, respectively. Since A is elliptic on M, it follows that

(a) $A_2(x) < 0$, $x \in W$;
(b) $a_1(x, \xi')^2 - 4A_2(x)a_0(x, \xi') < 0$, $x = (x', t) \in W$, $\xi' \in T^*_{x'}(\partial\Omega) \setminus \{0\}$.

Hence the principal symbol of A can be decomposed as follows:

$$A_2(x)\xi_n^2 + a_1(x, \xi')\xi_n + a_0(x, \xi') = A_2(x) \left(\xi_n - \xi_n^+(x, \xi')\right) \left(\xi_n - \xi_n^-(x, \xi')\right)$$

where

$$\xi_n^\pm(x, \xi') = \frac{-a_1(x, \xi') \mp \sqrt{-1}\left(4A_2(x)a_0(x, \xi') - a_1(x, \xi')^2\right)^{1/2}}{2A_2(x)} .$$

Since the principal symbol of $Q = A^{-1}$ is

$$\frac{1}{A_2(x)\left(\xi_n - \xi_n^+(x, \xi')\right)\left(\xi_n - \xi_n^-(x, \xi')\right)} ,$$

applying part (ii) of Theorem 5.16 to our situation we obtain that the homogeneous principal symbol $k(x', \xi')$ of K is given by the following formula:

$$\begin{aligned} k(x', \xi') &= \frac{1}{2\pi} \int_\Gamma \frac{d\xi_n}{A_2(x', 0)\left(\xi_n - \xi_n^+(x', 0, \xi')\right)\left(\xi_n - \xi_n^-(x', 0, \xi')\right)} \\ &= -\frac{1}{\left(4A_2(x', 0)a_0(x', 0, \xi') - a_1(x', 0, \xi')^2\right)^{1/2}} . \end{aligned} \tag{6.4}$$

This proves that $K \in L^{-1}_{cl}(\partial\Omega)$ is elliptic.

For example, if the operator A is the usual Laplacian Δ, then it follows that

$$k(x', \xi') = -\frac{1}{2|\xi'|}$$

where $|\xi'|$ is the length of ξ' with respect to the Riemannian metric of $\partial\Omega$ induced by the natural metric of $\mathbf{R}^n$.

Assertion (ii): First, since the homogeneous principal symbol $k(x', \xi')$ of K is real, we obtain that

$$\operatorname{ind} K = 0 .$$

Now we show that

$$\mathcal{N}(K) = \{v \in C^\infty(\partial\Omega): Kv = 0\} = \{0\} ;$$

then part (ii) of Theorem 6.2 follows.

Assume that

$$v \in C^\infty(\partial\Omega) \text{ and } Kv = 0 .$$

Then, applying part (i) of Theorem 5.16 to our situation, we obtain that

$$Q(v \otimes \delta)|_\Omega \in C^\infty(\overline{\Omega}) , \tag{6.5}$$

$$Q(v \otimes \delta)|_{M\setminus\overline{\Omega}} \in C^\infty(M \setminus \Omega) \tag{6.6}$$

and also

$$Q(v \otimes \delta)|_{\partial\Omega} = Kv = 0 . \tag{6.7}$$

However, we have

$$A\left(Q(v \otimes \delta)|_\Omega\right) = AQ(v \otimes \delta)|_\Omega = v \otimes \delta|_\Omega = 0 \quad \text{in } \Omega , \tag{6.8}$$

since A is a differential operator. Therefore, in view of assertions (6.5), (6.8) and (6.7) we can apply the weak maximum principle (Theorem C.1 in Appendix C) to obtain that

$$Q(v \otimes \delta) = 0 \quad \text{on } \overline{\Omega} . \tag{6.9}$$

This gives that

$$Q(v \otimes \delta) = \left(Q(v \otimes \delta)|_{M\setminus\overline{\Omega}}\right)^0 .$$

Thus it follows from an application of the jump formula (5.8) that

$$\begin{aligned} v \otimes \delta &= AQ(v \otimes \delta) \\ &= A\left(Q(v \otimes \delta)|_{|M\setminus\overline{\Omega}}\right)^0 \\ &= \left(AQ(v \otimes \delta)|_{M\setminus\overline{\Omega}}\right)^0 + \frac{1}{i}\left\{A_2(x)\left(D_t Q(v \otimes \delta)|_{\partial'\Omega}\right) \otimes \delta\right. \\ &\qquad \left. + A_2(x)\left(Q(v \otimes \delta)|_{\partial'\Omega}\right) \otimes D_t\delta + A_1(x, D_{x'})\left(Q(v \otimes \delta)|_{\partial'\Omega}\right) \otimes \delta\right\} \\ &= \frac{1}{i}\left\{A_2(x)\left(D_t Q(v \otimes \delta)|_{\partial'\Omega}\right) + A_1(x, D_{x'})\left(Q(v \otimes \delta)|_{\partial'\Omega}\right)\right\} \otimes \delta \\ &\qquad + \frac{1}{i} A_2(x)\left(Q(v \otimes \delta)|_{\partial'\Omega}\right) \otimes D_t\delta , \quad i = \sqrt{-1} , \end{aligned} \tag{6.10}$$

where

$$u|_{\partial'\Omega} = \text{the trace of } u \text{ on } \partial\Omega \text{ from } M \setminus \overline{\Omega} .$$

In order that formula (6.10) hold, the last term on the right-hand side must vanish; hence we have

$$Q(v \otimes \delta)|_{\partial' \Omega} = 0 \,, \tag{6.11}$$

since $A_2(x) < 0$ in W. However, we have

$$A\left(Q(v \otimes \delta)|_{M \setminus \overline{\Omega}}\right) = AQ(v \otimes \delta)|_{M \setminus \overline{\Omega}} = v \otimes \delta|_{M \setminus \overline{\Omega}} = 0 \,. \tag{6.12}$$

Therefore, in view of assertions (6.6), (6.12) and (6.11) we can apply the maximum principle to obtain that

$$Q(v \otimes \delta) = 0 \quad \text{on } M \setminus \Omega \,. \tag{6.13}$$

Consequently it follows from assertions (6.9) and (6.13) that

$$Q(v \otimes \delta) = 0 \quad \text{on } M \,.$$

Since the operator Q is invertible, this implies that

$$v \otimes \delta = 0 \quad \text{on } M \,,$$

so that

$$v = 0 \quad \text{on } \partial\Omega \,.$$

The proof of Theorem 6.2 is complete. □

Now we prove the Poisson integral formula for the solution of the Dirichlet problem

$$\begin{cases} Au = 0 & \text{in } \Omega, \\ u = \varphi & \text{on } \partial\Omega, \end{cases} \tag{6.14}$$

by generalizing formula (4.14). The candidate for a solution of problem (6.14) is given by the formula

$$P\varphi = Q(L\varphi \otimes \delta)|_{\Omega} \,, \quad \varphi \in C^\infty(\partial\Omega) \,.$$

Indeed, it follows that P maps $C^\infty(\partial\Omega)$ continuously into $C^\infty(\overline{\Omega})$ and extends to a continuous linear operator

$$P: B^{s-1/p,p}(\partial\Omega) \longrightarrow H^{s,p}(\Omega)$$

for all $s \in \mathbf{R}$ and $1 < p < \infty$. Moreover, we have, for all $\varphi \in B^{s-1/p,p}(\partial\Omega)$,

$$\begin{cases} AP\varphi = AQ(L\varphi \otimes \delta)|_{\Omega} = (L\varphi \otimes \delta)|_{\Omega} = 0 & \text{in } \Omega, \\ P\varphi|_{\partial\Omega} = KL\varphi = \varphi & \text{on } \partial\Omega. \end{cases}$$

The operator P is called the *Poisson operator*.

We let

$$\mathcal{N}(A, s, p) = \{u \in H^{s,p}(\Omega) : Au = 0 \text{ in } \Omega\} \,, \quad s \in \mathbf{R} \,.$$

Since the injection $H^{s,p}(\Omega) \to \mathcal{D}'(\Omega)$ is continuous, it follows that the space $\mathcal{N}(A, s, p)$ is a closed subspace of $H^{s,p}(\Omega)$; hence it is a Banach space.

Then we have the following (cf. Seeley [Se2, Theorems 5 and 6]):

Theorem 6.3. *The Poisson operator P maps $B^{s-1/p,p}(\partial\Omega)$ isomorphically onto $\mathcal{N}(A,s,p)$ for all $s \in \mathbf{R}$ and $1 < p < \infty$.*

It should be emphasized that the spaces $\mathcal{N}(A,s,p)$ and $B^{s-1/p,p}(\partial\Omega)$ are isomorphic in such a way that

$$\mathcal{N}(A,s,p) \xrightarrow{\gamma_0} B^{s-1/p,p}(\partial\Omega) ,$$
$$\mathcal{N}(A,s,p) \underset{P}{\longleftarrow} B^{s-1/p,p}(\partial\Omega) .$$

Combining Theorems 6.1 and 6.3, we can obtain the following uniqueness and existence theorem for the Dirichlet problem in the framework of Sobolev spaces of L^p style (cf. Agmon–Douglis–Nirenberg [ADN], Lions–Magenes [LM], Gilbarg–Trudinger [GT]):

Theorem 6.4. *Let $s > 1/p$ where $1 < p < \infty$. Then the Dirichlet problem (D) has a unique solution u in $H^{s,p}(\Omega)$ for any $f \in H^{s-2,p}(\Omega)$ and any $\varphi \in B^{s-1/p,p}(\partial\Omega)$.*

Next, we consider the Neumann problem

$$\begin{cases} Au = f & \text{in } \Omega, \\ \dfrac{\partial u}{\partial \mathbf{n}} = \varphi & \text{on } \partial\Omega \end{cases} \tag{N}$$

Then we have the following (cf. [ADN], [LM], [GT]):

Theorem 6.5. *Let $s > 1+1/p$ where $1 < p < \infty$. Then the Neumann problem (N) has a unique solution u in the space $H^{s,p}(\Omega)$ for any $f \in H^{s-2,p}(\Omega)$ and any $\varphi \in B^{s-1-1/p,p}(\partial\Omega)$.*

By Theorem 6.5, we can introduce a linear operator

$$G_N : H^{s-2,p}(\Omega) \longrightarrow H^{s,p}(\Omega)$$

as follows: For any $f \in H^{s-2,p}(\Omega)$, the function $G_N f$ is the unique solution of the problem

$$\begin{cases} Au = f & \text{in } \Omega, \\ \dfrac{\partial u}{\partial \mathbf{n}} = 0 & \text{on } \partial\Omega. \end{cases}$$

The operator G_N is called the *Green operator* for the Neumann problem.

6.2 Formulation of a Boundary Value Problem

Let $s > 1 + 1/p$ for $1 < p < \infty$. If $u \in H^{s,p}(\Omega)$, we define its traces $\gamma_0 u$ and $\gamma_1 u$ respectively by the formulas

$$\begin{cases} \gamma_0 u = u|_{\partial\Omega}, \\ \gamma_1 u = \dfrac{\partial u}{\partial \mathbf{n}}, \end{cases}$$

and let

$$\gamma u = \{\gamma_0 u, \gamma_1 u\} .$$

Then we have the following (see Bergh–Löfström [BL], Stein [Sn1], Taibleson [Tb], Triebel [Tr]):

Theorem 6.6 (the trace theorem). *Let $1 < p < \infty$. Then the trace map*

$$\gamma = (\gamma_0, \gamma_1): H^{s,p}(\Omega) \longrightarrow B^{s-1/p,p}(\partial\Omega) \bigoplus B^{s-1-1/p,p}(\partial\Omega)$$

is continuous for all $s > 1 + 1/p$.

Now we consider a boundary condition

$$B_0 u = a(x')\frac{\partial u}{\partial \mathbf{n}} + b(x')\,(u|_{\partial\Omega}) = a(x')(\gamma_1 u) + b(x')(\gamma_0 u) , \qquad u \in H^{s,p}(\Omega) , \tag{6.15}$$

where $a(x')$ and $b(x')$ are real-valued, smooth functions on $\partial\Omega$ such that

(H.1) $a(x') \geq 0$ and $b(x') \leq 0$ on $\partial\Omega$.
(H.2) $a(x') + |b(x')| > 0$ on $\partial\Omega$.

It is worth while pointing out here that the boundary condition B_0 and condition (H.2) are essentially the boundary condition L_0 and condition (H), respectively, introduced in Chap. 1.

We introduce a subspace of $B^{s-1-1/p,p}(\partial\Omega)$ which is associated with the boundary condition B_0 in the following way: We let

$$\begin{aligned} &B_{B_0}^{s-1-1/p,p}(\partial\Omega) \\ &= \left\{\varphi = a(x')\varphi_1 - b(x')\varphi_2 : \varphi_1 \in B^{s-1-1/p,p}(\partial\Omega),\ \ \varphi_2 \in B^{s-1/p,p}(\partial\Omega)\right\}, \end{aligned}$$

and define a norm

$$\begin{aligned} &|\varphi|_{B_{B_0}^{s-1-1/p,p}(\partial\Omega)} \\ &= \inf\left\{|\varphi_1|_{B^{s-1-1/p,p}(\partial\Omega)} + |\varphi_2|_{B^{s-1/p,p}(\partial\Omega)} : \varphi = a(x')\varphi_1 - b(x')\varphi_2\right\} . \end{aligned}$$

Then it is easy to verify that the space $B_{B_0}^{s-1-1/p,p}(\partial\Omega)$ is a Banach space with respect to the norm $|\cdot|_{B_{B_0}^{s-1-1/p,p}(\partial\Omega)}$. It should be noticed that

$$\begin{cases} B_{B_0}^{s-1-1/p,p}(\partial\Omega) = B^{s-1/p,p}(\partial\Omega) & \text{if } a(x') \equiv 0 \text{ on } \partial D, \\ B_{B_0}^{s-1-1/p,p}(\partial\Omega) = B^{s-1-1/p,p}(\partial\Omega) & \text{if } a(x') > 0 \text{ on } \partial D. \end{cases}$$

Moreover, using Theorem 6.6 we can prove the following:

Proposition 6.7. *If conditions (H.1) and (H.2) are satisfied, then the mapping*

$$B_0: H^{s,p}(\Omega) \longrightarrow B_{B_0}^{s-1-1/p,p}(\partial\Omega)$$

is continuous for $s > 1 + 1/p$ where $1 < p < \infty$.

Proof. Indeed, by formula (6.15) and the definition of $B_{B_0}^{s-1-1/p,p}(\partial\Omega)$ it follows that

$$\begin{aligned}|B_0 u|_{B_{B_0}^{s-1-1/p,p}(\partial\Omega)} &\leq |\gamma_1 u|_{B^{s-1-1/p,p}(\partial\Omega)} + |\gamma_0 u|_{B^{s-1/p,p}(\partial\Omega)} \\ &\leq C\|u\|_{s,p} \,.\end{aligned}$$

This proves the continuity of the mapping B_0. □

Now we can formulate our boundary value problem for (A, B_0) as follows: Given functions $f \in H^{s-2,p}(\Omega)$ and $\varphi \in B_{B_0}^{s-1-1/p,p}(\partial\Omega)$, find a function $u \in H^{s,p}(\Omega)$ such that

$$\begin{cases} Au = f & \text{in } \Omega, \\ B_0 u = \varphi & \text{on } \partial\Omega. \end{cases} \tag{$*$}$$

6.3 Reduction to the Boundary

In this section, using the Dirichlet and Neumann problems we show that problem $(*)$ can be reduced to the study of a pseudo-differential operator on the boundary.

Let f be an arbitrary element of $H^{s-2,p}(\Omega)$, and φ an arbitrary element of $B_{B_0}^{s-1-1/p,p}(\partial\Omega)$ such that

$$\varphi = a(x')\varphi_1 - b(x')\varphi_2 \,, \quad \varphi_1 \in B^{s-1-1/p,p}(\partial\Omega) \,, \quad \varphi_2 \in B^{s-1/p,p}(\partial\Omega) \,,$$

where $1 < p < \infty$ and $s > 1 + 1/p$. First, we consider the following Neumann problem:

$$\begin{cases} Av = f & \text{in } \Omega, \\ \dfrac{\partial v}{\partial \mathbf{n}} = \varphi_1 & \text{on } \partial\Omega. \end{cases} \tag{N}$$

By Theorem 6.5, we can find a unique solution $v \in H^{s,p}(\Omega)$ of problem (N). Then it is easy to see that a function u in $H^{s,p}(\Omega)$ is a solution of problem

$$\begin{cases} Au = f & \text{in } \Omega, \\ B_0 u = \varphi & \text{on } \partial\Omega \end{cases} \tag{$*$}$$

if and only if $w = u - v \in H^{s,p}(\Omega)$ is a solution of the problem

$$\begin{cases} Aw = 0 & \text{in } \Omega, \\ B_0 w = \varphi - B_0 v = -b(x')\left(v|_{\partial\Omega} + \varphi_2\right) & \text{on } \partial\Omega. \end{cases} \tag{6.16}$$

However, Theorem 6.3 tells us that the spaces $\mathcal{N}(A, s, p)$ and $B^{s-1/p,p}(\partial\Omega)$ are isomorphic in such a way that

$$\begin{aligned}\mathcal{N}(A, s, p) &\xrightarrow{\gamma_0} B^{s-1/p,p}(\partial\Omega) \,, \\ \mathcal{N}(A, s, p) &\underset{P}{\longleftarrow} B^{s-1/p,p}(\partial\Omega) \,.\end{aligned}$$

Therefore, we find that $w \in H^{s,p}(\Omega)$ is a solution of problem (6.16) if and only if $\psi \in B^{s-1/p,p}(\partial\Omega)$ is a solution of the equation

$$B_0 P\psi = -b(x')\,(\gamma_0 v + \varphi_2) \quad \text{on } \partial\Omega\,. \tag{$**$}$$

Here $\psi = \gamma_0 w$, or equivalently, $w = P\psi$.

Summing up, we obtain the following:

Proposition 6.8. *Let $1 < p < \infty$ and $s > 1 + 1/p$. For given functions $f \in H^{s-2,p}(\Omega)$ and $\varphi \in B_{B_0}^{s-1-1/p,p}(\partial\Omega)$, there exists a solution $u \in H^{s,p}(\Omega)$ of problem $(*)$ if and only if there exists a solution $\psi \in B^{s-1/p,p}(\partial\Omega)$ of equation $(**)$.*

We remark that equation $(**)$ is a generalization of the classical *Fredholm integral equation.*

We let

$$\begin{aligned} T\colon C^\infty(\partial\Omega) &\longrightarrow C^\infty(\partial\Omega) \\ \varphi &\longmapsto B_0 P\varphi\,. \end{aligned}$$

Then we have, by condition (6.15),

$$T = B_0 P = a(x')\Pi + b(x')\,,$$

where

$$\Pi\varphi = \gamma_1(P\varphi) = \frac{\partial}{\partial \mathbf{n}}(P\varphi)\,.$$

It is known (see [Ho3, Chap. XX], [Se2], [RS, Chap. 3]) that the operator Π is a first-order, classical pseudo-differential operator on $\partial\Omega$; hence the operator

$$T = B_0 P = a(x')\Pi + b(x')$$

is a first-order, classical pseudo-differential operator on the boundary $\partial\Omega$.

Consequently, Proposition 6.8 asserts that problem $(*)$ can be reduced to the study of the pseudo-differential operator T on the boundary $\partial\Omega$. We shall formulate this fact more precisely in terms of functional analysis.

First, it should be noticed that the operator $T\colon C^\infty(\partial\Omega) \to C^\infty(\partial\Omega)$ extends to a continuous linear operator $T\colon B^{\sigma,p}(\partial\Omega) \to B^{\sigma-1,p}(\partial\Omega)$ for all $\sigma \in \mathbf{R}$.

By Proposition 6.7, we can associate with problem $(*)$ a continuous linear operator

$$\mathcal{A} = (A, B_0)\colon H^{s,p}(\Omega) \longrightarrow H^{s-2,p}(\Omega) \bigoplus B_{B_0}^{s-1-1/p,p}(\partial\Omega)$$

as follows.

(a) The domain $\mathcal{D}(\mathcal{A})$ of $\mathcal{A}$ is the space $H^{s,p}(\Omega)$.
(b) $\mathcal{A}u = \{Au, B_0 u\}$, $u \in \mathcal{D}(\mathcal{A})$.

Similarly, we associate with equation (**) a linear operator

$$\mathcal{T}: B^{s-1/p,p}(\partial\Omega) \longrightarrow B^{s-1/p,p}(\partial\Omega)$$

as follows.

(α) The domain $\mathcal{D}(\mathcal{T})$ of $\mathcal{T}$ is the space

$$\mathcal{D}(\mathcal{T}) = \left\{\varphi \in B^{s-1/p,p}(\partial\Omega): T\varphi \in B^{s-1/p,p}(\partial\Omega)\right\}.$$

(β) $\mathcal{T}\varphi = T\varphi$, $\varphi \in \mathcal{D}(\mathcal{T})$.

It should be emphasized that the operator $\mathcal{T}$ is a densely defined, closed linear operator, since the operator $T: B^{s-1/p,p}(\partial\Omega) \to B^{s-1-1/p,p}(\partial\Omega)$ is continuous and since the domain $\mathcal{D}(\mathcal{T})$ contains the space $C^\infty(\partial\Omega)$.

Then Proposition 6.8 can be reformulated in the following form (see [Ta2, Sect. 8.3]):

Theorem 6.9. *(i) The null space $\mathcal{N}(\mathcal{A})$ of $\mathcal{A}$ has finite dimension if and only if the null space $\mathcal{N}(\mathcal{T})$ of $\mathcal{T}$ has finite dimension, and we have*

$$\dim \mathcal{N}(\mathcal{A}) = \dim \mathcal{N}(\mathcal{T}) .$$

(ii) The range $\mathcal{R}(\mathcal{A})$ of $\mathcal{A}$ is closed if and only if the range $\mathcal{R}(\mathcal{T})$ of $\mathcal{T}$ is closed; and $\mathcal{R}(\mathcal{A})$ has finite codimension if and only if $\mathcal{R}(\mathcal{T})$ has finite codimension, and we have

$$\operatorname{codim} \mathcal{R}(\mathcal{A}) = \operatorname{codim} \mathcal{R}(\mathcal{T}) .$$

(iii) The operator $\mathcal{A}$ is a Fredholm operator if and only if the operator $\mathcal{T}$ is a Fredholm operator, and we have

$$\operatorname{ind} \mathcal{A} = \operatorname{ind} \mathcal{T} .$$

Furthermore, the next theorem states that the operator $\mathcal{A}$ has the regularity property if and only if the operator $\mathcal{T}$ has:

Theorem 6.10. *Let $1 < p < \infty$ and $s > 1+1/p$. The following two conditions are equivalent:*

(i)
$$u \in L^p(\Omega)\ , \quad Au \in H^{s-2,p}(\Omega)\ , \quad B_0 u \in B_{B_0}^{s-1-1/p,p}(\partial\Omega) \Longrightarrow u \in H^{s,p}(\Omega)\ . \tag{6.17}$$

(ii) $$\varphi \in B^{-1/p,p}(\partial\Omega)\ , \quad T\varphi \in B^{s-1/p,p}(\partial\Omega) \Longrightarrow \varphi \in B^{s-1/p,p}(\partial\Omega)\ . \tag{6.18}$$

Proof. **Step 1**: The proof of the implication (6.17) $\Longrightarrow$ (6.18). Assume that

$$\varphi \in B^{-1/p,p}(\partial\Omega) \quad \text{and} \quad T\varphi \in B^{s-1/p,p}(\partial\Omega)\ .$$

Then, letting $u = P\varphi$ we obtain that

$$u \in L^p(\Omega) , \quad Au = 0 \quad \text{and} \quad B_0 u = T\varphi \in B^{s-1/p,p}(\partial\Omega) \subset B_{B_0}^{s-1-1/p,p}(\partial\Omega) .$$

Hence it follows from condition (6.17) that

$$u \in H^{s,p}(\Omega) ,$$

so that, by Theorem 6.6,

$$\varphi = \gamma_0 u \in B^{s-1/p,p}(\partial\Omega) .$$

Step 2: The proof of the implication (6.18) $\Longrightarrow$ (6.17). Conversely, assume that

$$u \in L^p(\Omega) , \quad Au \in H^{s-2,p}(\Omega) \quad \text{and} \quad B_0 u \in B_{B_0}^{s-1-1/p,p}(\partial\Omega) ,$$

where

$$B_0 u = a(x')\varphi_1 - b(x')\varphi_2 ,$$

with

$$\varphi_1 \in B^{s-1-1/p,p}(\partial\Omega) , \quad \varphi_2 \in B^{s-1/p,p}(\partial\Omega) .$$

Let $v \in H^{s,p}(\Omega)$ be a unique solution of the Neumann problem

$$\begin{cases} Av = Au & \text{in } \Omega, \\ \dfrac{\partial v}{\partial \mathbf{n}} = \varphi_1 & \text{on } \partial\Omega. \end{cases}$$

Then the function $u \in L^p(\Omega)$ can be decomposed in the form

$$u = v + w ,$$

where

$$w = u - v \in \mathcal{N}(A, 0, p) .$$

However, Theorem 6.3 tells us that the distribution w can be written as

$$w = P\varphi , \quad \varphi = \gamma_0 w \in B^{-1/p,p}(\partial\Omega) .$$

Hence we have, by Theorem 6.6,

$$T\varphi = B_0 P\varphi = B_0 u - B_0 v = -b(x')(\gamma_0 v + \varphi_2) \in B^{s-1/p,p}(\partial\Omega) ,$$

since $\gamma_1 v = \varphi_1$. Thus it follows from condition (5.18) that

$$\varphi \in B^{s-1/p,p}(\partial\Omega) ,$$

so that, again by Theorem 6.3,

$$w = P\varphi \in H^{s,p}(\Omega) .$$

This proves that

$$u = v + w \in H^{s,p}(\Omega) .$$

The proof of Theorem 6.10 is complete. $\square$

The next theorem states that *a priori* estimates for $\mathcal{A}$ are entirely equivalent to corresponding *a priori* estimates for $\mathcal{T}$:

Theorem 6.11. *The following two estimates are equivalent:*

$$(i) \qquad \|u\|_{s,p} \le C\left(\|Au\|_{s-2,p} + |B_0 u|_{B^{s-1-1/p,p}_{B_0}} + \|u\|_p\right), \quad u \in \mathcal{D}(\mathcal{A}) \,. \tag{6.19}$$

$$(ii) \qquad |\varphi|_{s-1/p,p} \le C\left(|\mathcal{T}\varphi|_{s-1/p,p} + |\varphi|_{-1/p,p}\right), \quad \varphi \in \mathcal{D}(\mathcal{T}) \,. \tag{6.20}$$

Here and in the following the letter C denotes a generic positive constant.

Proof. **Step 1**: The proof of the implication (6.19) $\Longrightarrow$ (6.20). Taking $u = P\varphi$ with $\varphi \in \mathcal{D}(\mathcal{T})$ in estimate (6.19), we obtain that

$$\|P\varphi\|_{s,p} \le C\left(|\mathcal{T}\varphi|_{B^{s-1-1/p,p}_{B_0}} + \|P\varphi\|_p\right) . \tag{6.21}$$

However, Theorem 6.3 tells us that the operator P maps $B^{\sigma-1/p,p}(\partial\Omega)$ isomorphically onto $\mathcal{N}(A,\sigma,p)$ for all $\sigma \in \mathbf{R}$. Thus estimate (6.20) follows from estimate (6.21).

Step 2: The proof of the implication (6.20) $\Longrightarrow$ (6.19). We decompose a function $u \in \mathcal{D}(\mathcal{A})$ as in the proof of Theorem 6.10

$$u = v + w \,,$$
$$B_0 u = a(x')\varphi_1 - b(x')\varphi_2 \,,$$

where $v \in H^{s,p}(\Omega)$ is a unique solution of the Neumann problem

$$\begin{cases} Av = Au & \text{in } \Omega, \\ \dfrac{\partial v}{\partial \mathbf{n}} = \varphi_1 & \text{on } \partial\Omega \end{cases}$$

and

$$w = u - v \in \mathcal{N}(A,s,p) \,.$$

Then we have

$$\|v\|_{s,p} \le C\left(\|Au\|_{s-2,p} + |\varphi_1|_{s-1-1/p,p}\right) . \tag{6.22}$$

Furthermore, applying estimate (6.20) to the function $\gamma_0 w$ we obtain that

$$\begin{aligned} |\gamma_0 w|_{s-1/p,p} &\le C\left(|\mathcal{T}(\gamma_0 w)|_{s-1/p,p} + |\gamma_0 w|_{-1/p,p}\right) \\ &= C\left(|B_0 w|_{s-1/p,p} + |\gamma_0 w|_{-1/p,p}\right) \\ &\le C\left(|\varphi_2|_{s-1/p,p} + |\gamma_0 v|_{s-1/p,p} + |\gamma_0 w|_{-1/p,p}\right) \\ &\le C\left(|\varphi_2|_{s-1/p,p} + \|v\|_{s,p} + |\gamma_0 w|_{-1/p,p}\right) . \end{aligned}$$

In view of Theorem 6.3, this gives that

$$\begin{aligned} \|w\|_{s,p} &\le C\left(|\varphi_2|_{s-1/p,p} + \|v\|_{s,p} + \|w\|_p\right) \\ &\le C\left(|\varphi_2|_{s-1/p,p} + \|v\|_{s,p} + \|u\|_p\right). \end{aligned} \tag{6.23}$$

Thus, it follows from estimates (6.22) and (6.23) that

$$\|w\|_{s,p} \le C\left(\|Au\|_{s-2,p} + |\varphi_1|_{s-1-1/p,p} + |\varphi_2|_{s-1/p,p} + \|u\|_p\right). \tag{6.24}$$

Therefore, the desired estimate (6.19) follows by combining estimates (6.22) and (6.24), since $u = v + w$. □

7

Elliptic Boundary Value Problems and Feller Semigroups

In the early 1950's, W. Feller characterized completely the analytic structure of one-dimensional diffusion processes; he gave an intrinsic representation of the infinitesimal generator $\mathfrak{A}$ of a one-dimensional diffusion process and determined all possible boundary conditions which describe the domain $D(\mathfrak{A})$. The probabilistic meaning of Feller's work was clarified by E. B. Dynkin, K. Itô, H. P. McKean, Jr., D. B. Ray and others. One-dimensional diffusion processes are completely studied both from analytic and probabilistic viewpoints.

The main purpose of this chapter is to generalize Feller's work to the multidimensional case. In 1959, A. D. Ventcel' studied the problem of determining all possible boundary conditions for multidimensional diffusion processes, which we formulate precisely. The results discussed here are adapted from Bony–Courrège–Priouret [BCP], Taira [Ta2] and [Ta4] (cf. Cancelier [Cn]).

7.1 Formulation of a Problem

Let D be a bounded domain of Euclidean space $\mathbf{R}^N$, with smooth boundary ∂D; its closure $\overline{D} = D \cup \partial D$ is an N-dimensional, compact smooth manifold with boundary.

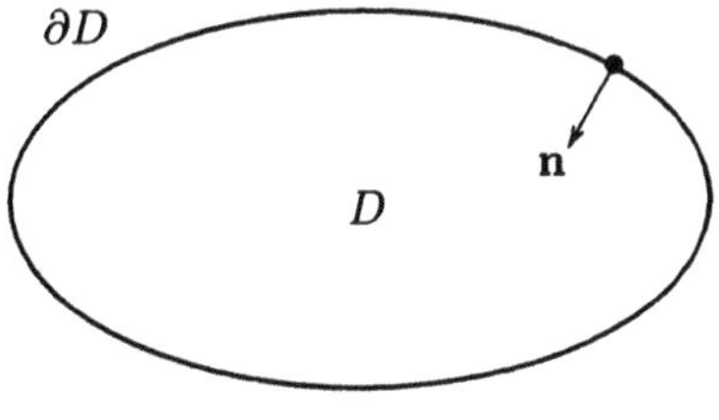

Fig. 7.1.

Let $C(\overline{D})$ be the space of real-valued, continuous functions f on $\overline{D}$. We equip the space $C(\overline{D})$ with the topology of uniform convergence on the whole $\overline{D}$; hence it is a Banach space with the maximum norm

$$\|f\|_\infty = \max_{x \in \overline{D}} |f(x)| .$$

A strongly continuous semigroup $\{T_t\}_{t \geq 0}$ on the space $C(\overline{D})$ is called a *Feller semigroup* on $\overline{D}$ if it is non-negative and contractive on $C(\overline{D})$:

$$f \in C(\overline{D}) , \; 0 \leq f(x) \leq 1 \quad \text{on } \overline{D} \Longrightarrow 0 \leq T_t f(x) \leq 1 \quad \text{on } \overline{D} .$$

It follows from an application of Theorem 3.2 that if T_t is a Feller semigroup on $\overline{D}$, then there exists a unique Markov transition function $p_t(x, \cdot)$ on $\overline{D}$ such that

$$T_t f(x) = \int_{\overline{D}} p_t(x, dy) f(y) , \quad f \in C(\overline{D}) .$$

It can be shown (see [Dy, Chap. III, Sect. 3]) that the function $p_t(x, \cdot)$ is the transition function of some strong *Markov process*; hence the value $p_t(x, E)$ expresses the transition probability that a Markovian particle starting at position x will be found in the set E at time t (see Fig. 3.2).

Furthermore, it is known (see [BCP], [SU], [Ta2], [Wa], [We]) that the infinitesimal generator of a Feller semigroup $\{T_t\}_{t \geq 0}$ is described analytically by a Waldenfels operator W and a Ventcel' boundary condition L, which we shall formulate precisely.

Let W be a second-order, *elliptic* integro-differential operator with real coefficients such that

$$\begin{aligned} Wu(x) &= Au(x) + S_r u(x) \\ &:= \sum_{i,j=1}^{N} a^{ij}(x) \frac{\partial^2 u}{\partial x_i \partial x_j}(x) + \sum_{i=1}^{N} b^i(x) \frac{\partial u}{\partial x_i}(x) + c(x)u(x) \\ &\quad + \int_D s(x,y) \left[u(y) - \sigma(x,y) \left(u(x) + \sum_{j=1}^{N} (y_j - x_j) \frac{\partial u}{\partial x_j}(x) \right) \right] dy , \end{aligned} \tag{7.1}$$

where:

(1) $a^{ij}(x) \in C^\infty(\mathbf{R}^N)$, $a^{ij}(x) = a^{ji}(x)$ and there exists a constant $a_0 > 0$ such that

$$\sum_{i,j=1}^{N} a^{ij}(x) \xi_i \xi_j \geq a_0 |\xi|^2 , \quad x \in \mathbf{R}^N , \; \xi \in \mathbf{R}^N .$$

(2) $b^i(x) \in C^\infty(\mathbf{R}^N)$.
(3) $c(x) \in C^\infty(\mathbf{R}^N)$ and $c(x) \leq 0$ in D.

(4) The integral kernel $s(x, y)$ is the distribution kernel of a properly supported, classical pseudo-differential operator $S \in L^{2-\kappa}_{1,0}(\mathbf{R}^N)$, $\kappa > 0$, which has the *transmission property* with respect to ∂D, and $s(x, y) \geq 0$ off the diagonal $\Delta_{\mathbf{R}^N} = \{(x, x) : x \in \mathbf{R}^N\}$ in $\mathbf{R}^N \times \mathbf{R}^N$. The measure dy is the Lebesgue measure on $\mathbf{R}^N$.

(5) The function $\sigma(x, y)$ is a local unity function on $\overline{D}$, that is, $\sigma(x, y)$ is a smooth function on $\overline{D} \times \overline{D}$ such that $\sigma(x, y) = 1$ in a neighborhood of the diagonal $\Delta_{\overline{D}} = \{(x, x) : x \in \overline{D}\}$ in $\overline{D} \times \overline{D}$ (see Sect. 3.4). The function $\sigma(x, y)$ depends on the shape of the domain D. More precisely, it depends on a family of local charts on D in each of which the Taylor expansion is valid for functions u. For example, if D is *convex*, we may take $\sigma(x, y) \equiv 1$ on $\overline{D} \times \overline{D}$.

(6) $W1(x) = c(x) + \int_D s(x, y)[1 - \sigma(x, y)]dy \leq 0$ in D.

The operator W is called a second-order *Waldenfels operator* (cf. [Wa]). The differential operator A is called a diffusion operator which describes analytically a strong Markov process with continuous paths in the interior D such as Brownian motion, and the functions $a^{ij}(x)$, $b^i(x)$ and $c(x)$ are called the diffusion coefficients, the drift coefficients and the termination coefficient, respectively. The operator S_r is called a second-order *Lévy operator* which is supposed to correspond to the jump phenomenon in the interior D; a Markovian particle moves by jumps to a random point, chosen with kernel $s(x, y)$, in the interior D. Therefore, the Waldenfels operator W is supposed to correspond to such a physical phenomenon that a Markovian particle moves both by jumps and continuously in the state space D (see Fig. 7.2).

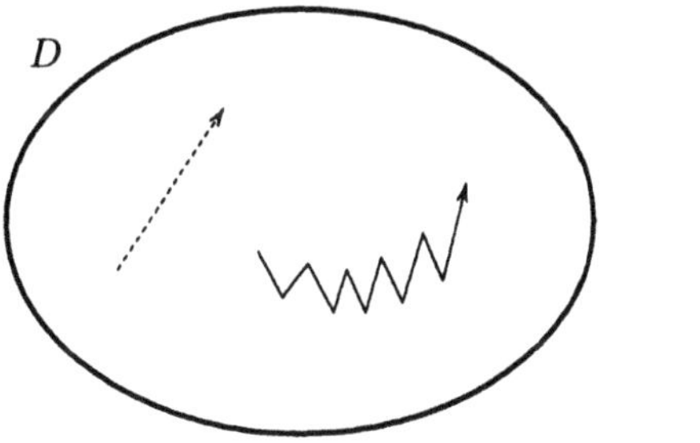

Fig. 7.2.

It should be noticed that the integral operator S_r is a "regularization" of S, since the integrand is absolutely convergent. Indeed, arguing as in the proof of Lemma A.10 in Appendix A we find that, for any compact $K \subset \mathbf{R}^N$ there exists a constant $C_K > 0$ such that the kernel $s(x, y)$ of $S \in L^{2-\kappa}_{1,0}(\mathbf{R}^N)$, $\kappa > 0$, satisfies the estimate

$$s(x, y) \leq \frac{C_K}{|x - y|^{N+2-\kappa}}, \quad x, y \in K, \ x \neq y.$$

Hence we have, with some constant $C > 0$,

$$\begin{aligned}
&\int_D s(x,y) \left| u(y) - \sigma(x,y) \left(u(x) + \sum_{j=1}^{N} (y_j - x_j) \frac{\partial u}{\partial x_j}(x) \right) \right| dy \\
&\le C \int_D \frac{1}{|x-y|^{N+2-\kappa}} \cdot |x-y|^2 \, dy \\
&= C \int_D \frac{1}{|x-y|^{N-\kappa}} \, dy \\
&< \infty .
\end{aligned}$$

Let L be a second-order boundary condition such that, in terms of local coordinates $(x_1, x_2, \ldots, x_{N-1})$,

$$\begin{aligned}
&Lu(x') \\
&= Qu(x') + \mu(x') \frac{\partial u}{\partial \mathbf{n}}(x') - \delta(x') W u(x') + \Gamma u(x') \\
&:= \sum_{i,j=1}^{N-1} \alpha^{ij}(x') \frac{\partial^2 u}{\partial x_i \partial x_j}(x') + \sum_{i=1}^{N-1} \beta^i(x') \frac{\partial u}{\partial x_i}(x') + \gamma(x') u(x') \\
&\quad + \mu(x') \frac{\partial u}{\partial \mathbf{n}}(x') - \delta(x') W u(x') \\
&\quad + \int_{\partial D} r(x',y') \left[u(y') - \tau(x',y') \left(u(x') + \sum_{j=1}^{N-1} (y_j - x_j) \frac{\partial u}{\partial x_j}(x') \right) \right] dy' \\
&\quad + \int_D t(x',y) \left[u(y) - \tau(x',y) \left(u(x') + \sum_{j=1}^{N-1} (y_j - x_j) \frac{\partial u}{\partial x_j}(x') \right) \right] dy, \qquad (7.2)
\end{aligned}$$

where:

(1) The operator Q is a second-order, *degenerate* elliptic differential operator on ∂D with non-positive principal symbol. In other words, the $\alpha^{ij}(x')$ are the components of a smooth symmetric contravariant tensor of type $\binom{2}{0}$ on ∂D satisfying the condition

$$\sum_{i,j=1}^{N-1} \alpha^{ij}(x') \xi_i \xi_j \ge 0 , \quad x' \in \partial D , \ \xi' = \sum_{j=1}^{N-1} \xi_j dx_j \in T^*_{x'}(\partial D) .$$

Here $T^*_{x'}(\partial D)$ is the cotangent space of ∂D at x'.

(2) $Q1(x') = \gamma(x') \in C^\infty(\partial D)$ and $\gamma(x') \le 0$ on ∂D.

(3) $\mu(x') \in C^\infty(\partial D)$ and $\mu(x') \ge 0$ on ∂D.

(4) $\delta(x') \in C^\infty(\partial D)$ and $\delta(x') \ge 0$ on ∂D.

(5) $\mathbf{n} = (n_1, n_2, \ldots, n_N)$ is the unit interior normal to the boundary ∂D (see Fig. 7.1).

(6) The integral kernel $r(x', y')$ is the distribution kernel of a classical pseudo-differential operator $R \in L^{2-\kappa_1}_{1,0}(\partial D)$, $\kappa_1 > 0$, and $r(x', y') \geq 0$ off the diagonal $\Delta_{\partial D} = \{(x', x') : x' \in \partial D\}$ in $\partial D \times \partial D$. The density dy' is a strictly positive density on ∂D.

(7) The integral kernel $t(x, y)$ is the distribution kernel of a properly supported, classical pseudo-differential operator $T \in L^{2-\kappa_2}_{1,0}(\mathbf{R}^N)$, $\kappa_2 > 0$, and which has the *transmission property* with respect to the boundary ∂D, and $t(x, y) \geq 0$ off the diagonal $\Delta_{\mathbf{R}^N}$.

(8) The function $\tau(x, y)$ is a local unity function on $\overline{D}$; more precisely, $\tau(x, y)$ is a smooth function on $\overline{D} \times \overline{D}$, with compact support in a neighborhood of the diagonal $\Delta_{\partial D}$, such that, at each point x' of ∂D, $\tau(x', y) = 1$ for y in a neighborhood of x' in $\overline{D}$. The function $\tau(x, y)$ depends on the shape of the boundary ∂D.

(9) The operator Γ is a boundary condition of order $2 - \min(\kappa_1; \kappa_2)$, and satisfies the condition

$$\Gamma 1(x') = \int_{\partial D} r(x', y')[1 - \tau(x', y')]dy' + \int_D t(x', y)[1 - \tau(x', y)]dy \leq 0 \quad \text{on } \partial D \,.$$

The boundary condition L is called a second-order *Ventcel' boundary condition* (cf. [We]). The six terms of L

$$\sum_{i,j=1}^{N-1} \alpha^{ij}(x') \frac{\partial^2 u}{\partial x_i \partial x_j}(x') + \sum_{i=1}^{N-1} \beta^i(x') \frac{\partial u}{\partial x_i}(x'),$$

$$\gamma(x')u(x'), \quad \mu(x') \frac{\partial u}{\partial \mathbf{n}}(x'), \quad \delta(x')Wu(x'),$$

$$\int_{\partial D} r(x', y') \left[u(y') - \tau(x', y') \left(u(x') + \sum_{j=1}^{N-1} (y_j - x_j) \frac{\partial u}{\partial x_j}(x') \right) \right] dy',$$

$$\int_D t(x', y) \left[u(y) - \tau(x', y) \left(u(x') + \sum_{j=1}^{N-1} (y_j - x_j) \frac{\partial u}{\partial x_j}(x') \right) \right] dy$$

are supposed to correspond to the diffusion phenomenon along the boundary (like Brownian motion on ∂D), the absorption phenomenon, the reflection phenomenon, the sticking (or viscosity) phenomenon and the jump phenomenon on the boundary and the inward jump phenomenon from the boundary, respectively (see Figs. 7.3 through 7.5).

This section is devoted to the functional analytic approach to the problem of construction of Markov processes with Ventcel' boundary conditions in probability theory. More precisely, we consider the following problem:

Problem. Conversely, given analytic data (W, L), can we construct a Feller semigroup $\{T_t\}_{t \geq 0}$ whose infinitesimal generator is characterized by (W, L)?

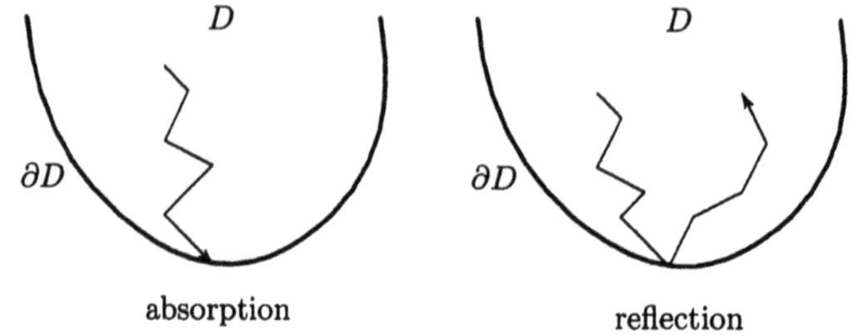

Fig. 7.3.

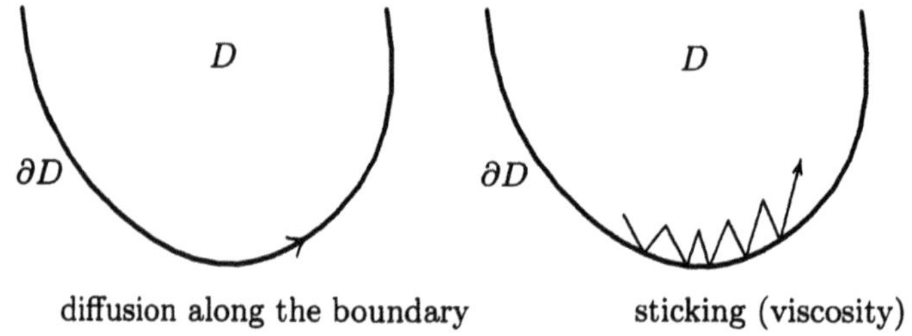

Fig. 7.4.

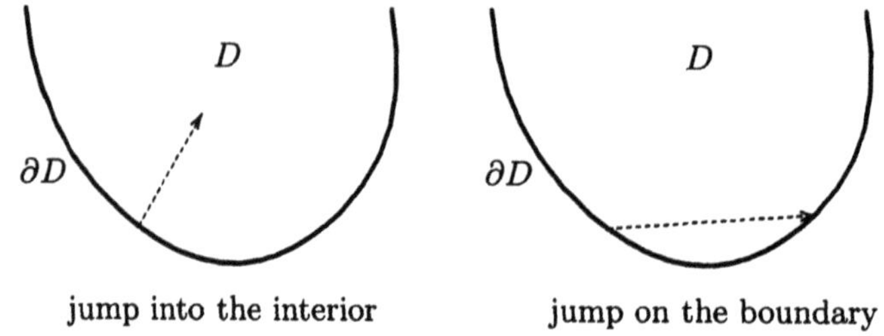

Fig. 7.5.

We shall only restrict ourselves to some aspects which have been discussed in our papers [Ta2] through [Ta6]. Our approach is distinguished by the extensive use of the ideas and techniques characteristic of the recent developments in the theory of partial differential equations. It focuses on the relationship between two interrelated subjects in analysis; Feller semigroups and elliptic boundary value problems, providing powerful methods for future research.

7.2 Transversal Case

First, we consider the transversal case. The boundary condition L is said to be *transversal* on the boundary ∂D if it satisfies the condition

$$\int_D t(x', y)\, dy = +\infty \quad \text{if } \mu(x') = \delta(x') = 0\,. \tag{7.3}$$

The intuitive meaning of condition (7.3) is that a Markovian particle jumps away "instantaneously" from the points $x' \in \partial D$ where neither reflection nor sticking phenomenon occurs (which is similar to the reflection phenomenon). Probabilistically, this means that every Markov process on the boundary ∂D is the "trace" on ∂D of trajectories of some Markov process on the closure $\overline{D} = D \cup \partial D$. $D \cup \partial D$ (see Remark 7.4 below).

7.2.1 Generation Theorem for Feller Semigroups

The next theorem asserts that there exists a Feller semigroup on $\overline{D}$ corresponding to such a physical phenomenon that one of the reflection phenomenon, the sticking phenomenon and the inward jump phenomenon from the boundary occurs at each point of the boundary ∂D (see Fig. 7.6 below):

Theorem 7.1. *We define a linear operator $\mathfrak{A}$ from the space $C(\overline{D})$ into itself as follows:*

(a) The domain of definition $D(\mathfrak{A})$ of $\mathfrak{A}$ is the set

$$D(\mathfrak{A}) = \{u \in C(\overline{D}) : Wu \in C(\overline{D}),\ \ Lu = 0\}\,. \tag{7.4}$$

(b) $\mathfrak{A}u = Wu$, $u \in D(\mathfrak{A})$.

Here Wu and Lu are taken in the sense of distributions.

Assume that the boundary condition L is transversal *on the boundary ∂D. Then the operator $\mathfrak{A}$ generates a Feller semigroup $\{T_t\}_{t\geq 0}$ on $\overline{D}$.*

Remark 7.1. Bony, Courrège and Priouret [BCP] proved Theorem 7.1 in the case where the differential operator Q in formula (7.2) is *elliptic* (see [BCP, Théorème XIX]). Theorem 7.1 is proved by Cancelier [Cn, Théorème 3.2] and also by Taira [Ta4, Theorem 1]. It should be emphasized that Takanobu and Watanabe give a probabilistic version of Theorem 7.1 in the case where the domain D is the half space $\mathbf{R}^N_+$ (see [TW, Corollary]).

7.2.2 Sketch of Proof of Theorem 7.1

The idea of our proof of Theorem 7.1 is stated as follows (see Bony–Courrège–Priouret [BCP], Sato–Ueno [SU], Taira [Ta2]). The proof is divided into three steps.

Step 1: First, we consider the following *Dirichlet problem*:

$$\begin{cases} (\alpha - W)v = f & \text{in } D, \\ v = 0 & \text{on } \partial D, \end{cases} \tag{D}$$

where $\alpha > 0$ is a parameter.

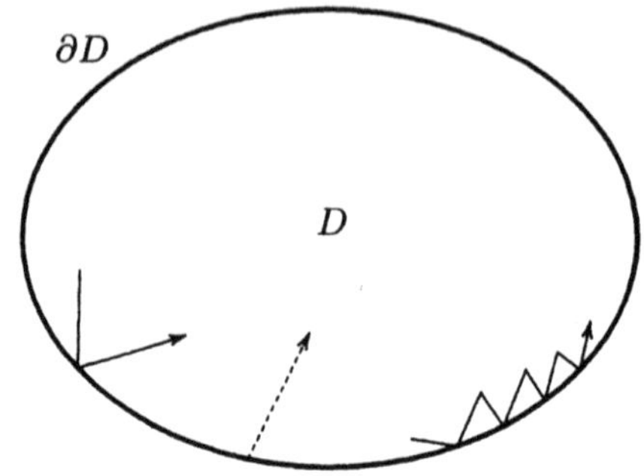

Fig. 7.6.

The existence and uniqueness theorem for this problem is well established in the framework of Hölder spaces. In fact, the next theorem summarizes the basic fact about the Dirichlet problem in the framework of *Hölder spaces* (see [BCP, Théorème XV]):

Theorem 7.2. *Let k be an arbitrary non-negative integer and $0 < \theta < 1$. For any $f \in C^{k+\theta}(\overline{D})$ and any $\varphi \in C^{k+2+\theta}(\partial D)$, problem (D) has a unique solution u in the space $C^{k+2+\theta}(\overline{D})$.*

Theorem 7.2 with $k := 0$ tells us that problem (D) has a unique solution u in $C^{2+\theta}(\overline{D})$ for any $f \in C^{\theta}(\overline{D})$ and any $\varphi \in C^{2+\theta}(\partial D)$, $0 < \theta < 1$. Therefore, we can introduce two linear operators

$$G_\alpha^0 : C^\theta(\overline{D}) \longrightarrow C^{2+\theta}(\overline{D}) \,,$$

and

$$H_\alpha : C^{2+\theta}(\partial D) \longrightarrow C^{2+\theta}(\overline{D})$$

as follows.

(a) For any $f \in C^\theta(\overline{D})$, the function $G_\alpha^0 f \in C^{2+\theta}(\overline{D})$ is the unique solution of the problem

$$\begin{cases} (\alpha - W)G_\alpha^0 f = f & \text{in } D, \\ G_\alpha^0 f = 0 & \text{on } \partial D. \end{cases} \tag{7.5}$$

(b) For any $\varphi \in C^{2+\theta}(\partial D)$, the function $H_\alpha \varphi \in C^{2+\theta}(\overline{D})$ is the unique solution of the problem

$$\begin{cases} (\alpha - W)H_\alpha \varphi = 0 & \text{in } D, \\ H_\alpha \varphi = \varphi & \text{on } \partial D. \end{cases} \tag{7.6}$$

The operator G_α^0 is called the *Green operator* and the operator H_α is called the *harmonic operator*, respectively.

Then we have the following results (cf. [BCP, Proposition III.1.6], [Ta2, Lemmas 9.6.2 and 9.6.3]):

Lemma 7.3. *The operator G_α^0, $\alpha > 0$, considered from $C(\overline{D})$ into itself, is non-negative and continuous with norm*

$$\|G_\alpha^0\| = \|G_\alpha^0 1\|_\infty = \max_{x \in \overline{D}} G_\alpha^0 1(x) .$$

Lemma 7.4. *The operator H_α, $\alpha > 0$, considered from $C(\partial D)$ into $C(\overline{D})$, is non-negative and continuous with norm*

$$\|H_\alpha\| = \|H_\alpha 1\|_\infty = \max_{x \in \overline{D}} H_\alpha 1(x) .$$

More precisely, we have the following (see [BCP, Proposition III.1.6]):

Theorem 7.5. *(i) (a) The operator G_α^0, $\alpha > 0$, can be uniquely extended to a non-negative, bounded linear operator on $C(\overline{D})$ into itself, denoted again by G_α^0, with norm*

$$\|G_\alpha^0\| = \|G_\alpha^0 1\|_\infty \leq \frac{1}{\alpha}. \tag{7.7}$$

(b) For any $f \in C(\overline{D})$, we have

$$G_\alpha^0 f \mid_{\partial D} = 0.$$

(c) For all α, $\beta > 0$, the resolvent equation *holds:*

$$G_\alpha^0 f - G_\beta^0 f + (\alpha - \beta) G_\alpha^0 G_\beta^0 f = 0 , \quad f \in C(\overline{D}) . \tag{7.8}$$

(d) For any $f \in C(\overline{D})$, we have

$$\lim_{\alpha \to +\infty} \alpha G_\alpha^0 f(x) = f(x) , \quad x \in D . \tag{7.9}$$

Furthermore, if $f|_{\partial D} = 0$, then this convergence is uniform in $x \in \overline{D}$, that is,

$$\lim_{\alpha \to +\infty} \alpha G_\alpha^0 f = f \quad in\ C(\overline{D}) . \tag{7.9'}$$

(e) The operator G_α^0 maps $C^{k+\theta}(\overline{D})$ into $C^{k+2+\theta}(\overline{D})$ for any non-negative integer k.

(ii) (a′) The operator H_α, $\alpha > 0$, can be uniquely extended to a non-negative, bounded linear operator on $C(\partial D)$ into $C(\overline{D})$, denoted again by H_α, with norm $\|H_\alpha\| \leq 1$.

(b′) For any $\varphi \in C(\partial D)$, we have

$$H_\alpha \varphi|_{\partial D} = \varphi .$$

(c′) For all α, $\beta > 0$, we have

$$H_\alpha \varphi - H_\beta \varphi + (\alpha - \beta) G_\alpha^0 H_\beta \varphi = 0 , \quad \varphi \in C(\partial D) . \tag{7.10}$$

(d′) For any $\varphi \in C(\partial D)$, we have

$$\lim_{\alpha \to +\infty} H_\alpha \varphi(x) = 0 , \quad x \in D .$$

(e′) The operator H_α maps $C^{k+2+\theta}(\partial D)$ into $C^{k+2+\theta}(\overline{D})$ for any non-negative integer k.

Step 2: Now we consider the following boundary value problem in the framework of the spaces of *continuous functions*:

$$\begin{cases} (\alpha - W)u = f & \text{in } D, \\ Lu = 0 & \text{on } \partial D, \end{cases} \tag{$*$}$$

where $\alpha > 0$ is a parameter.

To do this, we introduce three operators associated with problem $(*)$.

Step 2-1: First, we introduce a linear operator

$$W : C(\overline{D}) \longrightarrow C(\overline{D})$$

as follows.

(a) The domain $D(W)$ of W is the space $C^{2+\theta}(\overline{D})$.
(b) $Wu = Pu + S_r u$, $u \in D(W)$.

Then we have the following (cf. [Ta2, Lemma 9.6.5]):

Lemma 7.6. *The operator W has its minimal closed extension $\overline{W}$ in the space $C(\overline{D})$.*

Remark 7.2. Since the injection: $C(\overline{D}) \to \mathcal{D}'(D)$ is continuous, we have the formula

$$\overline{W}u(x) = \sum_{i,j=1}^{N} a^{ij}(x)\frac{\partial^2 u}{\partial x_i \partial x_j}(x) + \sum_{i=1}^{N} b^i(x)\frac{\partial u}{\partial x_i}(x) + c(x)u(x)$$
$$+ \int_D s(x,y)\left[u(y) - \sigma(x,y)\left(u(x) + \sum_{j=1}^{N}(y_j - x_j)\frac{\partial u}{\partial x_j}(x)\right)\right] dy,$$

where the right-hand side is taken in the sense of *distributions*.

The extended operators $G_\alpha^0 : C(\overline{D}) \to C(\overline{D})$ and $H_\alpha : C(\partial D) \to C(\overline{D})$, $\alpha > 0$, still satisfy formulas (6.5) and (6.6) respectively in the following sense (cf. [Ta2, Lemma 9.6.7 and Corollary 9.6.8]):

Lemma 7.7. *(i) For any $f \in C(\overline{D})$, we have*

$$\begin{cases} G_\alpha^0 f \in D(\overline{W}), \\ (\alpha I - \overline{W})G_\alpha^0 f = f \quad \text{in } D. \end{cases}$$

(ii) For any $\varphi \in C(\partial D)$, we have

$$\begin{cases} H_\alpha \varphi \in D(\overline{W}), \\ (\alpha I - \overline{W})H_\alpha \varphi = 0 \quad \text{in } D. \end{cases}$$

Here $D(\overline{W})$ is the domain of the closed extension $\overline{W}$.

Corollary 7.8. *Every u in $D(\overline{W})$ can be written in the following form:*

$$u = G_\alpha^0\left((\alpha I - \overline{W})u\right) + H_\alpha(u|_{\partial D}), \quad \alpha > 0\,. \tag{7.11}$$

Step 2-2: Secondly, we introduce a linear operator

$$LG_\alpha^0 : C(\overline{D}) \longrightarrow C(\partial D)$$

as follows.

(a) The domain $D\left(LG_\alpha^0\right)$ of LG_α^0 is the space $C^\theta(\overline{D})$.
(b) $LG_\alpha^0 f = L\left(G_\alpha^0 f\right)$, $f \in D\left(LG_\alpha^0\right)$.

Then we have the following (cf. [BCP, Lemme III.2.4], [Ta2, Lemma 9.6.9]):

Lemma 7.9. *The operator LG_α^0, $\alpha > 0$, can be uniquely extended to a non-negative, bounded linear operator $\overline{LG_\alpha^0} : C(\overline{D}) \to C(\partial D)$.*

The next lemma states a fundamental relationship between the operators $\overline{LG_\alpha^0}$ and $\overline{LG_\beta^0}$ for α, $\beta > 0$ (cf. [Ta2, Lemma 9.6.10]):

Lemma 7.10. *For any $f \in C(\overline{D})$, we have*

$$\overline{LG_\alpha^0} f - \overline{LG_\beta^0} f + (\alpha - \beta)\overline{LG_\alpha^0}\, G_\beta^0 f = 0\,, \quad \alpha, \beta > 0\,. \tag{7.12}$$

Step 2-3: Finally, we introduce a linear operator

$$LH_\alpha : C(\partial D) \longrightarrow C(\partial D)$$

as follows.

(a) The domain $D\left(LH_\alpha\right)$ of LH_α is the space $C^{2+\theta}(\partial D)$.
(b) $LH_\alpha \psi = L\left(H_\alpha \psi\right)$, $\psi \in D\left(LH_\alpha\right)$.

Then we have the following (cf. [Ta2, Lemma 9.6.11]):

Lemma 7.11. *The operator LH_α, $\alpha > 0$, has its minimal closed extension $\overline{LH_\alpha}$ in the space $C(\partial D)$.*

Remark 7.3. The operator $\overline{LH_\alpha}$ enjoys the following property:

> If a function ψ in the domain $D\left(\overline{LH_\alpha}\right)$ takes its *positive* maximum at some point x' of ∂D, then we have
> $$\overline{LH_\alpha}\psi(x') \le 0\,. \tag{7.13}$$

The next lemma states a fundamental relationship between the operators $\overline{LH_\alpha}$ and $\overline{LH_\beta}$ for α, $\beta > 0$ (cf. [Ta2, Lemma 9.6.13]):

Lemma 7.12. *The domain $D\left(\overline{LH_\alpha}\right)$ of $\overline{LH_\alpha}$ does* not *depend on $\alpha > 0$; so we denote by $\mathcal{D}$ the common domain. Then we have*

$$\overline{LH_\alpha}\psi - \overline{LH_\beta}\psi + (\alpha - \beta)\overline{LG_\alpha^0}\, H_\beta\psi = 0\,, \quad \alpha, \beta > 0\,, \quad \psi \in \mathcal{D}\,. \tag{7.14}$$

Step 2-4: Now we can state a general existence theorem for Feller semigroups on ∂D in terms of boundary value problem $(*)$.

The next theorem tells us that the operator $\overline{LH_\alpha}$ is the infinitesimal generator of some Feller semigroup on ∂D if and only if problem $(*)$ is solvable for sufficiently *many* functions φ in the space $C(\partial D)$ (cf. [BCP, Théorème XX], [Ta2, Theorem 9.6.15]):

Theorem 7.13. *(i) If the operator $\overline{LH_\alpha}$, $\alpha > 0$, is the infinitesimal generator of a Feller semigroup on ∂D, then, for each constant $\lambda > 0$, the boundary value problem*

$$\begin{cases} (\alpha - W)u = 0 & \text{in } D, \\ (\lambda - L)u = \varphi & \text{on } \partial D \end{cases} \tag{$*'$}$$

has a solution $u \in C^{2+\theta}(\overline{D})$ for any φ in some dense *subset of $C(\partial D)$.*

(ii) Conversely, if, for some constant $\lambda \geq 0$, problem $(*')$ has a solution $u \in C^{2+\theta}(\overline{D})$ for any φ in some dense *subset of $C(\partial D)$, then the operator $\overline{LH_\alpha}$ is the infinitesimal generator of some Feller semigroup on ∂D.*

We can give a precise meaning to the boundary conditions Lu for functions u in the domain $D(\overline{W})$.

We let

$$D(L) = \left\{u \in D(\overline{W}) : u|_{\partial D} \in \mathcal{D}\right\},$$

where $\mathcal{D}$ is the common domain of the operators $\overline{LH_\alpha}$, $\alpha > 0$. It should be noticed that the domain $D(L)$ contains the space $C^{2+\theta}(\overline{D})$, since $C^{2+\theta}(\partial D) = D\left(LH_\alpha\right) \subset \mathcal{D}$. Corollary 7.8 tells us that every function u in $D(L) \subset D(\overline{W})$ can be written in the form

$$u = G_\alpha^0\left((\alpha I - \overline{W})u\right) + H_\alpha\left(u|_{\partial D}\right), \quad \alpha > 0. \tag{7.11}$$

Then we define Lu by the formula

$$Lu = \overline{LG_\alpha^0}\left((\alpha I - \overline{W})u\right) + \overline{LH_\alpha}\left(u|_{\partial D}\right). \tag{7.15}$$

The next lemma justifies definition (7.15) of Lu for all $u \in D(L)$ (cf. [Ta2, Lemma 9.6.16]):

Lemma 7.14. *The right-hand side of formula (7.15) depends only on u,* not *on the choice of expression (7.11).*

Step 3: The next theorem proves Theorem 7.1:

Theorem 7.15. *We define a linear operator*

$$\mathfrak{A} : C(\overline{D}) \longrightarrow C(\overline{D})$$

as follows (see formula (7.4)).

(a) The domain $D(\mathfrak{A})$ of $\mathfrak{A}$ is the set

$$D(\mathfrak{A}) = \left\{u \in D(\overline{W}) : u|_{\partial D} \in \mathcal{D}\,,\ \ Lu = 0\right\}\,, \tag{7.16}$$

where $\mathcal{D}$ is the common domain of the operators $\overline{LH_\alpha}$, $\alpha > 0$.
(b) $\mathfrak{A}u = \overline{W}u$, $u \in D(\mathfrak{A})$.

If the boundary condition L is transversal *on the boundary ∂D, then the operator $\mathfrak{A}$ is the infinitesimal generator of some Feller semigroup on $\overline{D}$, and the Green operator $G_\alpha = (\alpha I - \mathfrak{A})^{-1}$, $\alpha > 0$, is given by the formula*

$$G_\alpha f = G_\alpha^0 f - H_\alpha\left(\overline{LH_\alpha}^{-1}\left(\overline{LG_\alpha^0 f}\right)\right), \quad f \in C(\overline{D})\,. \tag{7.17}$$

Remark 7.4. Intuitively, formula (7.17) asserts that if the boundary condition L is transversal on the boundary ∂D, then we can "piece together" a Markov process (Feller semigroup) on the boundary ∂D with W-process in the interior D to construct a Markov process (Feller semigroup) on the whole $\overline{D} = D \cup \partial D$. The situation may be represented schematically by Fig. 7.7.

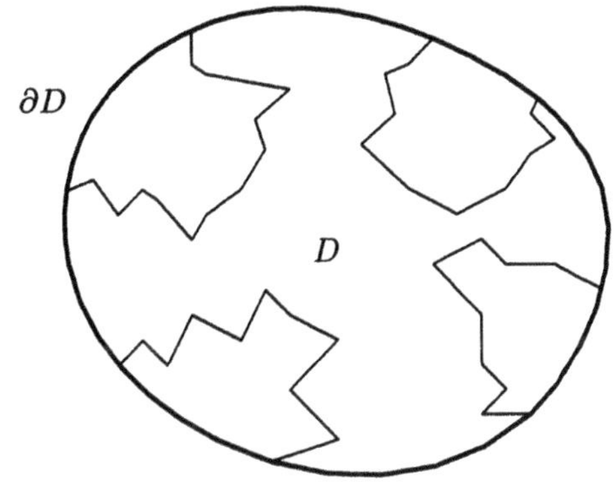

Fig. 7.7.

7.2.3 Proof of Theorem 7.15

We shall verify that the operator $\mathfrak{A}$, defined by formula (7.16), satisfies conditions (a) through (d) in Theorem 3.3. The proof is divided into several steps.

Step 1: First, by the Boutet de Monvel calculus [Bt] (see Sect. 5.6) we know that if $T \in L^{2-\kappa_2}_{1,0}(\mathbf{R}^N)$ has the *transmission property* with respect to the boundary ∂D, then the operator

$$\varphi(x') \longmapsto \int_D t(x', y) H_\alpha \varphi(y)\, dy$$

is a classical pseudo-differential operator of order $2-\kappa_2$ on the boundary ∂D. Therefore, we find that the operator LH_α is the sum of a second-order degenerate elliptic differential operator and a classical pseudo-differential operator of order $2-\min(\kappa_1,\kappa_2)$:

$$\begin{aligned} & LH_\alpha\varphi(x') \\ &= \sum_{i,j=1}^{N-1} \alpha^{ij}(x')\frac{\partial^2\varphi}{\partial x_i \partial x_j}(x') + \sum_{i=1}^{N-1} \beta^i(x')\frac{\partial\varphi}{\partial x_i}(x') + \gamma(x')\varphi(x') \\ &\quad - \alpha\delta(x')\varphi(x') + \mu(x')\frac{\partial}{\partial \mathbf{n}}(H_\alpha\varphi)(x') \\ &\quad + \int_D t(x',y)\left[H_\alpha\varphi(y) - \tau(x',y)\left(\varphi(x') + \sum_{j=1}^{N-1}(y_j - x_j)\frac{\partial\varphi}{\partial x_j}(x')\right)\right] dy \\ &\quad + \int_{\partial D} r(x',y')\left[\varphi(y') - \tau(x',y')\left(\varphi(x') + \sum_{j=1}^{N-1}(y_j - x_j)\frac{\partial\varphi}{\partial x_j}(x')\right)\right] dy' \end{aligned}$$

Now we prove that

For *all* $\alpha > 0$, the operator $\overline{LH_\alpha}$ generates a Feller semigroup on the boundary ∂D.

First, we have the following six assertions:

(i) The operator

$$Q\varphi = \sum_{i,j=1}^{N-1} \alpha^{ij}(x')\frac{\partial^2\varphi}{\partial x_i \partial x_j} + \sum_{i=1}^{N-1} \beta^i(x')\frac{\partial\varphi}{\partial x_i} + \gamma(x')\varphi$$

is a second-order, degenerate elliptic differential operator on ∂D with non-positive principal symbol, and $Q1(x') = \gamma(x') \leq 0$ on ∂D.

(ii) The operator

$$\Pi_\alpha\varphi = \frac{\partial}{\partial \mathbf{n}}(H_\alpha\varphi)$$

is a first-order, classical pseudo-differential operator on ∂D (see [Ho1], [RS]).

Moreover, it should be noticed that

$$\begin{aligned} & x_0' \in \partial D, \varphi \in C^2(\partial D), \varphi \geq 0 \text{ on } \partial D \text{ and } x_0' \notin \operatorname{supp}\varphi \\ & \Longrightarrow \Pi_\alpha\varphi(x_0') = \frac{\partial}{\partial \mathbf{n}}(H_\alpha\varphi)(x_0') \geq 0 . \end{aligned} \tag{7.18}$$

Indeed, if we let

$$u = H_\alpha\varphi ,$$

then we have

$$\begin{cases} (\alpha - W)u = 0 & \text{in } D, \\ u = \varphi & \text{on } \partial D. \end{cases}$$

Since $\varphi \geq 0$ on ∂D, it follows from an application of the weak maximum principle (see Theorem C.1 in Appendix C) that

$$H_\alpha \varphi = u \geq 0 \quad \text{in } D\,.$$

Hence this implies that

$$\Pi_\alpha \varphi(x_0') = \frac{\partial}{\partial \mathbf{n}}(H_\alpha \varphi)(x_0') \geq 0\,,$$

since $\varphi(x_0') = 0$.

(iii) The operator

$$\varphi(x') \longmapsto \int_{\partial D} r(x', y') \left[\varphi(y') - \tau(x', y') \left(\varphi(x') + \sum_{j=1}^{N-1} (y_j - x_j) \frac{\partial \varphi}{\partial x_j}(x') \right) \right] dy'$$

is a classical pseudo-differential operator of order $2 - \kappa_1$ on ∂D.

(iv) The operator

$$\varphi(x') \longmapsto \int_{D} t(x', y) \left[H_\alpha \varphi(y) - \tau(x', y) \left(\varphi(x') + \sum_{j=1}^{N-1} (y_j - x_j) \frac{\partial \varphi}{\partial x_j}(x') \right) \right] dy$$

is a classical pseudo-differential operator of order $2 - \kappa_2$ on the boundary ∂D, since $T \in L^{2-\kappa_2}_{1,0}(\mathbf{R}^N)$ has the *transmission property* with respect to the boundary ∂D (see [Bo], [RS]).

(v) We remark that

$$\begin{aligned} &x_0' \in \partial D, \varphi \in C^2(\partial D)\,, \varphi \geq 0 \text{ on } \partial D \text{ and } x_0' \notin \operatorname{supp} \varphi \\ &\Longrightarrow \; LH_\alpha \varphi(x_0') \geq 0\,. \end{aligned}$$

Indeed, we have, by assertion (7.18),

$$\begin{aligned} LH_\alpha \varphi(x_0') &= \mu(x_0') \frac{\partial}{\partial \mathbf{n}}(H_\alpha \varphi)(x_0') + \int_{\partial D} r(x_0', y') \varphi(y')\, dy' \\ &\quad + \int_D t(x_0', y) H_\alpha \varphi(y)\, dy \\ &\geq 0\,. \end{aligned}$$

Therefore, by applying Theorem 3.7 to our situation we obtain that the pseudo-differential operator LH_α may be written in the form

$$LH_\alpha \varphi = P_\alpha \varphi + S_\alpha \varphi\,, \quad \varphi \in C^2(\partial D)\,.$$

Here:

(a) P_α is a second-order, degenerate elliptic differential operator on ∂D with non-positive principal symbol and $P_\alpha 1 \leq 0$ on ∂D.
(b) S_α is given by the formula

$$S_\alpha \varphi(x') = \int_{\partial D} s(x', y') \Bigg[\varphi(y') - \sigma(x', y') \times \left(\varphi(x') + \sum_{i=1}^{N-1} \frac{\partial \varphi}{\partial x_i}(x') (y_i - x_i) \right) \Bigg] dy' .$$

It should be emphasized that $s(x', y') \geq 0$ off the diagonal $\Delta_{\partial D} = \{(x', x') : x' \in \partial D\}$.

(vi) Finally, since the function $H_\alpha 1$ takes its positive maximum 1 only on the boundary ∂D, it follows from an application of the boundary point lemma (see Lemma C.5 in Appendix C) that

$$\begin{aligned} H_\alpha 1(x) - 1 &< 0 \quad \text{in } D, \\ \Pi_\alpha 1(x') &= \frac{\partial}{\partial \mathbf{n}} (H_\alpha 1)(x') < 0 \quad \text{on } \partial D . \end{aligned}$$

Hence we have, by transversality condition (7.3),

$$\begin{aligned} & LH_\alpha 1(x') \\ &= \gamma(x') + \mu(x') \, \Pi_\alpha 1(x') - \alpha\delta(x') \\ &\quad + \int_{\partial D} r(x', y') \, [1 - \tau(x', y')] \, dy' + \int_D t(x', y) \, [H_\alpha 1(y) - \tau(x', y)] \, dy \\ &= (\gamma(x') - \alpha\delta(x')) + \mu(x') \Pi_\alpha 1(x') \\ &\quad + \left(\int_{\partial D} r(x', y') \, [1 - \tau(x', y')] \, dy' + \int_D t(x', y) \, [1 - \tau(x', y)] \, dy \right) \\ &\quad + \int_D t(x', y) \, [H_\alpha 1(y) - 1] \, dy \\ &< 0 \quad \text{on } \partial D . \end{aligned} \tag{7.19}$$

Step 2: The next unique solvability theorem for pseudo-differential operators will play an essential role in the construction of Feller semigroups (see [Cn, Théorème 4.5], [Ta4, Theorem 2.1]):

Theorem 7.16. *Let T be a second-order, classical pseudo-differential operator on an n-dimensional, compact smooth manifold M without boundary such that*

$$T = P + S ,$$

where:

(a) *The operator P is a second-order, degenerate* elliptic differential operator *on M with non-positive principal symbol, and $P1 \leq 0$ on M.*

(b) *The operator S is a classical pseudo-differential operator of order $2-\kappa$, $\kappa > 0$, on M and its distribution kernel $s(x,y)$ is* non-negative *off the diagonal $\Delta_M = \{(x,x) : x \in M\}$ in $M \times M$.*
(c) $T1 = P1 + S1 \leq 0$ *on M.*

Then, for each integer $k \geq 1$ there exists a constant $\lambda = \lambda(k) > 0$ such that, for any $f \in C^{k+\theta}(M)$, we can find a function $\varphi \in C^{k+\theta}(M)$ satisfying the equation

$$(T - \lambda I)\varphi = f \quad \text{on } M \,,$$

and the estimate

$$\|\varphi\|_{C^{k+\theta}(M)} \leq C(\lambda)\|f\|_{C^{k+\theta}(M)} \,.$$

Here $C(\lambda) > 0$ is a constant independent of f.

Theorem 7.16 will be proved in Appendix B (Theorem B.1) due to its length.

Step 3: By applying Theorem 7.16 to the operator LH_α, we find that

$$\begin{aligned}&\text{If } \lambda > 0 \text{ is sufficiently large, then the range } R(LH_\alpha - \lambda I)\\ &\text{contains the space } C^{2+\theta}(\partial D)\,. \end{aligned} \tag{7.20}$$

This implies that the range $R(LH_\alpha - \lambda I)$ is a *dense* subset of $C(\partial D)$. Therefore, applying part (ii) of Theorem 7.15 to the operator L we obtain that the operator $\overline{LH_\alpha}$ is the infinitesimal generator of some Feller semigroup on ∂D, for any $\alpha > 0$.

Step 4: Now we prove that

$$\begin{aligned}&\text{The equation}\\ &\quad \overline{LH_\alpha}\psi = \varphi\\ &\text{has a unique solution } \psi \text{ in } D(\overline{LH_\alpha}) \text{ for any } \varphi \in C(\partial D)\text{; hence}\\ &\text{the inverse } \overline{LH_\alpha}^{-1} \text{ of } \overline{LH_\alpha} \text{ can be defined on the whole space } C(\partial D).\\ &\text{Further the operator } -\overline{LH_\alpha}^{-1} \text{ is non-negative and bounded on } C(\partial D).\end{aligned} \tag{7.21}$$

We have, by inequality (7.19) and transversality condition (7.3),

$$\ell_\alpha = -\sup_{x' \in \partial D} LH_\alpha 1(x') > 0 \,.$$

Further, using Corollary 3.4 with $K := \partial D$, $A := \overline{LH_\alpha}$ and $c := \ell_\alpha$ we obtain that the operator $\overline{LH_\alpha} + \ell_\alpha I$ is the infinitesimal generator of some Feller semigroup on ∂D. Therefore, since $\ell_\alpha > 0$, it follows from an application of part (i) of Theorem 3.3 with $A := \overline{LH_\alpha} + \ell_\alpha I$ that the equation

$$-\overline{LH_\alpha}\,\psi = \left(\ell_\alpha I - (\overline{LH_\alpha} + \ell_\alpha I)\right)\psi = \varphi$$

has a unique solution $\psi \in D(\overline{LH_\alpha})$ for any $\varphi \in C(\partial D)$, and further that the operator $-\overline{LH_\alpha}^{-1} = \left(\ell_\alpha I - (\overline{LH_\alpha} + \ell_\alpha I)\right)^{-1}$ is non-negative and bounded on the space $C(\partial D)$ with norm

$$\left\| -\overline{LH_\alpha}^{-1} \right\| = \left\| \left(\ell_\alpha I - (\overline{LH_\alpha} + \ell_\alpha I)\right)^{-1} \right\|_\infty \leq \frac{1}{\ell_\alpha} .$$

Step 5: By assertion (7.21), we can define the formula

$$G_\alpha f = G_\alpha^0 f - H_\alpha \left(\overline{LH_\alpha}^{-1} \left(\overline{LG_\alpha^0 f} \right) \right) , \quad f \in C(\overline{D}) . \tag{7.17}$$

Step 5-1: First, we prove that

$$G_\alpha = (\alpha I - \mathfrak{A})^{-1} , \quad \alpha > 0 . \tag{7.22}$$

In view of Lemmas 7.7 and 7.14, it follows that we have, for all $f \in C(\overline{D})$,

$$\begin{cases} G_\alpha f = G_\alpha^0 f - H_\alpha \left(\overline{LH_\alpha}^{-1} \left(\overline{LG_\alpha^0 f} \right) \right) \in D(\overline{W}), \\ G_\alpha f|_{\partial D} = -\overline{LH_\alpha}^{-1} \left(\overline{LG_\alpha^0 f} \right) \in D\left(\overline{LH_\alpha}\right) = \mathcal{D}, \\ LG_\alpha f = \overline{LG_\alpha^0} f - \overline{LH_\alpha} \left(\overline{LH_\alpha}^{-1} \left(\overline{LG_\alpha^0 f} \right) \right) = 0, \end{cases}$$

and that

$$(\alpha I - \overline{W}) G_\alpha f = f .$$

This proves that

$$\begin{cases} G_\alpha f \in D(\mathfrak{A}), \\ (\alpha I - \mathfrak{A}) G_\alpha f = f, \end{cases}$$

that is,

$$(\alpha I - \mathfrak{A}) G_\alpha = I \quad \text{on } C(\overline{D}) .$$

Therefore, in order to prove formula (7.22) it suffices to show the injectivity of the operator $\alpha I - \mathfrak{A}$ for $\alpha > 0$.

Assume that

$$u \in D(\mathfrak{A}) \text{ and } (\alpha I - \mathfrak{A}) u = 0 .$$

Then, by Corollary 7.8, the function u can be written as

$$u = H_\alpha (u|_{\partial D}) , \quad u|_{\partial D} \in \mathcal{D} = D\left(\overline{LH_\alpha}\right) .$$

Thus we have

$$\overline{LH_\alpha} (u|_{\partial D}) = Lu = 0 .$$

In view of assertion (7.21), this implies that

$$u|_{\partial D} = 0 ,$$

so that

$$u = H_\alpha (u|_{\partial D}) = 0 \quad \text{in } D .$$

Step 5-2: The non-negativity of G_α, $\alpha > 0$, follows immediately from formula (7.17), since the operators G_α^0, H_α, $-\overline{LH_\alpha}^{-1}$ and $\overline{LG_\alpha^0}$ are all non-negative.

Step 5-3: We prove that the operator G_α is bounded on the space $C(\overline{D})$ with norm

$$\|G_\alpha\| \leq \frac{1}{\alpha}, \quad \alpha > 0 . \tag{7.23}$$

To do this, it suffices to show that

$$G_\alpha 1 \leq \frac{1}{\alpha} \quad \text{on } \overline{D} . \tag{7.23'}$$

since G_α is non-negative on $C(\overline{D})$.

First, it follows from the uniqueness property of solutions of problem (D′) that

$$\alpha G_\alpha^0 1 + H_\alpha 1 = 1 + G_\alpha^0(W1) \quad \text{on } \overline{D} . \tag{7.24}$$

In fact, the both sides have the same boundary value 1 and satisfy the same equation: $(\alpha - W)u = \alpha$ in D.

Applying the operator L to the both hand sides of equality (7.24), we obtain that

$$\begin{aligned}
&-LH_\alpha 1(x') \\
&= -L1(x') - LG_\alpha^0(W1)(x') + \alpha LG_\alpha^0 1(x') \\
&= -\gamma(x') - \mu(x')\frac{\partial}{\partial \mathbf{n}}(G_\alpha^0(W1))(x') - \int_D t(x',y)G_\alpha^0(W1)(y)dy + \alpha LG_\alpha^0 1(x') \\
&\geq \alpha LG_\alpha^0 1(x') \quad \text{on } \partial D ,
\end{aligned}$$

since $G_\alpha^0(W1)|_{\partial D} = 0$ and $G_\alpha^0(W1) \leq 0$ on $\overline{D}$. Hence we have, by the non-negativity of $-\overline{LH_\alpha}^{-1}$,

$$-\overline{LH_\alpha}^{-1}\left(LG_\alpha^0 1\right) \leq \frac{1}{\alpha} \quad \text{on } \partial D . \tag{7.25}$$

By using formula (7.17) with $f := 1$, inequality (7.25) and equality (7.24), we obtain that

$$\begin{aligned}
G_\alpha 1 = G_\alpha^0 1 + H_\alpha\left(-\overline{LH_\alpha}^{-1}\left(LG_\alpha^0 1\right)\right) &\leq G_\alpha^0 1 + \frac{1}{\alpha} H_\alpha 1 \\
&= \frac{1}{\alpha} + \frac{1}{\alpha} G_\alpha^0(W1) \\
&\leq \frac{1}{\alpha} \quad \text{on } \overline{D} ,
\end{aligned}$$

since the operators H_α and G_α^0 are non-negative and since $W1 \leq 0$ in D.

Step 5-4: Finally, we prove that

$$\text{The domain } D(\mathfrak{A}) \text{ is } dense \text{ in the space } C(\overline{D})\,. \tag{7.26}$$

(1) Before the proof, we need some lemmas on the behavior of G_α^0, H_α and $-\overline{LH_\alpha}^{-1}$ as $\alpha \to +\infty$ (see [BCP, Proposition III.1.6]; [Ta2, Lemmas 9.6.19 and 9.6.20]):

Lemma 7.17. *For all $f \in C(\overline{D})$, we have*

$$\lim_{\alpha \to +\infty} \left[\alpha G_\alpha^0 f + H_\alpha \left(f|_{\partial D}\right)\right] = f \quad \text{in } C(\overline{D})\,. \tag{7.27}$$

Lemma 7.18. *The function*

$$\Pi_\alpha 1 = \frac{\partial}{\partial \mathbf{n}} \left(H_\alpha 1\right)$$

diverges to $-\infty$ uniformly and monotonically as $\alpha \to +\infty$.

Corollary 7.19. *If the boundary condition L is* transversal *on the boundary ∂D, then we have*

$$\lim_{\alpha \to +\infty} \| -\overline{LH_\alpha}^{-1} \| = 0\,.$$

(2) **Proof of assertion (7.26)**
In view of formula (7.22) and inequality (7.23), it suffices to prove that

$$\lim_{\alpha \to +\infty} \|\alpha G_\alpha f - f\|_\infty = 0\,, \quad f \in C^{2+\theta}(\overline{D})\,, \tag{7.28}$$

since the space $C^{2+\theta}(\overline{D})$ is dense in $C(\overline{D})$.

First, we remark that

$$\begin{aligned}
\|\alpha G_\alpha f - f\|_\infty &= \left\|\alpha G_\alpha^0 f - \alpha H_\alpha \left(\overline{LH_\alpha}^{-1}\left(LG_\alpha^0 f\right)\right) - f\right\|_\infty \\
&\le \left\|\alpha G_\alpha^0 f + H_\alpha \left(f|_{\partial D}\right) - f\right\|_\infty \\
&\quad + \left\|-\alpha H_\alpha \left(\overline{LH_\alpha}^{-1}\left(LG_\alpha^0 f\right)\right) - H_\alpha \left(f|_{\partial D}\right)\right\|_\infty \\
&\le \left\|\alpha G_\alpha^0 f + H_\alpha \left(f|_{\partial D}\right) - f\right\|_\infty \\
&\quad + \left\|-\alpha \overline{LH_\alpha}^{-1}\left(LG_\alpha^0 f\right) - f|_{\partial D}\right\|_\infty\,.
\end{aligned}$$

Thus, in view of formula (7.27) it suffices to show that

$$\lim_{\alpha \to +\infty} \left[-\alpha \overline{LH_\alpha}^{-1}\left(LG_\alpha^0 f\right) - f|_{\partial D}\right] = 0 \quad \text{in } C(\partial D)\,. \tag{7.29}$$

Take a constant β such that $0 < \beta < \alpha$, and write

$$f = G_\beta^0 g + H_\beta \varphi\,,$$

where (cf. formula (7.11)):

$$\begin{cases} g = (\beta - W)f \in C^{\theta}(\overline{D}), \\ \varphi = f|_{\partial D} \in C^{2+\theta}(\partial D). \end{cases}$$

Then, using equations (7.8) (with $f := g$) and (7.10) we obtain that

$$G^0_\alpha f = G^0_\alpha G^0_\beta g + G^0_\alpha H_\beta \varphi = \frac{1}{\alpha - \beta}\left(G^0_\beta g - G^0_\alpha g + H_\beta \varphi - H_\alpha \varphi\right).$$

Hence we have

$$\begin{aligned} &\left\| -\alpha \overline{LH_\alpha}^{-1}\left(LG^0_\alpha f\right) - f|_{\partial D}\right\|_\infty \\ &= \left\| \frac{\alpha}{\alpha - \beta}\left(-\overline{LH_\alpha}^{-1}\right)\left(LG^0_\beta g - LG^0_\alpha g + LH_\beta \varphi\right) + \frac{\alpha}{\alpha - \beta}\varphi - \varphi\right\|_\infty \\ &\leq \frac{\alpha}{\alpha - \beta}\left\| -\overline{LH_\alpha}^{-1}\right\| \cdot \left\| LG^0_\beta g + LH_\beta \varphi\right\|_\infty \\ &\quad + \frac{\alpha}{\alpha - \beta}\left\| -\overline{LH_\alpha}^{-1}\right\| \cdot \left\| LG^0_\alpha\right\|_\infty \cdot \|g\|_\infty + \frac{\beta}{\alpha - \beta}\|\varphi\|_\infty . \end{aligned}$$

By Corollary 7.19, it follows that the first term on the last inequality converges to zero as $\alpha \to +\infty$. For the second term, using formula (7.8) with $f := 1$ and the non-negativity of G^0_β and LG^0_α we find that

$$\begin{aligned} \|LG^0_\alpha\| &= \|LG^0_\alpha 1\|_\infty \\ &= \|LG^0_\beta 1 - (\alpha - \beta) LG^0_\alpha G^0_\beta 1\|_\infty \\ &\leq \|LG^0_\beta 1\|_\infty . \end{aligned}$$

Hence the second term also converges to zero as $\alpha \to +\infty$. It is clear that the third term converges to zero as $\alpha \to +\infty$. This completes the proof of assertion (7.29) and hence that of assertion (7.28). □

Step 6: Summing up, we have proved that the operator $\mathfrak{A}$, defined by formula (7.16), satisfies conditions (a) through (d) in Theorem 3.3. Hence it follows from an application of the same theorem that the operator $\mathfrak{A}$ is the infinitesimal generator of some Feller semigroup on $\overline{D}$.

The proof of Theorem 7.15 is now complete. □

7.3 Non-Transversal Case

Secondly, we consider the non-transversal case. To do this, we assume that

(A) There exists a second-order Ventcel' boundary condition L_ν such that

$$Lu = m(x')\, L_\nu u + \gamma(x') u \quad \text{on } \partial D,$$

where

(3′) $m(x') \in C^\infty(\partial D)$ and $m(x') \geq 0$ on ∂D, and L_ν is given in terms of local coordinates $(x_1, x_2, \ldots, x_{N-1})$ by the formula

$$\begin{aligned}
&L_\nu u(x') \\
&= \overline{Q}u(x') + \overline{\mu}(x')\frac{\partial u}{\partial \mathbf{n}}(x') - \overline{\delta}(x')Wu(x') + \overline{\Gamma}u(x') \\
&:= \sum_{i,j=1}^{N-1} \overline{\alpha}^{ij}(x')\frac{\partial^2 u}{\partial x_i \partial x_j}(x') + \sum_{i=1}^{N-1} \overline{\beta}^i(x')\frac{\partial u}{\partial x_i}(x') \\
&\quad + \overline{\mu}(x')\frac{\partial u}{\partial \mathbf{n}}(x') - \overline{\delta}(x')\,Wu(x') \\
&\quad + \int_{\partial D} \overline{r}(x',y')\left[u(y') - \overline{\tau}(x',y')\left(u(x') + \sum_{j=1}^{N-1}(y_j - x_j)\,\frac{\partial u}{\partial x_j}(x')\right)\right] dy' \\
&\quad + \int_D \overline{t}(x',y)\left[u(y) - \overline{\tau}(x',y)\left(u(x') + \sum_{j=1}^{N-1}(y_j - x_j)\,\frac{\partial u}{\partial x_j}(x')\right)\right] dy\,,
\end{aligned}$$

and satisfies the *transversality* condition

$$\int_D \overline{t}(x',y)\,dy = +\infty \quad \text{if } \overline{\mu}(x') = \overline{\delta}(x') = 0\,. \tag{7.3$'$}$$

We let

$$M = \{x' \in \partial D : \mu(x') = \delta(x') = 0\,,\ \int_D t(x',y)\,dy < \infty\}\,.$$

Then, by condition (7.3′) it follows that

$$M = \{x' \in \partial D : m(x') = 0\}\,,$$

since we have

$$\mu(x') = m(x')\,\overline{\mu}(x')\,, \quad \delta(x') = m(x')\,\overline{\delta}(x')\,, \quad t(x',y) = m(x')\,\overline{t}(x',y)\,.$$

Hence we find that the boundary condition L is *not* transversal on ∂D.

Furthermore, we assume that (see Remark 7.5 below)

(H) $m(x') + |\gamma(x')| > 0$ on ∂D.

The intuitive meaning of conditions (A) and (H) is that a Markovian particle does not stay on ∂D for any period of time until it "dies" at the time when it reaches the set M where the particle is definitely absorbed.

7.3.1 The Space $C_0(\overline{D} \setminus M)$

Now we introduce a subspace of $C(\overline{D})$ which is associated with the boundary condition L. To do this, we consider a one-point compactification $K_\partial = K \cup \{\partial\}$ of the space $K = \overline{D} \setminus M$.

We say that two points x and y of $\overline{D}$ are equivalent modulo M if $x = y$ or x, $y \in M$. It is easy to verify that this relation $\sim$ enjoys the so-called equivalence laws. We denote by $\overline{D}/M$ the totality of equivalence classes modulo M. On the set $\overline{D}/M$ we define the quotient topology induced by the projection

$$q : \overline{D} \longrightarrow \overline{D}/M \,.$$

Namely, a subset O of $\overline{D}/M$ is defined to be open if and only if the inverse image $q^{-1}(O)$ of O is open in $\overline{D}$. It is easy to see that the topological space $\overline{D}/M$ is a *one-point compactification* of the space $\overline{D} \setminus M$ and that the *point at infinity* ∂ corresponds to the set M (see Fig. 7.8):

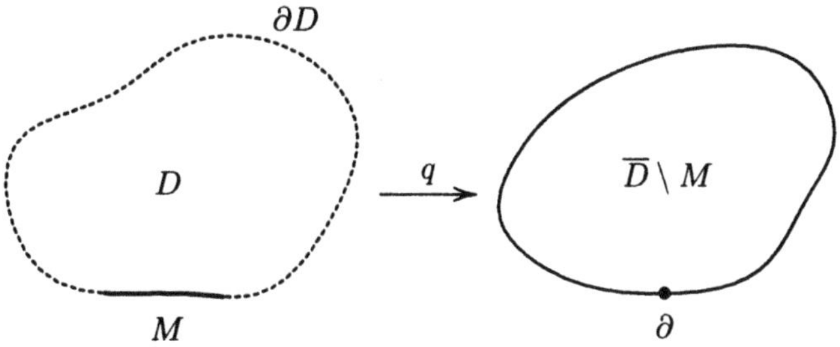

Fig. 7.8.

Furthermore, we have the following two assertions:

(i) If $\tilde{u}$ is a continuous function defined on K_∂, then the function $\tilde{u} \circ q$ is continuous on $\overline{D}$ and constant on M.
(ii) Conversely, if u is a continuous function defined on $\overline{D}$ and constant on M, then it defines a continuous function $\tilde{u}$ on K_∂.

In other words we have the isomorphism

$$C(K_\partial) \cong \left\{ u \in C(\overline{D}) : u \text{ is constant on } M \right\} . \tag{7.30}$$

Now we introduce a closed subspace of $C(K_\partial)$ as in Sect. 3.1:

$$C_0(K) = \{ u \in C(K_\partial) : u(\partial) = 0 \} .$$

Then we have, by assertion (7.30),

$$C_0(K) \cong C_0(\overline{D} \setminus M) = \left\{ u \in C(\overline{D}) : u = 0 \text{ on } M \right\} . \tag{7.31}$$

7.3.2 Generation Theorem for Feller Semigroups

A strongly continuous semigroup $\{T_t\}_{t \geq 0}$ on the space $C_0(\overline{D} \setminus M)$ is called a *Feller semigroup* on $\overline{D} \setminus M$ if it is non-negative and contractive on $C_0(\overline{D} \setminus M)$:

$$f \in C_0(\overline{D} \setminus M) \,, 0 \leq f(x) \leq 1 \text{ on } \overline{D} \setminus M \implies 0 \leq T_t f(x) \leq 1 \text{ on } \overline{D} \setminus M \,.$$

We define a linear operator $\mathfrak{W}$ from $C_0(\overline{D} \setminus M)$ into itself as follows:

(a) The domain of definition $D(\mathfrak{W})$ of $\mathfrak{W}$ is the set

$$D(\mathfrak{W}) = \left\{u \in C_0(\overline{D} \setminus M) : Wu \in C_0(\overline{D} \setminus M),\ Lu = 0\right\}. \tag{7.32}$$

(b) $\mathfrak{W}u = Wu$, $u \in D(\mathfrak{W})$.

The next theorem is a generalization of Theorem 7.1 to the non-transversal case:

Theorem 7.20. *If conditions (A) and (H) are satisfied, then the operator $\mathfrak{W}$, defined by formula (7.32), generates a Feller semigroup $\{T_t\}_{t \geq 0}$ on $\overline{D} \setminus M$.*

Remark 7.5. Theorem 1.5 is a special case of Theorem 7.20 if we take $m(x') \equiv \mu(x')$ and $L_\nu = \partial/\partial\mathbf{n}$ (i.e., $\overline{Q} \equiv 0$, $\overline{\mu}(x') \equiv 1$, $\overline{\delta}(x') \equiv 0$ and $\overline{\Gamma} \equiv 0$). If $m(x') \equiv 0$ and $\gamma(x') \equiv -1$ on ∂D, then Theorem 7.20 was proved by Bony–Courrège–Priouret [BCP, Théorème XVI]. Moreover, it was extended to the degenerate (but non-characteristic) case by Cancelier [Cn, Théorème 7.2].

If T_t is a Feller semigroup on $\overline{D} \setminus M$, then there exists a unique Markov transition function $p_t(x, \cdot)$ on $\overline{D} \setminus M$ such that

$$T_t f(x) = \int_{\overline{D} \setminus M} p_t(x, dy) f(y), \quad f \in C_0(\overline{D} \setminus M),$$

and further that $p_t(x, \cdot)$ is the transition function of some strong Markov process. On the other hand, the intuitive meaning of conditions (A) and (H) is that the absorption phenomenon occurs at each point of the set $M = \{x' \in \partial D : m(x') = 0\}$. Therefore, Theorem 7.20 asserts that there exists a Feller semigroup on $\overline{D} \setminus M$ corresponding to such a physical phenomenon that a Markovian particle moves both by jumps and continuously in the state space $\overline{D} \setminus M$ until it "dies" at which time it reaches the set M. The situation may be represented schematically by Fig. 7.9.

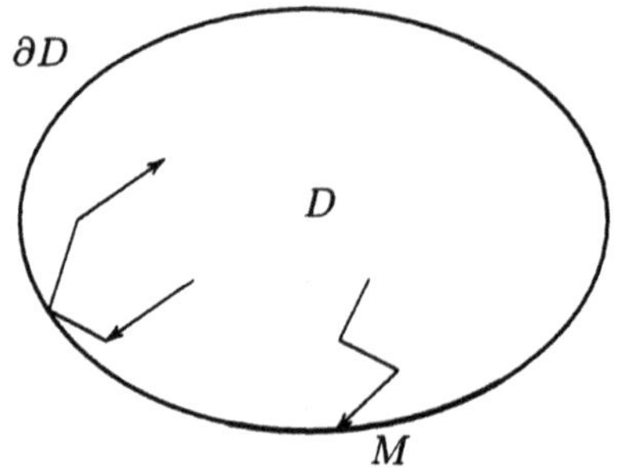

Fig. 7.9.

7.3.3 Sketch of Proof of Theorem 7.20

We explain the idea of proof of Theorem 7.20.

First, it should be noticed that if condition (A) is satisfied, then the boundary condition L can be written in the form

$$Lu = m(x')\, L_\nu u + \gamma(x')u \quad \text{on } \partial D\,,$$

where the boundary condition L_ν is *transversal* on ∂D. Hence, applying Theorem 7.1 to the boundary condition L_ν we can solve uniquely the following boundary value problem:

$$\begin{cases} (\alpha - W)v = f & \text{in } D, \\ L_\nu v = 0 & \text{on } \partial D. \end{cases}$$

We let

$$v = G_\alpha^\nu f\,.$$

The operator G_α^ν is the *Green operator* for the boundary condition L_ν. Then it follows that a function u is a solution of the problem

$$\begin{cases} (\alpha - W)u = f & \text{in } D, \\ Lu = m(x')\, L_\nu u + \gamma(x')u = 0 & \text{on } \partial D \end{cases} \tag{7.33}$$

if and only if the function

$$w = u - v$$

is a solution of the problem

$$\begin{cases} (\alpha - W)w = 0 & \text{in } D, \\ Lw = -Lv = -\gamma(x')v & \text{on } \partial D. \end{cases}$$

Thus, as in the proof of Theorem 7.1 we can reduce the study of problem (7.33) to that of the equation

$$LH_\alpha\psi = -LG_\alpha^\nu f = -\gamma(x')G_\alpha^\nu f \quad \text{on } \partial D\,.$$

We shall apply part (ii) of Theorem 3.3 to the operator $\mathfrak{W}$ defined by formula (7.32).

First, we simplify the boundary condition

$$Lu = 0 \quad \text{on } \partial D\,.$$

If conditions (A) and (H) are satisfied, then we may assume that the boundary condition L is of the form

$$Lu = m(x')L_\nu u + (m(x') - 1)u\,, \tag{7.34}$$

with

$$0 \leq m(x') \leq 1 \quad \text{on } \partial D\,.$$

Indeed, it suffices to note that the boundary condition

$$Lu = m(x')L_\nu u + \gamma(x')\,(u|_{\partial D}) = 0 \quad \text{on } \partial D$$

is equivalent to the condition

$$\left(\frac{m(x')}{m(x') - \gamma(x')}\right) L_\nu u + \left(\frac{\gamma(x')}{m(x') - \gamma(x')}\right) (u|_{\partial D}) = 0 \quad \text{on } \partial D\,.$$

Furthermore, we remark that

$$\overline{LG_\alpha^0} f = m(x')\overline{L_\nu G_\alpha^0} f\,,$$

and

$$\overline{LH_\alpha}\varphi = m(x')\overline{L_\nu H_\alpha}\varphi + (m(x') - 1)\varphi\,.$$

Hence, in view of definition (7.15) it follows that

$$Lu = m(x')L_\nu u + (m(x') - 1)\,(u|_{\partial D})\,, \quad u \in D(L)\,. \tag{7.34'}$$

Therefore, the next theorem proves Theorem 7.20:

Theorem 7.21. *We define a linear operator*

$$\mathfrak{W} : C_0(\overline{D} \setminus M) \longrightarrow C_0(\overline{D} \setminus M)$$

as follows (see formula (7.32)).

(a) The domain $D(\mathfrak{W})$ of $\mathfrak{W}$ is the set

$$\begin{aligned} D(\mathfrak{W}) = \{u \in C_0(\overline{D} \setminus M) : \overline{W}u \in C_0(\overline{D} \setminus M)\,, \\ Lu = m(x')L_\nu u + (m(x') - 1)\,(u|_{\partial D}) = 0\}\,. \end{aligned} \tag{7.35}$$

(b) $\mathfrak{W}u = \overline{W}u$, $u \in D(\mathfrak{W})$.

Assume that the following condition (A′) *is satisfied:*

(A′) $0 \le m(x') \le 1$ on ∂D.

Then the operator $\mathfrak{W}$ is the infinitesimal generator of some Feller semigroup $\{T_t\}_{t\ge0}$ on $\overline{D} \setminus M$, and the Green operator $G_\alpha = (\alpha I - \mathfrak{W})^{-1}$, $\alpha > 0$, is given by the formula

$$G_\alpha f = G_\alpha^\nu f - H_\alpha \left(\overline{LH_\alpha}^{-1} (LG_\alpha^\nu f)\right), \quad f \in C_0(\overline{D} \setminus M)\,. \tag{7.36}$$

Here G_α^ν is the Green operator for the boundary condition L_ν given by formula (7.17) with $L := L_\nu$:

$$G_\alpha^\nu f = G_\alpha^0 f - H_\alpha \left(\overline{L_\nu H_\alpha}^{-1} \left(\overline{L_\nu G_\alpha^0} f\right)\right), \quad f \in C(\overline{D})\,.$$

Proof. We apply part (ii) of Theorem 3.3 to the operator $\mathfrak{W}$ defined by formula (7.35), just as in the proof of Theorem 7.1. The proof is divided into several steps.

Step 1: First, we prove that

For *all* $\alpha > 0$, the operator $\overline{LH_\alpha}$ generates a Feller semigroup on the boundary ∂D.

By virtue of the *transmission property* of $\overline{T} \in L^{2-\kappa_2}_{1,0}(\mathbf{R}^N)$, it follows (see [Bo], [RS, Chap. 3]) that the operator LH_α is the sum of a second-order degenerate elliptic differential operator and a classical pseudo-differential operator of order $2 - \min(\kappa_1, \kappa_2)$:

$$\begin{aligned}
&LH_\alpha\varphi(x') \\
&= m(x')\, L_\nu H_\alpha\varphi(x') + (m(x') - 1)\varphi(x') \\
&= m(x') \left(\sum_{i,j=1}^{N-1} \overline{\alpha}^{ij}(x') \frac{\partial^2\varphi}{\partial x_i \partial x_j}(x') + \sum_{i=1}^{N-1} \overline{\beta}^{i}(x') \frac{\partial\varphi}{\partial x_i}(x') \right) \\
&\quad + \left((m(x') - 1) - \alpha m(x')\, \overline{\delta}(x') \right) \varphi(x') \\
&\quad + m(x')\, \overline{\mu}(x') \frac{\partial}{\partial \mathbf{n}} \left(H_\alpha\varphi \right)(x') \\
&\quad + \int_{\partial D} \overline{r}(x', y') \left[\varphi(y') - \overline{\tau}(x', y') \left(\varphi(x') + \sum_{j=1}^{N-1} (y_j - x_j) \frac{\partial\varphi}{\partial x_j}(x') \right) \right] dy' \\
&\quad + \int_{D} \overline{t}(x', y) \left[H_\alpha\varphi(y) - \overline{\tau}(x', y) \left(\varphi(x') + \sum_{j=1}^{N-1} (y_j - x_j) \frac{\partial\varphi}{\partial x_j}(x') \right) \right] dy\,.
\end{aligned}$$

Furthermore, it follows from an application of the boundary point lemma (see Lemma C.5 in Appendix C) that

$$\Pi_\alpha 1 = \frac{\partial}{\partial \mathbf{n}}(H_\alpha 1) < 0 \quad \text{on } \partial D\,.$$

This implies that

$$LH_\alpha 1(x') = m(x') L_\nu H_\alpha 1(x') + (m(x') - 1) < 0 \quad \text{on } \partial D\,.$$

Indeed, it suffices to note (see inequality (7.19)) that

$$\begin{aligned}
&L_\nu H_\alpha 1(x') \\
&= -\alpha\overline{\delta}(x') + \overline{\mu}(x')\, \Pi_\alpha 1(x') \\
&\quad + \left(\int_{\partial D} \overline{r}(x', y') \left[1 - \overline{\tau}(x', y') \right] dy' + \int_D \overline{t}(x', y) \left[1 - \overline{\tau}(x', y) \right] dy \right) \\
&\quad + \int_D \overline{t}(x', y) \left[H_\alpha 1(y) - 1 \right] dy \\
&< 0 \quad \text{on } \partial D. \qquad (7.19')
\end{aligned}$$

Thus, applying Theorem 7.16 to the operator LH_α (cf. the proof of assertion (7.20)) we obtain that

If $\lambda > 0$ is sufficiently large, then the range $R(LH_\alpha - \lambda I)$ contains the space $C^{2+\theta}(\partial D)$. (7.37)

This implies that the range $R(LH_\alpha - \lambda I)$ is a *dense* subset of $C(\partial D)$. Therefore, applying part (ii) of Theorem 7.15 to the operator L we obtain that the operator $\overline{LH_\alpha}$ is the infinitesimal generator of some Feller semigroup on ∂D, for all $\alpha > 0$.

Step 2: Now we prove that

If condition (A′) is satisfied, then the equation
$\overline{LH_\alpha}\psi = \varphi$
has a unique solution ψ in $D\left(\overline{LH_\alpha}\right)$ for any $\varphi \in C(\partial D)$; hence the inverse $\overline{LH_\alpha}^{-1}$ of $\overline{LH_\alpha}$ can be defined on the whole space $C(\partial D)$. Further the operator $-\overline{LH_\alpha}^{-1}$ is non-negative and bounded on $C(\partial D)$. (7.38)

Since we have, by inequality (7.19′) and condition (A′),

$$LH_\alpha 1(x') = m(x')L_\nu H_\alpha 1(x') + (m(x') - 1) < 0 \quad \text{on } \partial D\,,$$

it follows that

$$k_\alpha = -\sup_{x' \in \partial D} LH_\alpha 1(x') > 0\,,$$

and further that the constants k_α are increasing in $\alpha > 0$:

$$\alpha \geq \beta > 0 \implies k_\alpha \geq k_\beta\,.$$

Indeed, it suffices to note (see Lemma 7.18) that $H_\alpha 1$ converges to zero and $\Pi_\alpha 1$ diverges to $-\infty$ *monotonically* as $\alpha \to +\infty$, respectively. Moreover, using Corollary 3.4 with $K := \partial D$, $A := \overline{LH_\alpha}$ and $c := k_\alpha$ we obtain that the operator $\overline{LH_\alpha} + k_\alpha I$ generates a Feller semigroup on ∂D. Therefore, since $k_\alpha > 0$, it follows from an application of part (i) of Theorem 3.3 with $A := \overline{LH_\alpha} + k_\alpha I$ that the equation

$$-\overline{LH_\alpha}\psi = \left(k_\alpha I - (\overline{LH_\alpha} + k_\alpha I)\right)\psi = \varphi$$

has a unique solution $\psi \in D\left(\overline{LH_\alpha}\right)$ for any $\varphi \in C(\partial D)$, and further that the operator $-\overline{LH_\alpha}^{-1} = \left(k_\alpha I - (\overline{LH_\alpha} + k_\alpha I)\right)^{-1}$ is non-negative and bounded on the space $C(\partial D)$ with norm

$$\left\|-\overline{LH_\alpha}^{-1}\right\| = \left\|\left(k_\alpha I - (\overline{LH_\alpha} + k_\alpha I)\right)^{-1}\right\| \leq \frac{1}{k_\alpha}\,. \tag{7.39}$$

Step 3: By assertion (7.38), we can define the operator G_α by formula (7.36) for all $\alpha > 0$. We prove that

$$G_\alpha = (\alpha I - \mathfrak{W})^{-1}, \quad \alpha > 0. \tag{7.40}$$

By virtue of Lemma 7.7 and Theorem 7.15, it follows that we have, for all $f \in C_0(\overline{D} \setminus M)$,

$$G_\alpha f \in D(\overline{W}),$$

and

$$\overline{W} G_\alpha f = \alpha G_\alpha f - f.$$

Furthermore, we have

$$LG_\alpha f = LG_\alpha^\nu f - \overline{LH_\alpha}\left(\overline{LH_\alpha}^{-1}(LG_\alpha^\nu f)\right) = 0 \quad \text{on } \partial D. \tag{7.41}$$

However we recall that

$$Lu = m(x')\, L_\nu u + (m(x') - 1)\,(u|_{\partial D}), \quad u \in D(L). \tag{7.34$'$}$$

Hence we find that formula (7.41) is equivalent to the following:

$$m(x')\, L_\nu(G_\alpha f) + (m(x') - 1)\,(G_\alpha f|_{\partial D}) = 0 \quad \text{on } \partial D. \tag{7.41$'$}$$

This implies that

$$G_\alpha f = 0 \quad \text{on } M = \{x' \in \partial D : m(x') = 0\},$$

and so

$$\overline{W} G_\alpha f = \alpha G_\alpha f - f = 0 \quad \text{on } M.$$

Summing up, we have proved that

$$G_\alpha f \in D(\mathfrak{W}) = \left\{u \in C_0(\overline{D} \setminus M) : \overline{W} u \in C_0(\overline{D} \setminus M), \ Lu = 0\right\},$$

and

$$(\alpha I - \mathfrak{W}) G_\alpha f = f, \quad f \in C_0(\overline{D} \setminus M),$$

that is,

$$(\alpha I - \mathfrak{W}) G_\alpha = I \quad \text{on } C_0(\overline{D} \setminus M).$$

Therefore, in order to prove formula (7.40) it suffices to show the injectivity of the operator $\alpha I - \mathfrak{W}$ for $\alpha > 0$.

Assume that

$$u \in D(\mathfrak{W}) \text{ and } (\alpha I - \mathfrak{W}) u = 0.$$

Then, by Corollary 7.8 it follows that the function u can be written in the form

$$u = H_\alpha (u|_{\partial D}), \quad u|_{\partial D} \in \mathcal{D} = D\left(\overline{LH_\alpha}\right).$$

Thus we have

$$\overline{LH_\alpha}\,(u|_{\partial D}) = Lu = 0.$$

In view of assertion (7.38), this implies that

$$u|_{\partial D} = 0 \,,$$

so that

$$u = H_\alpha \left(u|_{\partial D} \right) = 0 \quad \text{in } D \,.$$

Step 4: Now we prove the following three assertions:

(i) The operator G_α is non-negative on the space $C_0(\overline{D} \setminus M)$:

$$f \in C_0(\overline{D} \setminus M),\ f(x) \geq 0 \quad \text{on } \overline{D} \setminus M \implies G_\alpha f(x) \geq 0 \quad \text{on } \overline{D} \setminus M \,.$$

(ii) The operator G_α is bounded on the space $C_0(\overline{D} \setminus M)$ with norm

$$\|G_\alpha\| \leq \frac{1}{\alpha} \,, \quad \alpha > 0 \,.$$

(iii) The domain $D(\mathfrak{W})$ is dense in the space $C_0(\overline{D} \setminus M)$.

Step 4-1: First, we show the non-negativity of G_α on the space $C(\overline{D})$:

$$f \in C(\overline{D}) \,, f(x) \geq 0 \quad \text{on } \overline{D} \Longrightarrow G_\alpha f(x) \geq 0 \quad \text{on } \overline{D} \,.$$

Recall that the Dirichlet problem

$$\begin{cases} (\alpha - W)u = f & \text{in } D, \\ u = \varphi & \text{on } \partial D \end{cases} \tag{D$'$}$$

is uniquely solvable. Hence it follows that

$$G_\alpha^\nu f = H_\alpha \left(G_\alpha^\nu f|_{\partial D} \right) + G_\alpha^0 f \quad \text{on } \overline{D}. \tag{7.42}$$

Indeed, the both sides have the same boundary values $G_\alpha^\nu f|_{\partial D}$ and satisfy the same equation: $(\alpha - W)u = f$ in D.

Thus, applying the operator L to the both sides of formula (7.42) we obtain that

$$LG_\alpha^\nu f = \overline{LH_\alpha} \left(G_\alpha^\nu f|_{\partial D} \right) + \overline{LG_\alpha^0} f \,.$$

Since the operators $-\overline{LH_\alpha}^{-1}$ and $\overline{LG_\alpha^0}$ are non-negative, it follows that

$$\begin{aligned} \left(-\overline{LH_\alpha}^{-1} \right) \left(LG_\alpha^\nu f \right) &= -G_\alpha^\nu f|_{\partial D} + \left(-\overline{LH_\alpha}^{-1} \right) \left(\overline{LG_\alpha^0} f \right) \\ &\geq -G_\alpha^\nu f|_{\partial D} \quad \text{on } \partial D \,. \end{aligned}$$

Therefore, by the non-negativity of H_α and G_α^0 we find that

$$\begin{aligned} G_\alpha f &= G_\alpha^\nu f + H_\alpha \left(-\overline{LH_\alpha}^{-1} \left(LG_\alpha^\nu f \right) \right) \\ &\geq G_\alpha^\nu f - H_\alpha \left(G_\alpha^\nu f|_{\partial D} \right) \\ &= G_\alpha^0 f \\ &\geq 0 \quad \text{on } \overline{D} \,. \end{aligned}$$

Step 4-2: Next we prove the boundedness of G_α on the space $C_0(\overline{D} \setminus M)$ with norm

$$\|G_\alpha\| \le \frac{1}{\alpha}, \quad \alpha > 0. \tag{7.43}$$

To do this, it suffices to show that

$$f \in C_0(\overline{D} \setminus M),\, f(x) \ge 0 \quad \text{on } \overline{D} \Longrightarrow \alpha G_\alpha f(x) \le \max_{\overline{D}} f \quad \text{on } \overline{D}, \tag{7.43$'$}$$

since G_α is non-negative on the space $C(\overline{D})$.

We remark (cf. formula (7.34′)) that

$$LG_\alpha^\nu f = m(x')\, L_\nu G_\alpha^\nu f + (m(x') - 1)\,(G_\alpha^\nu f|_{\partial D}) = (m(x') - 1)\,(G_\alpha^\nu f|_{\partial D}),$$

so that

$$\begin{aligned} G_\alpha f &= G_\alpha^\nu f - H_\alpha \left(\overline{LH_\alpha}^{-1} (LG_\alpha^\nu f)\right) \\ &= G_\alpha^\nu f + H_\alpha \left(-\overline{LH_\alpha}^{-1} ((m(x') - 1)G_\alpha^\nu f|_{\partial D})\right). \end{aligned} \tag{7.36$'$}$$

Therefore, by the non-negativity of H_α and $-\overline{LH_\alpha}^{-1}$ it follows that

$$\begin{aligned} G_\alpha f &= G_\alpha^\nu f + H_\alpha \left(-\overline{LH_\alpha}^{-1} ((m(x') - 1)G_\alpha^\nu f|_{\partial D})\right) \\ &\le G_\alpha^\nu f \\ &\le \frac{1}{\alpha} \max_{\overline{D}} f \quad \text{on } \overline{D}, \end{aligned}$$

since $(m(x') - 1)G_\alpha^\nu f|_{\partial D} \le 0$ on ∂D and $\|G_\alpha^\nu\| \le 1/\alpha$. This proves assertion (7.43′) and hence assertion (7.43).

Step 4-3: Finally, we prove the density of $D(\mathfrak{W})$ in the space $C_0(\overline{D} \setminus M)$. In view of formula (7.40), it suffices to show that

$$\lim_{\alpha \to +\infty} \|\alpha G_\alpha f - f\|_\infty = 0, \quad f \in C_0(\overline{D} \setminus M) \cap C^\infty(\overline{D}). \tag{7.44}$$

We recall (cf. formula (5.5′)) that

$$\begin{aligned} \alpha G_\alpha f - f &= \alpha G_\alpha^\nu f - f - \alpha H_\alpha \left(\overline{LH_\alpha}^{-1} (LG_\alpha^\nu f)\right) \\ &= (\alpha G_\alpha^\nu f - f) + H_\alpha \left(\overline{LH_\alpha}^{-1} (\alpha(1 - m(x'))G_\alpha^\nu f|_{\partial D})\right). \end{aligned} \tag{7.45}$$

We estimate each term on the right of formula (7.45).

Step 4-3-1: First, applying Theorem 7.1 to the boundary condition L_ν we find from assertion (7.28) that the first term on the right of formula (7.45) tends to zero:

$$\lim_{\alpha \to +\infty} \|\alpha G_\alpha^\nu f - f\|_\infty = 0. \tag{7.46}$$

Step 4-3-2: To estimate the second term on the right of formula (7.45), we remark that

$$\begin{aligned}
&H_\alpha\left(\overline{LH_\alpha}^{-1}(\alpha(1-m(x'))G_\alpha^\nu f|_{\partial D})\right) \\
&= H_\alpha\left(\overline{LH_\alpha}^{-1}((1-m(x'))f|_{\partial D})\right) \\
&\quad + H_\alpha\left(\overline{LH_\alpha}^{-1}((1-m(x'))(\alpha G_\alpha^\nu f - f)|_{\partial D})\right).
\end{aligned}$$

However we have, by assertion (7.39),

$$\begin{aligned}
&\left\|H_\alpha\left(\overline{LH_\alpha}^{-1}((1-m(x'))(\alpha G_\alpha^\nu f - f)|_{\partial D})\right)\right\|_\infty \\
&\leq \left\|-\overline{LH_\alpha}^{-1}\right\| \cdot \|(1-m(x'))(\alpha G_\alpha^\nu f - f)|_{\partial D}\|_\infty \\
&\leq \frac{1}{k_\alpha}\|(1-m(x'))(\alpha G_\alpha^\nu f - f)|_{\partial D}\|_\infty \\
&\leq \frac{1}{k_1}\|\alpha G_\alpha^\nu f - f\|_\infty \longrightarrow 0 \quad \text{as } \alpha \to +\infty.
\end{aligned} \tag{7.47}$$

Here we have used the fact that

$$k_1 = -\sup_{x' \in \partial D} LH_1 1(x') \leq k_\alpha = -\sup_{x' \in \partial D} LH_\alpha 1(x') \quad \text{for all } \alpha \geq 1,$$

since the constants k_α are increasing in $\alpha > 0$. Thus we are reduced to the study of the term

$$H_\alpha\left(\overline{LH_\alpha}^{-1}((1-m(x'))f|_{\partial D})\right).$$

Now, for any given $\varepsilon > 0$, we can find a function $h(x') \in C^\infty(\partial D)$ such that

$$\begin{cases} h = 0 & \text{near } M = \{x' \in \partial D : m(x') = 0\}, \\ \|(1-m(x'))f|_{\partial D} - h\|_\infty < \varepsilon. \end{cases}$$

Then we have, for all $\alpha \geq 1$,

$$\begin{aligned}
&\left\|H_\alpha\left(\overline{LH_\alpha}^{-1}((1-m(x'))f|_{\partial D})\right) - H_\alpha\left(\overline{LH_\alpha}^{-1}h\right)\right\|_\infty \\
&\leq \left\|-\overline{LH_\alpha}^{-1}\right\| \cdot \|(1-m(x'))f|_{\partial D} - h\|_\infty \\
&\leq \frac{\varepsilon}{k_\alpha} \\
&\leq \frac{\varepsilon}{k_1}.
\end{aligned} \tag{7.48}$$

Furthermore, we can find a function $\theta(x') \in C_0^\infty(\partial D)$ such that

$$\begin{cases} \theta = 1 & \text{near } M, \\ (1-\theta)h = h & \text{on } \partial D. \end{cases}$$

Then we have

$$\begin{aligned} h(x') &= (1-\theta(x'))\, h(x') \\ &= (-LH_\alpha 1(x'))\left(\frac{1-\theta(x')}{-LH_\alpha 1(x')}\right) h(x') \\ &\le \left[\sup_{x'\in\partial D}\left(\frac{1-\theta(x')}{-LH_\alpha 1(x')}\right)\right] \|h\|_\infty\,(-LH_\alpha 1(x')) . \end{aligned}$$

Since the operator $-\overline{LH_\alpha}^{-1}$ is non-negative on the space $C(\partial D)$, it follows that

$$-\overline{LH_\alpha}^{-1} h \le \sup_{x'\in\partial D}\left(\frac{1-\theta(x')}{-LH_\alpha 1(x')}\right)\cdot \|h\|_\infty \quad \text{on } \partial D\,,$$

so that

$$\left\| H_\alpha\left(\overline{LH_\alpha}^{-1} h\right)\right\|_\infty \le \left\| -\overline{LH_\alpha}^{-1} h\right\|_\infty \le \sup_{x'\in\partial D}\left(\frac{1-\theta(x')}{-LH_\alpha 1(x')}\right)\cdot \|h\|_\infty\,. \tag{7.49}$$

However there exists a constant $c_0 > 0$ such that

$$0 \le \frac{1-\theta(x')}{m(x')} \le c_0\,, \quad x' \in \partial D\,.$$

Hence we have

$$\begin{aligned} \frac{1-\theta(x')}{-LH_\alpha 1(x')} &\le \left(\frac{1-\theta(x')}{m(x')\,(-L_\nu H_\alpha 1(x')) + (1-m(x'))}\right) \\ &\le c_0 \left\|\frac{1}{-L_\nu H_\alpha 1}\right\|_\infty . \end{aligned}$$

In view of Lemma 7.18, this implies that

$$\lim_{\alpha\to+\infty}\left[\sup_{x'\in\partial D}\left(\frac{1-\theta(x')}{-LH_\alpha 1(x')}\right)\right] = 0\,.$$

Summing up, we obtain from inequalities (7.48) and (7.49) that

$$\begin{aligned} &\limsup_{\alpha\to+\infty}\left\| H_\alpha\left(\overline{LH_\alpha}^{-1}((1-m(x'))f|_{\partial D})\right)\right\|_\infty \\ &\le \limsup_{\alpha\to+\infty}\Bigg[\left\| H_\alpha\left(\overline{LH_\alpha}^{-1} h\right)\right\|_\infty \\ &\qquad + \left\| H_\alpha\left(\overline{LH_\alpha}^{-1}((1-m(x'))f|_{\partial D})\right) - H_\alpha\left(\overline{LH_\alpha}^{-1} h\right)\right\|_\infty\Bigg] \\ &\le \lim_{\alpha\to+\infty}\left[\sup_{x'\in\partial D}\left(\frac{1-\theta(x')}{-LH_\alpha 1(x')}\right)\right]\|h\|_\infty + \frac{\varepsilon}{k_1} \\ &\le \frac{\varepsilon}{k_1}\,. \end{aligned}$$

Since ε is arbitrary, this proves that

$$\lim_{\alpha \to +\infty} \left\| H_\alpha \left(\overline{LH_\alpha}^{-1} ((1 - m(x'))f|_{\partial D}) \right) \right\|_\infty = 0 . \tag{7.50}$$

Therefore, combining assertions (7.47) and (7.50) we find that the second term on the right of formula (7.45) also tends to zero:

$$\lim_{\alpha \to +\infty} \left\| H_\alpha \left(\overline{LH_\alpha}^{-1} (\alpha(1 - m(x'))G_\alpha^\nu f|_{\partial D}) \right) \right\|_\infty = 0 .$$

This completes the proof of assertion (7.44) and hence that of assertion (iii).

Step 5: Summing up, we have proved that the operator $\mathfrak{W}$, defined by formula (7.35), satisfies conditions (a) through (d) in Theorem 3.3. Therefore, in view of assertion (7.31) it follows from an application of part (ii) of the same theorem that the operator $\mathfrak{W}$ is the infinitesimal generator of some Feller semigroup $\{T_t\}_{t \geq 0}$ on $\overline{D} \setminus M$.

The proof of Theorem 7.21 and hence that of Theorem 7.20 is now complete. □

Remark 7.6. It is worth while pointing out that if we use instead of G_α^ν the Green operator G_α^0 for the Dirichlet problem as in the proof of Theorem 7.1, our proof would break down.

8

Proof of Theorem 1.1

In this chapter we consider the boundary value problem

$$\begin{cases} Au = f & \text{in } D, \\ L_0 u = \mu(x')\dfrac{\partial u}{\partial \mathbf{n}} + \gamma(x')u = \varphi & \text{on } \partial D \end{cases} \tag{1.1}$$

in the framework of Sobolev spaces of L^p style, and prove Theorem 1.1. If $1 < p < \infty$ and $s > 1+1/p$, then we associate with problem (1.1) a continuous linear operator

$$\mathcal{A} = (A, L_0) \colon H^{s,p}(D) \longrightarrow H^{s-2,p}(D) \bigoplus B_{L_0}^{s-1-1/p,p}(\partial D) .$$

To prove Theorem 1.1 for $s = 2$, it suffices to show that the operator $\mathcal{A}$ is *bijective*. Indeed, since the inverse $\mathcal{A}^{-1}$ is a closed operator, we obtain that $\mathcal{A}^{-1}$ is continuous if we apply the following Banach's closed graph theorem (see [Yo, Chap. II, Sect. 6, Theorem 1], [Fr2, Theorem 4.6.4]) with

$$X := L^p(D) \bigoplus B_{L_0}^{1-1/p,p}(\partial D) , \quad Y := H^{2,p}(D) ,$$
$$T := \mathcal{A}^{-1} .$$

Theorem 8.1 (Banach's closed graph theorem). *Let X and Y be Banach spaces. Then every closed linear operator* on *X into Y is continuous.*

Theorem 1.1 will be proved in a series of theorems in the subsequent sections (Theorems 8.2, 8.6 and 8.8). In the proof of surjectivity of the operator T we make use of Agmon's method (Proposition 8.11). This is a technique of treating a spectral parameter as a second-order differential operator of an extra variable and relating the old problem to a new one with the additional variable. More precisely, we introduce an auxiliary variable y of the unit circle $S = \mathbf{R}/2\pi\mathbf{Z}$, and replace a spectral parameter λ by the second-order differential operator $-\partial^2/\partial y^2$.

8.1 Regularity Theorem for Problem (1.1)

The next theorem states that the operator $\mathcal{A}$ has the *regularity property*:

Theorem 8.2. *Assume that conditions (H.1) and (H.2) are satisfied:*

(H.1) $\mu(x') \geq 0$ *and* $\gamma(x') \leq 0$ *on* ∂D.
(H.2) $\gamma(x') < 0$ *on* $M = \{x' \in \partial D: \mu(x') = 0\}$.

Then we have, for all $s > 1 + 1/p$ *where* $1 < p < \infty$,

$$u \in L^p(D)\ , Au \in H^{s-2,p}(D)\ , L_0 u \in B_{L_0}^{s-1-1/p,p}(\partial D) \Longrightarrow u \in H^{s,p}(D)\ .$$

Proof. The proof is divided into several steps.

Step 1: By Theorem 6.10 with $B_0 := L_0$, we know that the question of regularity for solutions of problem (1.1) is reduced to the corresponding question for the operator T, where

$$T = L_0 P = \mu(x')\Pi + \gamma(x')\ ,$$

and

$$\Pi\varphi = \frac{\partial}{\partial \mathbf{n}}(P\varphi)\ .$$

The operator Π is a first-order, classical pseudo-differential operator on the boundary ∂D, and its complete symbol is given by the following (cf. [Ta2, Sect. 10.2]):

$$\begin{aligned}&\left(p_1(x',\xi') + \sqrt{-1}\, q_1(x',\xi')\right) + \left(p_0(x',\xi') + \sqrt{-1}\, q_0(x',\xi')\right)\\ &+ \text{terms of order} \leq -1\ ,\end{aligned}$$

where

$$\begin{aligned} p_1(x',\xi') < 0 \quad &\text{on the bundle } T^*(\partial D) \setminus \{0\}\\ &\text{of non-zero cotangent vectors}\ . \end{aligned} \tag{8.1}$$

For example, if A is the usual Laplacian $\Delta = \partial^2/\partial x_1^2 + \partial^2/\partial x_2^2 + \ldots + \partial^2/\partial x_N^2$, then we have

$$p_1(x',\xi') = -|\xi'|\ ,$$

where $|\xi'|$ is the length of ξ' with respect to the Riemannian metric of ∂D induced by the natural metric of $\mathbf{R}^N$.

Thus, we obtain that the operator

$$T = L_0 P = \mu(x')\Pi + \gamma(x')$$

is a first-order, classical pseudo-differential operator on the boundary ∂D, and its complete symbol $t(x',\xi')$ is given by the following:

$$\begin{aligned} t(x',\xi') &= \mu(x')\left(p_1(x',\xi') + \sqrt{-1}\, q_1(x',\xi')\right)\\ &\quad + \left([\gamma(x') + \mu(x')p_0(x',\xi)] + \sqrt{-1}\, \mu(x')q_0(x',\xi)\right)\\ &\quad + \text{terms of order} \leq -1\ . \end{aligned} \tag{8.2}$$

Step 2: To prove Theorem 8.2, it suffices to show the following:

Lemma 8.3. *If conditions (H.1) and (H.2) are satisfied, then we have, for all* $s \in \mathbf{R}$,

$$\varphi \in \mathcal{D}'(\partial D)\ , T\varphi \in B^{s,p}(\partial D) \Longrightarrow \varphi \in B^{s,p}(\partial D)\ . \tag{8.3}$$

Furthermore, for any $t < s$, *there exists a constant* $C_{s,t} > 0$ *such that*

$$|\varphi|_{s,p} \leq C_{s,t}\left(|T\varphi|_{s,p} + |\varphi|_{t,p}\right)\ . \tag{8.4}$$

Proof. (1) The proof of Lemma 8.3 is based on the following lemma (cf. Kannai [Ka]):

Lemma 8.4. *If conditions (H.1) and (H.2) are satisfied, then, for each point* x' *of* ∂D, *we can find a neighborhood* $U(x')$ *of* x' *such that: For any compact* $K \subset U(x')$ *and any multi-indices* α, β, *there exist constants* $C_{K,\alpha,\beta} > 0$ *and* $C_K > 0$ *such that we have, for all* $x' \in K$ *and* $|\xi'| \geq C_K$,

$$\left|D_{\xi'}^{\alpha} D_{x'}^{\beta} t(x',\xi')\right| \leq C_{K,\alpha,\beta}\, |t(x',\xi')|\, (1+|\xi'|)^{-|\alpha|+(1/2)|\beta|}\ , \tag{8.5a}$$

$$|t(x',\xi')|^{-1} \leq C_K\ . \tag{8.5b}$$

Granting Lemma 8.4 for the moment, we shall prove Lemma 8.3.

(2) First, we cover ∂D by a finite number of local charts $\{(U_j,\chi_j)\}_{j=1}^{m}$ in each of which inequalities (8.5a) and (8.5b) hold. Since the operator T satisfies conditions (5.7a) and (5.7b) of Theorem 5.15 with $\mu := 0$, $\rho := 1$ and $\delta := 1/2$, it follows from an application of the same theorem that there exists a parametrix S in the class $L^0_{1,1/2}(U_j)$ for T. Let $\{\varphi_j\}_{j=1}^{m}$ be a partition of unity subordinate to the covering $\{U_j\}_{j=1}^{m}$, and choose a function $\psi_j \in C_0^\infty(U_j)$ such that $\psi_j = 1$ on $\operatorname{supp}\varphi_j$, so that $\varphi_j\psi_j = \varphi_j$.

Now we may assume that $\varphi \in B^{t,p}(\partial D)$ for some $t < s$ and that $T\varphi \in B^{s,p}(\partial D)$. It should be noticed that the operator T can be written in the following form:

$$T = \sum_{j=1}^{m} \varphi_j T\psi_j + \sum_{j=1}^{m} \varphi_j T(1-\psi_j)\ .$$

However, the second terms $\varphi_j T(1-\psi_j)$ are in $L^{-\infty}(\partial D)$, since $\varphi_j(1-\psi_j) = 0$. Hence we are reduced to the study of the first terms $\varphi_j T\psi_j$. This implies that we have only to prove the following *local* version of assertions (8.3) and (8.4):

$$\psi_j\varphi \in B^{t,p}(U_j),\ T\psi_j\varphi \in B^{s,p}(U_j) \Longrightarrow \psi_j\varphi \in B^{s,p}(U_j); \tag{8.3$'$}$$

$$|\psi_j\varphi|_{s,p} \leq C'_{s,t}\left(|T\psi_j\varphi|_{s,p} + |\psi_j\varphi|_{t,p}\right). \tag{8.4$'$}$$

However, applying the Besov-space boundedness theorem (Theorem 5.14) to our situation we find that the parametrix S maps the space $B^{\sigma,p}_{\mathrm{loc}}(U_j)$ continuously into itself for all $\sigma \in \mathbf{R}$. This proves assertions (8.3′) and (8.4′), since we have $ST \equiv I \bmod L^{-\infty}(U_j)$.

Lemma 8.3 is proved, apart from the proof of Lemma 8.4. □

(3) **Proof of Lemma 8.4**

(3-1) First, we verify condition (8.5b): By assertions (8.2) and (8.1), it follows that we have, for $|\xi'|$ sufficiently large,

$$\begin{aligned}|t(x',\xi')| &\geq \mu(x')\,|p_1(x',\xi')+p_0(x',\xi')|-\gamma(x')\\ &\geq \begin{cases}\frac{c_0}{2}\,\mu(x')|\xi'|-\gamma(x') & \text{if } \mu(x')>0,\\ -\gamma(x') & \text{if } \mu(x')=0,\end{cases}\end{aligned}$$

so that

$$|t(x',\xi')| \geq C\left(\mu(x')|\xi'|+1\right), \tag{8.6}$$

since $\gamma(x')<0$ on $M=\{x'\in\partial D\colon \mu(x')=0\}$. Here and in the following the letter C denotes a generic positive constant.

Inequality (8.6) implies condition (8.5b):

$$|t(x',\xi')| \geq C\,. \tag{8.7}$$

(3-2) Next we verify condition (8.5a) for $|\alpha|=1$ and $|\beta|=0$:
Since we have, for $|\xi'|$ sufficiently large,

$$\left|D^{\alpha}_{\xi'}t(x',\xi')\right| \leq C\left(\mu(x')+|\xi'|^{-1}\right),$$

it follows from inequality (8.6) that

$$\begin{aligned}\left|D^{\alpha}_{\xi'}t(x',\xi')\right| &\leq C(1+|\xi'|)^{-1}\left(\mu(x')|\xi'|+1\right)\\ &\leq C(1+|\xi'|)^{-1}\,|t(x',\xi')|\;.\end{aligned}$$

This inequality proves condition (8.5a) for $|\alpha|=1$ and $|\beta|=0$.

(3-3) We verify condition (8.5a) for $|\beta|=1$ and $|\alpha|=0$: To do this, we need the following elementary lemma on non-negative functions:

Lemma 8.5. *Let $f(x)$ be a non-negative, C^2 function on $\mathbf{R}$ such that*

$$\sup_{x\in\mathbf{R}}|f''(x)| \leq c \tag{8.8}$$

for some constant $c>0$. Then we have the inequality

$$|f'(x)| \leq \sqrt{2c}\sqrt{f(x)} \quad \text{on } \mathbf{R}\,. \tag{8.9}$$

Proof. In view of Taylor's formula, it follows that

$$0 \leq f(y) = f(x)+f'(x)(y-x)+\frac{f''(\xi)}{2}\,(y-x)^2\;,$$

where ξ is between x and y. Thus, letting $z=x-y$, we obtain from estimate (8.8) that

$$\begin{aligned} 0 &\le f(x) + f'(x)z + \frac{f''(\xi)}{2} z^2 \\ &\le f(x) + f'(x)z + \frac{c}{2} z^2 \quad \text{for all } z \in \mathbf{R} . \end{aligned}$$

This implies inequality (8.9). □

Since we have, for $|\xi'|$ sufficiently large,

$$\left| D^\beta_{x'} t(x', \xi') \right| \le C \left(\left| D^\beta_{x'} \mu(x') \right| \cdot |\xi'| + \mu(x')|\xi'| + 1 \right),$$

it follows from an application of Lemma 8.5 and inequalities (8.6) and (8.7) that

$$\begin{aligned} \left| D^\beta_{x'} t(x', \xi') \right| &\le C \left[\left(\sqrt{\mu(x')}\, |\xi'| + 1 \right) + (\mu(x')|\xi'| + 1) \right] \\ &\le C \left[|\xi'|^{1/2} (\mu(x')|\xi'| + 1)^{1/2} + (\mu(x')|\xi'| + 1) \right] \\ &\le C\, |t(x', \xi')| \left(|\xi'|^{1/2} \, |t(x', \xi')|^{-1/2} + 1 \right) \\ &\le C\, |t(x', \xi')| \, (1 + |\xi'|)^{1/2} . \end{aligned}$$

This inequality proves condition (8.5a) for $|\beta| = 1$ and $|\alpha| = 0$.

(3-4) Similarly, we can verify condition (8.5a) for the general case: $|\alpha| + |\beta| = k$, $k \in \mathbf{N}$. □

Now the proof of Lemma 8.3, and hence that of Theorem 8.2, is complete. □

8.2 Uniqueness Theorem for Problem (1.1)

The next theorem asserts that the operator $\mathcal{A}$ is *injective*:

Theorem 8.6. *Assume that conditions (H.1) and (H.2) are satisfied:*

(H.1) $\mu(x') \ge 0$ *and* $\gamma(x') \le 0$ *on* ∂D.
(H.2) $\gamma(x') < 0$ *on* $M = \{x' \in \partial D \colon \mu(x') = 0\}$.

Then we have, for $s > 1 + 1/p$ *where* $1 < p < \infty$,

$$u \in H^{s,p}(D) , \quad Au = 0 \ \ in\ D , \quad L_0 u = 0 \ \ on\ \partial D \Longrightarrow u = 0 \ \ in\ D . \tag{8.10}$$

Proof. In view of Theorem 8.2, it follows that

$$u \in H^{s,p}(D) , \quad Au = 0 \text{ in } D , \quad L_0 u = 0 \text{ on } \partial D \Longrightarrow u \in C^\infty(\overline{D}) .$$

Therefore, the uniqueness result (8.10) is an immediate consequence of the following *maximum principle*:

Proposition 8.7. *Assume that conditions (H.1) and (H.2) are satisfied. Then we have*

$$v \in C^2(\overline{D}) \,, \quad Av \geq 0 \ \ in \ D \,, \quad L_0 v \leq 0 \ \ on \ \partial D \Longrightarrow v \leq 0 \ \ on \ \overline{D} \,.$$

Proof. If $v(x)$ is a constant m, then we have

$$0 \leq Av(x) = mc(x) \quad \text{in } D \,.$$

This implies that m is non-positive, since $c(x) \leq 0$ and $c(x) \not\equiv 0$ in D.

Now we consider the case when $v(x)$ is not a constant. Assume, to the contrary, that

$$m = \max_{x \in \overline{D}} v(x) > 0 \,.$$

Then, applying the weak maximum principle (see Theorem C.1 in Appendix C) to the operator A, we obtain that there exists a point x_0' of ∂D such that

$$\begin{cases} v(x_0') = m, \\ v(x) < v(x_0') \quad \text{for all } x \in D. \end{cases}$$

Furthermore, it follows from an application of the boundary point lemma (see Lemma C.5 in Appendix C) that

$$\frac{\partial v}{\partial \mathbf{n}}(x_0') < 0 \,.$$

Since we have

$$L_0 v(x_0') = \mu(x_0') \frac{\partial v}{\partial \mathbf{n}}(x_0') + \gamma(x_0') v(x_0') \leq 0 \,,$$

it follows from condition (H.1) that

$$\mu(x_0') = 0 \,, \quad \gamma(x_0') = 0 \,.$$

This contradicts condition (H.2). □

8.3 Existence Theorem for Problem (1.1)

The next theorem asserts that the operator $\mathcal{A}$ is *surjective*:

Theorem 8.8. *Let $s > 1 + 1/p$ where $1 < p < \infty$. Assume that conditions (H.1) and (H.2) are satisfied:*

(H.1) $\mu(x') \geq 0$ and $\gamma(x') \leq 0$ on ∂D.
(H.2) $\gamma(x') < 0$ on $M = \{x' \in \partial D \colon \mu(x') = 0\}$.

Then, for any $f \in H^{s-2,p}(D)$ and any $\varphi \in B_{L_0}^{s-1-1/p,p}(\partial D)$, the boundary value problem

$$\begin{cases} Au = f & \text{in } D, \\ L_0 u = \mu(x') \dfrac{\partial u}{\partial \mathbf{n}} + \gamma(x') u = \varphi & \text{on } \partial D \end{cases} \tag{1.1}$$

has a (unique) solution u in the space $H^{s,p}(D)$.

8.3.1 Proof of Theorem 8.8

First, by Theorem 6.9 we know that

$$\operatorname{ind} \mathcal{A} = \operatorname{ind} \mathcal{T} ,$$

where the operator

$$\mathcal{T}: B^{s-1/p,p}(\partial D) \longrightarrow B^{s-1/p,p}(\partial D)$$

is defined as follows:

(a) The domain $\mathcal{D}(\mathcal{T})$ of $\mathcal{T}$ is the space

$$\mathcal{D}(\mathcal{T}) = \left\{ \varphi \in B^{s-1/p,p}(\partial D) : T\varphi \in B^{s-1/p,p}(\partial D) \right\} .$$

(b) $\mathcal{T}\varphi = T\varphi$, $\varphi \in \mathcal{D}(\mathcal{T})$.

However, Theorem 8.6 tells us that the operator $\mathcal{A}$ (or equivalently the operator $\mathcal{T}$) is injective. Hence, to prove the surjectivity it suffices to show the following:

Proposition 8.9. *The index of the operator* $\mathcal{T}$ *is equal to* zero.

Proof. The proof is divided into several steps.

Step 1: We replace the differential operator A by the differential operator $A-\lambda$ with $\lambda \geq 0$, and consider instead of problem (1.1) the following boundary value problem:

$$\begin{cases} (A-\lambda)u = f & \text{in } D, \\ L_0 u = \mu(x')\dfrac{\partial u}{\partial \mathbf{n}} + \gamma(x')u = \varphi & \text{on } \partial D. \end{cases} \tag{1.1$_\lambda$}$$

We associate with problem $(1.1)_\lambda$ a linear operator

$$\mathcal{A}(\lambda) = (A - \lambda I, L_0): H^{s,p}(D) \longrightarrow H^{s-2,p}(D) \bigoplus B_{L_0}^{s-1-1/p,p}(\partial D) .$$

It should be noticed that the operator $\mathcal{A}(\lambda)$ coincides with the operator $\mathcal{A}$ when $\lambda = 0$.

We reduce the study of problem $(1.1)_\lambda$ to that of a pseudo-differential operator on the boundary, just as in the proof of Theorem 8.2.

We can prove that Theorems 6.3 and 6.4 remain valid for the operator $A - \lambda$. Namely, we have the following two assertions:

(a) The Dirichlet problem

$$\begin{cases} (A-\lambda)w = 0 & \text{in } D, \\ w = \varphi & \text{on } \partial D \end{cases}$$

has a unique solution w in $H^{t,p}(D)$ for any $\varphi \in B^{t-1/p,p}(\partial D)$, where $t \in \mathbf{R}$.

(b) The Poisson operator

$$P(\lambda): B^{t-1/p,p}(\partial D) \longrightarrow H^{t,p}(D) ,$$

defined by $w = P(\lambda)\varphi$, is an isomorphism of the space $B^{t-1/p,p}(\partial D)$ onto the space $\mathcal{N}(A - \lambda I, t, p) = \{u \in H^{t,p}(D): (A - \lambda)u = 0 \text{ in } D\}$ for all $t \in \mathbf{R}$; and its inverse is the trace operator on ∂D.

Let $T(\lambda)$ be a first-order, classical pseudo-differential operator on the boundary ∂D defined by the formula

$$T(\lambda) = L_0 P(\lambda) = \mu(x')\,\Pi(\lambda) + \gamma(x') , \quad \lambda \geq 0 ,$$

where

$$\begin{aligned} \Pi(\lambda): C^\infty(\partial D) &\longrightarrow C^\infty(\partial D) \\ \varphi &\longmapsto \frac{\partial}{\partial \mathbf{n}}(P(\lambda)\varphi) . \end{aligned}$$

Since the operator $T(\lambda): C^\infty(\partial D) \to C^\infty(\partial D)$ extends to a continuous linear operator $T(\lambda): B^{t,p}(\partial D) \to B^{t-1,p}(\partial D)$ for all $t \in \mathbf{R}$, we can introduce a densely defined, closed linear operator

$$\mathcal{T}(\lambda): B^{s-1/p,p}(\partial D) \longrightarrow B^{s-1/p,p}(\partial D)$$

as follows.

(α) The domain $\mathcal{D}(\mathcal{T}(\lambda))$ of $\mathcal{T}(\lambda)$ is the space

$$\mathcal{D}(\mathcal{T}(\lambda)) = \left\{\varphi \in B^{s-1/p,p}(\partial D): T(\lambda)\varphi \in B^{s-1/p,p}(\partial D)\right\} .$$

(β) $\mathcal{T}(\lambda)\varphi = T(\lambda)\varphi$, $\varphi \in \mathcal{D}(\mathcal{T}(\lambda))$.

It should be noticed that the operator $\mathcal{T}(\lambda)$ coincides with the operator $\mathcal{T}$ when $\lambda = 0$.

Then we can obtain the following results (see Theorem 6.9):

(I) The null space $\mathcal{N}(\mathcal{A}(\lambda))$ of $\mathcal{A}(\lambda)$ has finite dimension if and only if the null space $\mathcal{N}(\mathcal{T}(\lambda))$ of $\mathcal{T}(\lambda)$ has finite dimension, and we have

$$\dim \mathcal{N}(\mathcal{A}(\lambda)) = \dim \mathcal{N}(\mathcal{T}(\lambda)) .$$

(II) The range $\mathcal{R}(\mathcal{A}(\lambda))$ of $\mathcal{A}(\lambda)$ is closed if and only if the range $\mathcal{R}(\mathcal{T}(\lambda))$ of $\mathcal{T}(\lambda)$ is closed; and $\mathcal{R}(\mathcal{A}(\lambda))$ has finite codimension if and only if $\mathcal{R}(\mathcal{T}(\lambda))$ has finite codimension, and we have

$$\operatorname{codim} \mathcal{R}(\mathcal{A}(\lambda)) = \operatorname{codim} \mathcal{R}(\mathcal{T}(\lambda)) .$$

(III) The operator $\mathcal{A}(\lambda)$ is a Fredholm operator if and only if the operator $\mathcal{T}(\lambda)$ is a Fredholm operator, and we have

$$\operatorname{ind} \mathcal{A}(\lambda) = \operatorname{ind} \mathcal{T}(\lambda) .$$

Step 2: To study problem $(1.1)_\lambda$, we shall make use of a method essentially due to Agmon (cf. Agmon [Ag], Fujiwara [Fu], Lions–Megenes [LM] and also Taira [Ta2, Sect. 8.4]).

We introduce an auxiliary variable y of the unit circle

$$S = \mathbf{R}/2\pi\mathbf{Z} ,$$

and replace the parameter λ by the second-order differential operator

$$-\frac{\partial^2}{\partial y^2} .$$

Namely, we replace the differential operator $A - \lambda$ by the differential operator

$$A + \frac{\partial^2}{\partial y^2} ,$$

and consider instead of problem $(1.1)_\lambda$ the following boundary value problem:

$$\begin{cases} \widetilde{\Lambda}\tilde{u} := \left(A + \dfrac{\partial^2}{\partial y^2}\right)\tilde{u} = \tilde{f} & \text{in } D \times S, \\ L_0\tilde{u} = \mu(x')\dfrac{\partial \tilde{u}}{\partial \mathbf{n}} + \gamma(x')\tilde{u} = \tilde{\varphi} & \text{on } \partial D \times S. \end{cases}$$

We can prove that Theorems 6.3 and 6.4 remain valid for the operator $\widetilde{\Lambda} = A + \partial^2/\partial y^2$:

($\tilde{a}$) The Dirichlet problem

$$\begin{cases} \widetilde{\Lambda}\tilde{w} = 0 & \text{in } D \times S, \\ \tilde{w} = \tilde{\varphi} & \text{on } \partial D \times S \end{cases}$$

has a unique solution $\tilde{w}$ in $H^{t,p}(D \times S)$ for any $\tilde{\varphi} \in B^{t-1/p,p}(\partial D \times S)$, where $t \in \mathbf{R}$.

($\tilde{b}$) The Poisson operator

$$\widetilde{P}: B^{t-1/p,p}(\partial D \times S) \longrightarrow H^{t,p}(D \times S) ,$$

defined by $\tilde{w} = \widetilde{P}\tilde{\varphi}$, is an isomorphism of the space $B^{t-1/p,p}(\partial D \times S)$ onto the space $\mathcal{N}(\widetilde{\Lambda}, t, p) = \{\tilde{u} \in H^{t,p}(D \times S): \widetilde{\Lambda}\tilde{u} = 0 \text{ in } D \times S\}$ for all $t \in \mathbf{R}$; and its inverse is the trace operator on $\partial D \times S$.

We let

$$\begin{aligned} \widetilde{T}: C^\infty(\partial D \times S) &\longrightarrow C^\infty(\partial D \times S) \\ \tilde{\varphi} &\longmapsto L_0\widetilde{P}\tilde{\varphi} . \end{aligned}$$

Then the operator $\widetilde{T}$ can be decomposed as follows:

$$\widetilde{T} = \mu(x')\widetilde{\Pi} + \gamma(x') ,$$

where

$$\widetilde{\Pi}\tilde{\varphi} = \frac{\partial}{\partial \mathbf{n}} \left(\widetilde{P}\tilde{\varphi} \right) .$$

The operator $\widetilde{\Pi}$ is a first-order, classical pseudo-differential operator on the boundary $\partial D \times S$, and its complete symbol is given by the following:

$$\begin{aligned}&\left[\tilde{p}_1(x',\xi',y,\eta) + \sqrt{-1}\ \tilde{q}_1(x',\xi',y,\eta)\right] \\ &+ \left[\tilde{p}_0(x',\xi',y,\eta) + \sqrt{-1}\ \tilde{q}_0(x',\xi',y,\eta)\right] + \text{terms of order} \leq -1 ,\end{aligned}$$

where

$$\tilde{p}_1(x',\xi',y,\eta) < 0 \quad \text{on } T^*(\partial D \times S) \setminus \{0\} . \tag{8.11}$$

For example, if A is the usual Laplacian $\Delta = \partial^2/\partial x_1^2 + \ldots + \partial^2/\partial x_N^2$, then we have

$$\tilde{p}_1(x',\xi',y,\eta) = -\sqrt{|\xi'|^2 + \eta^2} .$$

Thus we find that the operator

$$\widetilde{T} = L_0\widetilde{P} = \mu(x')\widetilde{\Pi} + \gamma(x')$$

is a first-order, classical pseudo-differential operator on the boundary $\partial D \times S$ and its complete symbol is given by the following:

$$\begin{aligned}&\mu(x')\left[\tilde{p}_1(x',\xi',y,\eta) + \sqrt{-1}\ \tilde{q}_1(x',\xi',y,\eta)\right] \\ &+ \left[(\gamma(x') + \mu(x')\tilde{p}_0(x',\xi',y,\eta)) + \sqrt{-1}\,\mu(x')\tilde{q}_0(x',\xi',y,\eta)\right] \\ &+ \text{terms of order} \leq -1 .\end{aligned} \tag{8.12}$$

Then, by virtue of assertions (8.12) and (8.11) it is easy to verify that the operator $\widetilde{T}$ satisfies conditions (5.7a) and (5.7b) of Theorem 5.15 with $\mu := 0$, $\rho := 1$ and $\delta := 1/2$, just as in the proof of Lemma 8.3. Hence there exists a parametrix $\widetilde{S}$ in the class $L^0_{1,1/2}(\partial D \times S)$ for the operator $\widetilde{T}$.

Therefore, we obtain the following result, analogous to Lemma 8.3:

Lemma 8.10. *If conditions (H.1) and (H.2) are satisfied, then we have, for all $s \in \mathbf{R}$,*

$$\tilde{\varphi} \in \mathcal{D}'(\partial D \times S),\ \ \widetilde{T}\tilde{\varphi} \in B^{s,p}(\partial D \times S) \Longrightarrow \tilde{\varphi} \in B^{s,p}(\partial D \times S) .$$

Furthermore, for any $t < s$, there exists a constant $\widetilde{C}_{s,t} > 0$ such that

$$|\tilde{\varphi}|_{s,p} \leq \widetilde{C}_{s,t} \left(|\widetilde{T}\tilde{\varphi}|_{s,p} + |\tilde{\varphi}|_{t,p} \right) . \tag{8.13}$$

We introduce a densely defined, closed linear operator

$$\widetilde{\mathcal{T}}: B^{s-1/p,p}(\partial D \times S) \longrightarrow B^{s-1/p,p}(\partial D \times S)$$

as follows.

($\tilde{\alpha}$) The domain $\mathcal{D}(\widetilde{\mathcal{T}})$ of $\widetilde{\mathcal{T}}$ is the space

$$\mathcal{D}(\widetilde{\mathcal{T}}) = \left\{\tilde{\varphi} \in B^{s-1/p,p}(\partial D \times S): \widetilde{\mathcal{T}}\tilde{\varphi} \in B^{s-1/p,p}(\partial D \times S)\right\}.$$

($\tilde{\beta}$) $\widetilde{\mathcal{T}}\tilde{\varphi} = \widetilde{\mathcal{T}}\tilde{\varphi}$, $\tilde{\varphi} \in \mathcal{D}(\widetilde{\mathcal{T}})$.

Then the most fundamental relationship between $\widetilde{\mathcal{T}}$ and $\mathcal{T}(\lambda)$ ($\lambda \geq 0$) is the following:

Proposition 8.11. *If* $\operatorname{ind} \widetilde{\mathcal{T}}$ *is finite, then there exists* a finite *subset* K *of* $\mathbf{Z}$ *such that the operator* $\mathcal{T}(\lambda')$ *is bijective for all* $\lambda' = \ell^2$ *satisfying* $\ell \in \mathbf{Z} \setminus K$.

Granting Proposition 8.11 for the moment, we shall prove Theorem 8.8.

Step 3: End of Proof of Theorem 8.8

(3-1) We show that if conditions (H.1) and (H.2) are satisfied, then we have the assertion

$$\operatorname{ind} \widetilde{\mathcal{T}} := \dim \mathcal{N}(\widetilde{\mathcal{T}}) - \operatorname{codim} \mathcal{R}(\widetilde{\mathcal{T}}) < \infty . \tag{8.14}$$

To this end, we need a useful criterion for Fredholm operators:

Lemma 8.12 (Peetre). *Let* X, Y, Z *be Banach spaces such that* $X \subset Z$ *is a* compact *injection, and let* T *be a closed linear operator from* X *into* Y *with domain* $\mathcal{D}(T)$. *Then the following two conditions are equivalent:*

(i) The null space $\mathcal{N}(T)$ *of* T *has finite dimension and the range* $\mathcal{R}(T)$ *of* T *is closed in* Y.

(ii) There is a constant $C > 0$ *such that*

$$\|x\|_X \leq C \left(\|Tx\|_Y + \|x\|_Z\right), \quad x \in \mathcal{D}(T) . \tag{8.15}$$

Proof. (i) $\Longrightarrow$ (ii): Since the null space $\mathcal{N}(T)$ has finite dimension, we can find a closed topological complement X_0 in X

$$X = \mathcal{N}(T) \bigoplus X_0 . \tag{8.16}$$

This gives that

$$\mathcal{D}(T) = \mathcal{N}(T) \bigoplus (\mathcal{D}(T) \cap X_0) .$$

Namely, every element x of $\mathcal{D}(T)$ can be written in the form

$$x = x_0 + x_1 , \quad x_0 \in \mathcal{D}(T) \cap X_0 , \quad x_1 \in \mathcal{N}(T) .$$

Moreover, since the range $\mathcal{R}(T)$ is closed in Y, there exists a constant C such that

$$\|x_0\|_X \leq C\|Tx_0\|_Y . \tag{8.17}$$

Here and in the following the letter C denotes a generic positive constant independent of x.

On the other hand, it should be noticed that all norms on a finite dimensional linear space are equivalent. This gives that

$$\|x_1\|_X \leq C\|x_1\|_Z . \tag{8.18}$$

Moreover, since the injection $X \to Z$ is compact and hence is continuous, we obtain that

$$\|x_1\|_Z \leq \|x\|_Z + \|x_0\|_Z \leq \|x\|_Z + C\|x_0\|_X . \tag{8.19}$$

Thus we have, by inequalities (8.18) and (8.19),

$$\|x_1\|_X \leq C\left(\|x\|_Z + \|x_0\|_X\right) . \tag{8.20}$$

Therefore, combining inequalities (8.17) and (8.20) we obtain the desired inequality (8.15)

$$\|x\|_X \leq \|x_0\|_X + \|x_1\|_X \leq C\left(\|Tx\|_Y + \|x\|_Z\right),$$

since $Tx_0 = Tx$.

(ii) $\Longrightarrow$ (i): By inequality (8.15), it follows that

$$\|x\|_X \leq C\|x\|_Z , \quad x \in \mathcal{N}(T) . \tag{8.21}$$

However, the null space $\mathcal{N}(T)$ is closed in X, and so it is a Banach space. Since the injection $X \to Z$ is compact, we obtain from inequality (8.21) that the closed unit ball $\{x \in \mathcal{N}(T) \colon \|x\|_X \leq 1\}$ of $\mathcal{N}(T)$ is compact. This implies (see [Fr2, Theorem 4.3.3]) that

$$\dim \mathcal{N}(T) < +\infty .$$

Let X_0 be a closed topological complement of $\mathcal{N}(T)$ as in decomposition (8.16).

To prove the closedness of $\mathcal{R}(T)$, it suffices to show that

$$\|x\|_X \leq C\|Tx\|_Y , \quad x \in \mathcal{D}(T) \cap X_0 .$$

Assume, to the contrary, that

For every $n \in \mathbf{N}$, there is an element x_n of $\mathcal{D}(T) \cap X_0$ such that

$$\|x_n\|_X > n\|Tx_n\|_Y .$$

If we let

$$x_n' = \frac{x_n}{\|x_n\|_X},$$

then we have

$$x'_n \in \mathcal{D}(T) \cap X_0\,, \quad \|x'_n\|_X = 1\,, \tag{8.22}$$

$$\|Tx'_n\|_Y < \frac{1}{n}\,. \tag{8.23}$$

Since the injection $X \to Z$ is compact, by passing to a subsequence we may assume that the sequence $\{x'_n\}$ is a Cauchy sequence in Z. Then, combining inequalities (8.23) and (8.15) we find that the sequence $\{x'_n\}$ is a Cauchy sequence in X, and hence converges to an element x' of X. Thus, by assertion (8.22) it follows that

$$\|x'\|_X = \lim_n \|x'_n\|_X = 1\,,$$

and further that

$$x' \in \mathcal{D}(T)\,, \quad Tx' = 0\,,$$

since the operator T is closed.

Summing up, we have proved that

$$x' \in \mathcal{N}(T)\,,$$
$$\|x'\|_X = 1\,.$$

However, this is a contradiction. Indeed, we then have

$$x' \in \mathcal{N}(T) \cap X_0 = \{0\}\,.$$

The proof of Lemma 8.12 is complete. □

Now, estimate (8.13) gives that

$$|\tilde{\varphi}|_{s-1/p,p} \le \widetilde{C}_{s,t}\left(|\widetilde{T}\tilde{\varphi}|_{s-1/p,p} + |\tilde{\varphi}|_{t,p}\right), \quad \tilde{\varphi} \in \mathcal{D}(\widetilde{T})\,, \tag{8.24}$$

where $t < s - 1/p$. However, it follows from an application of Rellich's theorem that the injection $B^{s-1/p,p}(\partial D \times S) \to B^{t,p}(\partial D \times S)$ is *compact* (or completely continuous) for $t < s - 1/p$. Thus, applying Peetre's Lemma (Lemma 8.12) with

$$X = Y := B^{s-1/p,p}(\partial D \times S)\,,$$
$$Z := B^{t,p}(\partial D \times S)\,,$$
$$T := \widetilde{T}\,,$$

we obtain that the range $\mathcal{R}(\widetilde{T})$ is closed in $B^{s-1/p,p}(\partial D \times S)$ and

$$\dim \mathcal{N}(\widetilde{T}) < \infty\,. \tag{8.25}$$

On the other hand, by formula (8.12) we find that the symbol of the adjoint $\widetilde{T}^*$ is given by the following (cf. Theorem 5.10):

$$\mu(x')\left(\tilde{p}_1(x',\xi',y,\eta) - \sqrt{-1}\ \tilde{q}_1(x',\xi',y,\eta)\right)$$
$$+ \left(\left[-\gamma(x') + \mu(x')\tilde{p}_0(x',\xi',y,\eta) - \sum_{j=1}^{N-1} \partial_{x_j}\left(\mu(x')\cdot \partial_{\xi_j}\tilde{q}_1(x',\xi',y,\eta)\right)\right]\right.$$
$$\left. - \sqrt{-1}\left[\mu(x')\tilde{q}_0(x',\xi',y,\eta) + \sum_{j=1}^{N-1} \partial_{x_j}\left(\mu(x')\cdot \partial_{\xi_j}\tilde{p}_1(x',\xi',y,\eta)\right)\right]\right)$$
$$+ \text{terms of order} \leq -1\,.$$

However, by virtue of Lemma 8.5 it follows that

$$\partial_{x_j}\mu(x') = 0 \text{ on } M = \{x' \in \partial D\colon \mu(x') = 0\}\,.$$

Thus we can easily verify that the pseudo-differential operator $\widetilde{T}^*$ satisfies conditions (5.7a) and (5.7b) of Theorem 5.15 with $\mu := 0$, $\rho := 1$ and $\delta := 1/2$. This implies that estimate (8.24) remains valid for the adjoint operator $\widetilde{T}^*$ of $\widetilde{T}$:

$$|\tilde{\psi}|_{-s+1/p,p'} \leq \widetilde{C}_{s,\tau}\left(|\widetilde{T}^*\tilde{\psi}|_{-s+1/p,p'} + |\tilde{\psi}|_{\tau,p'}\right), \quad \tilde{\psi} \in \mathcal{D}(\widetilde{T}^*)\,,$$

where

$$\widetilde{T^*}\colon B^{-s+1/p,p'}(\partial D \times S) \longrightarrow B^{-s+1/p,p'}(\partial D \times S)\,,$$

and $\tau < -s + 1/p$ and $p' = p/(p-1)$, the exponent conjugate to p (see Lemma 8.15 below). Hence we have, by Peetre's Lemma (Lemma 8.12),

$$\dim \mathcal{N}(\widetilde{T}^*) < \infty\,, \tag{8.26}$$

since the injection $B^{-s+1/p,p'}(\partial D \times S) \to B^{\tau,p'}(\partial D \times S)$ is compact for $\tau < -s + 1/p$.

Moreover, we obtain that

$$\operatorname{codim} \mathcal{R}(\widetilde{T}) = \dim \mathcal{N}(\widetilde{T}^*)\,, \tag{8.27}$$

if we apply the following Banach's closed range theorem (see [Yo, Chap. VII, Sect. 5, Theorem]) with

$$X = Y := B^{s-1/p,p}(\partial D \times S)\,, \quad T := \widetilde{T}\,.$$

Theorem 8.13 (Banach's closed range theorem). *Let X and Y be Banach spaces and T a densely defined, closed linear operator from X into Y. Then the following four conditions are equivalent:*

(i) The range $\mathcal{R}(T)$ of T is closed in Y.
(ii) The range $\mathcal{R}(T')$ of the dual operator T' is closed in the dual space X'.
(iii) $\mathcal{R}(T) = {}^0\mathcal{N}(T') := \{x \in X\colon \langle x, x'\rangle = 0$ for all $x' \in \mathcal{N}(T')\}$.
(iv) $\mathcal{R}(T') = {}^0\mathcal{N}(T) := \{x' \in X'\colon \langle x', x\rangle = 0$ for all $x \in \mathcal{N}(T)\}$.

Therefore, the desired assertion (8.14) follows from assertions (8.25), (8.26) and (8.27).

(3-2) By assertion (8.14), we can apply Proposition 8.11 to obtain that the operator $\mathcal{T}(\ell^2)\colon B^{s-1/p,p}(\partial D) \to B^{s-1/p,p}(\partial D)$ is bijective if $\ell \in \mathbf{Z} \setminus K$ for some finite subset K of $\mathbf{Z}$. In particular, we have

$$\operatorname{ind} \mathcal{T}(\lambda_0) = 0 \quad \text{if } \lambda_0 = \ell^2,\ \ell \in \mathbf{Z} \setminus K\,. \tag{8.28}$$

However, it is easy to see that the symbol $t(x', \xi'; \lambda)$ of the operator

$$T(\lambda) = \mu(x')\Pi(\lambda) + \gamma(x')\,, \quad \lambda \geq 0\,,$$

has the following asymptotic expansion:

$$\begin{aligned} t(x', \xi'; \lambda) = {} & \mu(x')\left[p_1(x', \xi') + \sqrt{-1}\, q_1(x', \xi')\right] \\ & + \left[(-\gamma(x') + \mu(x')p_0(x', \xi)) + \sqrt{-1}\, \mu(x')q_0(x', \xi')\right] \\ & + \text{terms of order} \leq -1 \text{ depending on } \lambda\,. \end{aligned} \tag{8.29}$$

Thus we can find a classical pseudo-differential operator $K(\lambda_0)$ of order -1 on the boundary ∂D such that

$$T = T(\lambda_0) + K(\lambda_0)\,.$$

Furthermore, Rellich's theorem tells us that the operator

$$K(\lambda_0) : B^{s-1/p,p}(\partial D) \longrightarrow B^{s-1/p,p}(\partial D)$$

is *compact*. Hence we have

$$\operatorname{ind} \mathcal{T} = \operatorname{ind} \mathcal{T}(\lambda_0)\,. \tag{8.30}$$

Therefore, Proposition 8.9 follows from assertions (8.28) and (8.30). □

Theorem 8.8 is proved, apart from the proof of Proposition 8.11.

8.3.2 Proof of Proposition 8.11

The proof is divided into three steps.

Step 1: First, we study the null spaces $\mathcal{N}(\widetilde{\mathcal{T}})$ and $\mathcal{N}(\mathcal{T}(\lambda'))$ when $\lambda' = \ell^2$ with $\ell \in \mathbf{Z}$:

$$\begin{aligned} \mathcal{N}(\widetilde{\mathcal{T}}) &= \left\{\tilde{\varphi} \in B^{s-1/p,p}(\partial D \times S) \colon \widetilde{\mathcal{T}}\tilde{\varphi} = 0\right\}, \\ \mathcal{N}(\mathcal{T}(\lambda')) &= \left\{\varphi \in B^{s-1/p,p}(\partial D) \colon T(\lambda')\varphi = 0\right\}\,. \end{aligned}$$

Since the pseudo-differential operators $\widetilde{T}$ and $T(\lambda')$ are both *hypoelliptic*, it follows that

$$\mathcal{N}(\widetilde{\mathcal{T}}) = \left\{ \tilde{\varphi} \in C^\infty(\partial D \times S) : \widetilde{\mathcal{T}}\tilde{\varphi} = 0 \right\},$$
$$\mathcal{N}(\mathcal{T}(\lambda')) = \{ \varphi \in C^\infty(\partial D) : \mathcal{T}(\lambda')\varphi = 0 \} .$$

Therefore, we can apply [Ta2, Proposition 8.4.6] to obtain the following most important relationship between $\mathcal{N}(\widetilde{\mathcal{T}})$ and $\mathcal{N}(\mathcal{T}(\lambda'))$ when $\lambda' = \ell^2$ with $\ell \in \mathbf{Z}$:

Lemma 8.14. *The following two conditions are equivalent:*

(1) $\dim \mathcal{N}(\widetilde{\mathcal{T}}) < \infty$.
(2) There exists a finite *subset I of $\mathbf{Z}$ such that*

$$\begin{cases} \dim \mathcal{N}\left(\mathcal{T}(\ell^2)\right) < \infty & \text{if } \ell \in I, \\ \dim \mathcal{N}\left(\mathcal{T}(\ell^2)\right) = 0 & \text{if } \ell \notin I. \end{cases}$$

Moreover, in this case, we have

$$\mathcal{N}(\widetilde{\mathcal{T}}) = \bigoplus_{\ell \in I} \mathcal{N}\left(\mathcal{T}(\ell^2)\right) \otimes e^{i\ell y} ,$$
$$\dim \mathcal{N}(\widetilde{\mathcal{T}}) = \sum_{\ell \in I} \dim \mathcal{N}(\mathcal{T}(\ell^2)) .$$

Step 2: Next, we study the ranges $\mathcal{R}(\widetilde{\mathcal{T}})$ and $\mathcal{R}(\mathcal{T}(\lambda'))$ when $\lambda' = \ell^2$ with $\ell \in \mathbf{Z}$. To do this, we consider the adjoint operators $\widetilde{\mathcal{T}}^*$ and $\mathcal{T}(\lambda')^*$ of $\widetilde{\mathcal{T}}$ and $\mathcal{T}(\lambda')$, respectively.

The next lemma allows us to give a characterization of the adjoint operators $\widetilde{\mathcal{T}}^*$ and $\mathcal{T}(\lambda')^*$ in terms of pseudo-differential operators (cf. [Ta2, Lemma 8.4.8]):

Lemma 8.15. *Let M be an n-dimensional, compact smooth manifold without boundary. If T is a classical pseudo-differential operator of order m on M, we define a densely defined, closed linear operator*

$$\mathcal{T} : B^{s,p}(M) \longrightarrow B^{s-m+1,p}(M) \quad (s \in \mathbf{R})$$

as follows.

(a) The domain $\mathcal{D}(\mathcal{T})$ of $\mathcal{T}$ is the space

$$\mathcal{D}(\mathcal{T}) = \{ \varphi \in H^{s,p}(M) : T\varphi \in H^{s-m+1,p}(M) \} .$$

(b) $\mathcal{T}\varphi = T\varphi$, $\varphi \in \mathcal{D}(\mathcal{T})$.
Then the adjoint operator $\mathcal{T}^*$ of $\mathcal{T}$ is characterized as follows:
(c) The domain $\mathcal{D}(\mathcal{T}^*)$ of $\mathcal{T}^*$ is contained in the space

$$\left\{ \psi \in B^{-s+m-1,p'}(M) : T^*\psi \in B^{-s,p'}(M) \right\},$$

where $p' = p/(p-1)$ and $T^* \in L^m_{\mathrm{cl}}(M)$ is the adjoint of T.

(d) $\widetilde{T}^*\psi = T^*\psi$, $\psi \in \mathcal{D}(T^*)$.

It should be noticed that the pseudo-differential operators $T(\lambda)^*$ and $\widetilde{T}^*$ also satisfy conditions (5.7a) and (5.7b) of Theorem 5.15 with $\mu := 0$, $\rho := 1$ and $\delta := 1/2$; hence they are *hypoelliptic.*

Therefore, applying Lemma 8.15 to the operators $\widetilde{T}$ and $T(\lambda')$ we obtain the following:

Lemma 8.16. *The null spaces* $\mathcal{N}(\widetilde{T}^*)$ *and* $\mathcal{N}(T(\lambda')^*)$ *are characterized respectively as follows:*

$$\mathcal{N}(\widetilde{T}^*) = \left\{\tilde{\psi} \in C^\infty(\partial D \times S): \widetilde{T}^*\tilde{\psi} = 0\right\},$$
$$\mathcal{N}(T(\lambda')^*) = \{\psi \in C^\infty(\partial D): T(\lambda')^*\psi = 0\}.$$

By Lemma 8.16, we find that Lemma 8.14 remains valid for the adjoint operators $\widetilde{T}^*$ and $T(\lambda')^*$ (cf. [Ta2, Lemma 8.4.10]):

Lemma 8.17. *The following two conditions are equivalent:*

(1) $\dim \mathcal{N}(\widetilde{T}^*) < \infty$.

(2) There exists a finite *subset* J *of* $\mathbf{Z}$ *such that*

$$\begin{cases} \dim \mathcal{N}\left(T(\ell^2)^*\right) < \infty & \text{if } \ell \in J, \\ \dim \mathcal{N}\left(T(\ell^2)^*\right) = 0 & \text{if } \ell \notin J. \end{cases}$$

Moreover, in this case, we have

$$\dim \mathcal{N}(\widetilde{T^*}) = \sum_{\ell \in J} \dim \mathcal{N}\left(T(\ell^2)^*\right).$$

Hence, combining Lemma 8.17 and Banach's closed range theorem (Theorem 8.13) we obtain the most important relationship between $\operatorname{codim} \mathcal{R}(\widetilde{T})$ and $\operatorname{codim} \mathcal{R}(T(\lambda'))$ when $\lambda' = \ell^2$, $\ell \in \mathbf{Z}$ (cf. [Ta2, Proposition 8.4.11]):

Lemma 8.18. *The following two conditions are equivalent:*

(1) $\operatorname{codim} \mathcal{R}(\widetilde{T}) < \infty$.

(2) There exists a finite *subset* J *of* $\mathbf{Z}$ *such that*

$$\begin{cases} \operatorname{codim} \mathcal{R}\left(T(\ell^2)\right) < \infty & \text{if } \ell \in J, \\ \operatorname{codim} \mathcal{N}\left(T(\ell^2)\right) = 0 & \text{if } \ell \notin J. \end{cases}$$

Moreover, in this case, we have

$$\operatorname{codim} \mathcal{R}(\widetilde{T}) = \sum_{\ell \in J} \operatorname{codim} \mathcal{R}\left(T(\ell^2)\right).$$

Step 3: Proposition 8.11 is an immediate consequence of Lemmas 8.14 and 8.18, with $K = I \cup J$.

Now the proof of Proposition 8.9, and hence that of Theorem 8.8, is complete. □

9

Proof of Theorem 1.2

In this chapter we consider the boundary value problem

$$\begin{cases} Au = f & \text{in } D\,, \\ L_0 u = \mu(x')\dfrac{\partial u}{\partial \mathbf{n}} + \gamma(x')u = \varphi & \text{on } \partial D\,, \end{cases} \tag{1.1}$$

in the framework of Hölder spaces, and prove Theorem 1.2 (Theorem 9.1). Theorem 1.2 follows from Theorem 1.1 by using the Hölder space theory of pseudo-differential operators (Proposition 9.3).

The next theorem proves Theorem 1.2:

Theorem 9.1. *Let* $0 < \theta < 1$. *Assume that conditions (H.1) and (H.2) are satisfied:*

(H.1) $\mu(x') \geq 0$ *and* $\gamma(x') \leq 0$ *on* ∂D.
(H.2) $\gamma(x') < 0$ *on* $M = \{x' \in \partial D : \mu(x') = 0\}$.

Then the mapping

$$(A, L_0)\colon C^{2+\theta}(\overline{D}) \longrightarrow C^{\theta}(\overline{D}) \bigoplus C^{1+\theta}_{L_0}(\partial D)$$

is an algebraic and topological isomorphism.

Proof. The proof is divided into three steps.

Step 1: Let (f, φ) be an arbitrary element of the space $C^{\theta}(\overline{D}) \oplus C^{1+\theta}_{L_0}(\partial D)$ with

$$\varphi = \mu(x')\varphi_1 - \gamma(x')\varphi_2, \quad \varphi_1 \in C^{1+\theta}(\partial D),\ \varphi_2 \in C^{2+\theta}(\partial D)\,.$$

First, we show that the boundary value problem

$$\begin{cases} Au = f & \text{in } D\,, \\ L_0 u = \varphi & \text{on } \partial D \end{cases} \tag{1.1}$$

can be reduced to the study of an operator on the boundary, just as in Sect. 6.3.

To do this, we consider the following Neumann problem:

$$\begin{cases} Av = f & \text{in } D\,, \\ \dfrac{\partial v}{\partial \mathbf{n}} = \varphi_1 & \text{on } \partial D\,. \end{cases} \tag{9.1}$$

Recall that the existence and uniqueness theorem for problem (9.1) is well established in the framework of Hölder spaces (see Gilbarg-Trudinger [GT, Theorem 6.31]). Thus we find that a function $u \in C^{2+\theta}(\overline{D})$ is a solution of problem (1.1) if and only if the function $w = u - v \in C^{2+\theta}(\overline{D})$ is a solution of the problem

$$\begin{cases} Aw = 0 & \text{in } D\,, \\ L_0 w = \varphi - L_0 v = \mu(x')\varphi_1 - \gamma(x')\varphi_2 - L_0 v & \text{on } \partial D\,. \end{cases}$$

Here it should be noticed that

$$L_0 v = \mu(x')\frac{\partial v}{\partial \mathbf{n}} + \gamma(x')v = \mu(x')\varphi_1 + \gamma(x')v\,,$$

so that

$$L_0 w = -\gamma(x')\left(\varphi_2 + (v|_{\partial D})\right) \in C^{2+\theta}(\partial D)\,.$$

However, we know that every solution $w \in C^{2+\theta}(\overline{D})$ of the homogeneous equation: $Aw = 0$ in D can be expressed as follows (see Gilbarg-Trudinger [GT, Theorem 6.14]):

$$w = P\psi, \quad \psi \in C^{2+\theta}(\partial D)\,.$$

Thus we can reduce the study of problem (1.1) to that of the equation

$$T\psi = L_0 P\psi = -\gamma(x')\left(\varphi_2 + (v|_{\partial D})\right) \quad \text{on } \partial D\,. \tag{9.2}$$

Summing up, we have proved the following:

Proposition 9.2. *For given functions $f \in C^{\theta}(\overline{D})$ and $\varphi \in C^{1+\theta}_{L_0}(\partial D)$, there exists a solution $u \in C^{2+\theta}(\overline{D})$ of problem (1.1) if and only if there exists a solution $\psi \in C^{2+\theta}(\partial D)$ of equation (9.2).*

Step 2: We study the operator $T = L_0 P$ in question. The next proposition is an essential step in the proof of Theorem 1.2:

Proposition 9.3. *If conditions (H.1) and (H.2) are satisfied, then there exists a* parametrix *E in the Hörmander class $L^0_{1,1/2}(\partial D)$ for T which maps the Hölder space $C^{k+\theta}(\partial D)$ continuously into itself, for any nonnegative integer k.*

Proof. Indeed, by virtue of Lemma 8.4 we can construct a parametrix E in the class $L^0_{1,1/2}(\partial D)$ for T. Furthermore, it follows from an application of the Besov-space boundedness theorem (Theorem 5.14) that the parametrix E maps the Hölder space $C^{k+\theta}(\partial D)$ continuously into itself, for any nonnegative integer k, since we have (see [Tr, Theorem 2.5.7])

$$C^{k+\theta}(\partial D) = B^{k+\theta}_{\infty,\infty}(\partial D) . \quad \square$$

Step 3: Now we remark that

$$\begin{cases} C^{\theta}(\overline{D}) \subset L^p(D) , \\ C^{1+\theta}_{L_0}(\partial D) \subset B^{1-1/p,p}_{L_0}(\partial D) \end{cases}$$

for $1 < p < \infty$. Thus we find from Theorem 1.1 that problem (1.1) has a unique solution $u \in H^{2,p}(D)$ for any $f \in C^{\theta}(\overline{D})$ and any $\varphi \in C^{1+\theta}_{L_0}(\partial D)$. Furthermore, by virtue of Proposition 9.2 it follows that the solution u can be written in the form

$$u = v + P\psi, \quad v \in C^{2+\theta}(\overline{D}),\ \psi \in B^{2-1/p,p}(\partial D) . \tag{9.3}$$

However, we have, by equation (9.2) and Proposition 9.3,

$$\psi \in C^{2+\theta}(\partial D) .$$

Indeed, it suffices to note that

$$\psi \equiv E(T\psi) = -E\left(\gamma(x')(\varphi_2 + (v|_{\partial D})\right) \in C^{2+\theta}(\partial D) \operatorname{mod} C^{\infty}(\partial D) .$$

Therefore, we obtain from formula (9.3) that

$$u = v + P\psi \in C^{2+\theta}(\overline{D}) .$$

The proof of Theorem 9.1 (and hence that of Theorem 1.2) is complete.
$\square$

10

A Priori Estimates

This Chap. 10 and the next Chap. 11 are devoted to the proof of Theorem 1.3. In this chapter we study the operator A_p, and prove estimate (1.2) for the operator $A_p - \lambda I$ (Theorem 10.3). Once again Agmon's method plays an important role in the proof of estimate (1.2) (Proposition 10.4) just as in Chap. 8, but we replace the differential operator $A-\lambda$, $\lambda = r^2 e^{i\vartheta}$, $-\pi < \vartheta < \pi$, by the differential operator $\widetilde{\Lambda}(\vartheta) = A + e^{i\vartheta}\, \partial^2/\partial y^2$.

Recall that the operator A_p is a unbounded linear operator from $L^p(D)$ into itself given by the following:

(a) The domain of definition $\mathcal{D}(A_p)$ of A_p is the space

$$\mathcal{D}(A_p) = \left\{ u \in H^{2,p}(D) : L_0 u = \mu(x')\frac{\partial u}{\partial \mathbf{n}} + \gamma(x')u = 0 \text{ on } \partial D \right\} .$$

(b) $A_p u = Au$, $u \in \mathcal{D}(A_p)$.

It should be emphasized that the operator A_p is densely defined, since the domain $\mathcal{D}(A_p)$ contains the space $C_0^\infty(D)$.

First, we have the following:

Lemma 10.1. *Let $1 < p < \infty$. Assume that conditions (H.1) and (H.2) are satisfied:*

(H.1) $\mu(x') \geq 0$ *and* $\gamma(x') \leq 0$ *on* ∂D.
(H.2) $\gamma(x') < 0$ *on* $M = \{x' \in \partial D : \mu(x') = 0\}$.

Then we have the a priori *estimate*

$$\|u\|_{2,p} \leq C\|Au\|_p, \quad u \in \mathcal{D}(A_p) . \tag{10.1}$$

Proof. Estimate (10.1) follows immediately from Theorem 1.1 with $\varphi := 0$. □

Corollary 10.2. *The operator A_p is a closed operator.*

Proof. Let $\{u_j\}$ be an arbitrary sequence in the domain $\mathcal{D}(A_p)$ such that:

$$\begin{cases} u_j \longrightarrow u & \text{in } L^p(D) , \\ Au_j \longrightarrow v & \text{in } L^p(D) . \end{cases}$$

Then, applying estimate (10.1) to the sequence $\{u_j\}$ we find that $\{u_j\}$ is a Cauchy sequence in the space $H^{2,p}(D)$, so that

$$u \in H^{2,p}(D) ,$$

and

$$u_j \longrightarrow u \quad \text{in } H^{2,p}(D) .$$

Hence we have

$$Au = \lim_{j\to\infty} Au_j = v \quad \text{in } L^p(D) ,$$

and also, by Proposition 6.7 with $s := 2$,

$$L_0 u = \lim_{j\to\infty} L_0 u_j = 0 \quad \text{in } B_{L_0}^{1-1/p,p}(\partial D) .$$

This proves that $u \in \mathcal{D}(A_p)$ and $A_p u = v$. □

The next theorem is the essential step in the proof of Theorem 1.3:

Theorem 10.3. *Assume that conditions (H.1) and (H.2) are satisfied. Then, for every $-\pi < \vartheta < \pi$ there exists a constant $R(\vartheta) > 0$ depending on ϑ such that if $\lambda = r^2 e^{i\vartheta}$ and $|\lambda| = r^2 \geq R(\vartheta)$, we have, for all $u \in H^{2,p}(D)$ satisfying $L_0 u = 0$ on ∂D (i.e., $u \in \mathcal{D}(A_p)$),*

$$|u|_{2,p} + |\lambda|^{1/2} |u|_{1,p} + |\lambda| \, \|u\|_p \leq C(\vartheta) \, \|(A - \lambda)u\|_p , \tag{10.2}$$

with a constant $C(\vartheta) > 0$ depending on ϑ. Here $|\cdot|_{j,p}$ $(j = 1, 2)$ is the seminorm on the space $H^{2,p}(D)$ defined by the formula

$$|u|_{j,p} = \left(\int_D \sum_{|\alpha|=j} |D^\alpha u(x)|^p \, dx \right)^{1/p} .$$

Proof. The proof is divided into two steps.

Step 1: We shall make use of a method essentially due to Agmon [Ag] just as in Chap. 8. We introduce an auxiliary variable y of the unit circle

$$S = \mathbf{R}/2\pi\mathbf{Z} ,$$

and replace the parameter λ by the second-order differential operator

$$-e^{i\vartheta} \frac{\partial^2}{\partial y^2}, \quad -\pi < \vartheta < \pi .$$

Namely, we replace the differential operator $A - \lambda$ by the differential operator

$$\widetilde{\Lambda}(\vartheta) := A + e^{i\vartheta}\,\frac{\partial^2}{\partial y^2}, \quad -\pi < \vartheta < \pi\,,$$

and consider instead of the problem

$$\begin{cases} (A-\lambda)u = f & \text{in } D\,, \\ L_0 u = \mu(x')\dfrac{\partial u}{\partial \mathbf{n}} + \gamma(x')u = 0 & \text{on } \partial D \end{cases} \tag{10.3}$$

the following boundary value problem:

$$\begin{cases} \widetilde{\Lambda}(\vartheta)\tilde{u} := \left(A + e^{i\vartheta}\,\dfrac{\partial^2}{\partial y^2}\right)\tilde{u} = \tilde{f} & \text{in } D\times S\,, \\ L_0\tilde{u} := \mu(x')\dfrac{\partial \tilde{u}}{\partial \mathbf{n}} + \gamma(x')\tilde{u} = 0 & \text{on } \partial D\times S\,. \end{cases} \tag{10.4}$$

It should be noticed that the operator $\widetilde{\Lambda}(\vartheta)$ is *elliptic* for $-\pi < \vartheta < \pi$.

Then we have the following result, analogous to Lemma 10.1:

Proposition 10.4. *Assume that conditions (H.1) and (H.2) are satisfied. Then we have, for all $\tilde{u} \in H^{2,p}(D\times S)$ satisfying the condition $L_0\tilde{u} = 0$ on $\partial D \times S$,*

$$\|\tilde{u}\|_{2,p} \leq \widetilde{C}(\vartheta)\left(\left\|\widetilde{\Lambda}(\vartheta)\tilde{u}\right\|_p + \|\tilde{u}\|_p\right)\,, \tag{10.5}$$

with a constant $\widetilde{C}(\vartheta) > 0$ depending on ϑ.

Proof. We reduce the study of problem (10.4) to that of a pseudo-differential operator on the boundary, just as in problem (10.3).

We can prove that Theorems 6.3 and 6.4 remain valid for the operator $\widetilde{\Lambda}(\vartheta) = A + e^{i\vartheta}\partial^2/\partial y^2$, $-\pi < \vartheta < \pi$:

($\tilde{a}$) The Dirichlet problem

$$\begin{cases} \widetilde{\Lambda}(\vartheta)\tilde{w} = 0 & \text{in } D\times S\,, \\ \tilde{w} = \tilde{\varphi} & \text{on } \partial D\times S \end{cases}$$

has a unique solution $\tilde{w}$ in $H^{t,p}(D\times S)$ for any $\tilde{\varphi} \in B^{t-1/p,p}(\partial D\times S)$ ($t \in \mathbf{R}$).

($\tilde{b}$) The Poisson operator

$$\widetilde{P}(\vartheta) : B^{t-1/p,p}(\partial D\times S) \longrightarrow H^{t,p}(D\times S)\,,$$

defined by $\tilde{w} = \widetilde{P}(\vartheta)\tilde{\varphi}$, is an isomorphism of $B^{t-1/p,p}(\partial D\times S)$ onto the space $\mathcal{N}(\widetilde{\Lambda}(\vartheta), t, p) = \{\tilde{u} \in H^{t,p}(D\times S) : \widetilde{\Lambda}(\vartheta)\tilde{u} = 0 \text{ in } D\times S\}$ for all $t \in \mathbf{R}$; and its inverse is the trace operator on $\partial D \times S$.

We let

$$\begin{aligned}\widetilde{T}(\vartheta): C^\infty(\partial D \times S) &\longrightarrow C^\infty(\partial D \times S)\\ \tilde{\varphi} &\longmapsto L_0 \widetilde{P}(\vartheta)\tilde{\varphi}\,.\end{aligned}$$

Then the operator $\widetilde{T}(\vartheta)$ can be decomposed as follows:

$$\widetilde{T}(\vartheta) = \mu(x')\,\widetilde{\Pi}(\vartheta) + \gamma(x')\,,$$

where

$$\widetilde{\Pi}(\vartheta)\tilde{\varphi} = \frac{\partial}{\partial \mathbf{n}}\left(\widetilde{P}(\vartheta)\tilde{\varphi}\right)\,.$$

The operator $\widetilde{\Pi}(\vartheta)$ is a first-order, classical pseudo-differential operator on the boundary $\partial D \times S$, and its complete symbol is given by the following (cf. [Ta2, Sect. 10.2]):

$$\begin{aligned}&\left(\tilde{p}_1(x',\xi',y,\eta;\vartheta) + \sqrt{-1}\,\tilde{q}_1(x',\xi',y,\eta;\vartheta)\right)\\ &\quad + \left(\tilde{p}_0(x',\xi',y,\eta;\vartheta) + \sqrt{-1}\,\tilde{q}_0(x',\xi',y,\eta;\vartheta)\right) + \text{terms of order} \le -1\,,\end{aligned}$$

where

$$\begin{aligned}&\tilde{p}_1(x',\xi',y,\eta;\vartheta) < 0 \quad \text{on the bundle } T^*(\partial D \times S) \setminus \{0\}\\ &\text{of non-zero cotangent vectors, for } -\pi < \vartheta < \pi\,. \end{aligned} \tag{10.6}$$

For example, if A is the usual Laplacian $\Delta = \partial^2/\partial x_1^2 + \cdots + \partial^2/\partial x_N^2$, then we have

$$\begin{aligned}&\tilde{p}_1(x',\xi',y,\eta;\vartheta)\\ &= -\left[\frac{\left[\left(|\xi'|^2 + \cos\vartheta\cdot\eta^2\right)^2 + \sin^2\vartheta\cdot\eta^4\right]^{1/2} + \left(|\xi'|^2 + \cos\vartheta\cdot\eta^2\right)}{2}\right]^{1/2}.\end{aligned}$$

Hence it follows that the operator

$$\widetilde{T}(\vartheta) = L_0\widetilde{P}(\vartheta) = \mu(x')\,\widetilde{\Pi}(\vartheta) + \gamma(x')$$

is a first-order, classical pseudo-differential operator on the boundary $\partial D \times S$ and that its complete symbol is given by the following:

$$\begin{aligned}&\mu(x')\left(\tilde{p}_1(x',\xi',y,\eta;\vartheta) + \sqrt{-1}\,\tilde{q}_1(x',\xi',y,\eta;\vartheta)\right)\\ &\quad + \left([\gamma(x') + \mu(x')\tilde{p}_0(x',\xi',y,\eta;\vartheta)] + \sqrt{-1}\,\mu(x')\tilde{q}_0(x',\xi',y,\eta;\vartheta)\right)\\ &\quad + \text{terms of order} \le -1\,.\end{aligned} \tag{10.7}$$

Then, by virtue of assertions (10.7) and (10.6) we can verify that the operator $\widetilde{T}(\vartheta)$ satisfies conditions (5.7a) and (5.7b) of Theorem 5.15 with $\mu := 0$, $\rho := 1$ and $\delta := 1/2$, just as in the proof of Lemma 8.4.

Therefore, we obtain the following result, analogous to Lemma 8.3:

Lemma 10.5. *conditions (H.1) and (H.2) are satisfied, then we have, for all* $s \in \mathbf{R}$,

$$\tilde{\varphi} \in \mathcal{D}'(\partial D \times S),\ \widetilde{T}(\vartheta)\tilde{\varphi} \in B^{s,p}(\partial D \times S) \Longrightarrow \tilde{\varphi} \in B^{s,p}(\partial D \times S)\,.$$

Furthermore, for any $t < s$, *there exists a constant* $\widetilde{C}_{s,t} > 0$ *such that*

$$|\tilde{\varphi}|_{s,p} \le \widetilde{C}_{s,t}\left(|\widetilde{T}(\vartheta)\tilde{\varphi}|_{s,p} + |\tilde{\varphi}|_{t,p}\right)\,. \tag{10.8}$$

The desired estimate (10.5) follows from estimate (10.8) with $s := 2 - 1/p$ and $t := -1/p$, just as in the proof of Theorem 6.11. □

Step 2: Now let u be an arbitrary function in the domain $\mathcal{D}(A_p)$:

$$u \in H^{2,p}(D) \quad \text{and} \quad L_0 u = 0 \text{ on } \partial D\,.$$

We choose a function $\zeta(y) \in C^\infty(S)$ such that

$$\begin{cases} 0 \le \zeta(y) \le 1 \quad \text{on } S\,, \\ \operatorname{supp}\zeta \subset \left[\dfrac{\pi}{3}, \dfrac{5\pi}{3}\right]\,, \\ \zeta(y) = 1 \quad \text{for} \quad \dfrac{\pi}{2} \le y \le \dfrac{3\pi}{2}\,, \end{cases}$$

and let

$$\tilde{v}_\eta(x, y) = u(x) \otimes \zeta(y) e^{i\eta y}, \quad \eta \ge 0\,.$$

Then we have

$$\begin{aligned} &\tilde{v}_\eta \in H^{2,p}(D \times S)\,, \\ &\widetilde{\Lambda}(\vartheta)\tilde{v}_\eta = \left(A + e^{i\vartheta}\frac{\partial^2}{\partial y^2}\right)\tilde{v}_\eta \\ &\qquad = (A - \eta^2 e^{i\vartheta}) u \otimes \zeta e^{i\eta y} + 2(i\eta) e^{i\vartheta} u \otimes \zeta' e^{i\eta y} + e^{i\vartheta} u \otimes \zeta'' e^{i\eta y}\,, \end{aligned}$$

and also

$$L_0 \tilde{v}_\eta = L_0 u \otimes \zeta e^{i\eta y} = 0 \quad \text{on } \partial D \times S\,.$$

Thus, applying inequality (10.5) to the functions $\tilde{v}_\eta = u \otimes \zeta e^{i\eta y}$ we obtain that

$$\left\| u \otimes \zeta e^{i\eta y} \right\|_{2,p} \le \widetilde{C}(\vartheta)\left(\left\| \widetilde{\Lambda}(\vartheta)(u \otimes \zeta e^{i\eta y}) \right\|_p + \left\| u \otimes \zeta e^{i\eta y} \right\|_p\right)\,. \tag{10.9}$$

We can estimate each term of inequality (10.9) as follows:

$$\left\|u \otimes \zeta e^{i\eta y}\right\|_p = \left(\int_{D \times S} |u(x)|^p |\zeta(y)|^p \, dx \, dy\right)^{1/p} = \|\zeta\|_p \cdot \|u\|_p . \tag{10.10}$$

$$\begin{aligned} \left\|\widetilde{\Lambda}(\vartheta)(u \otimes \zeta e^{i\eta y})\right\|_p &\leq \left\|(A - \eta^2 e^{i\vartheta}) u \otimes \zeta e^{i\eta y}\right\|_p \\ &\quad + 2\eta \left\|u \otimes \zeta' e^{i\eta y}\right\|_p + \left\|u \otimes \zeta'' e^{i\eta y}\right\|_p \\ &\leq \|\zeta\|_p \left\|(A - \eta^2 e^{i\vartheta}) u\right\|_p + (2\eta \|\zeta'\|_p + \|\zeta''\|_p) \|u\|_p . \end{aligned} \tag{10.11}$$

$$\begin{aligned} \left\|u \otimes \zeta e^{i\eta y}\right\|_{2,p}^p &= \sum_{|\alpha| \leq 2} \int_{D \times S} \left|D_{x,y}^\alpha (u(x) \otimes \zeta(y) e^{i\eta y})\right|^p \, dx \, dy \\ &\geq \sum_{|\alpha| \leq 2} \int_D \int_{\pi/2}^{3\pi/2} \left|D_{x,y}^\alpha (u(x) \otimes e^{i\eta y})\right|^p \, dx \, dy \\ &= \sum_{k + |\beta| \leq 2} \int_D \int_{\pi/2}^{3\pi/2} \left|\eta^k D^\beta u(x)\right|^p \, dx \, dy \\ &\geq \pi \left(\sum_{|\beta| = 2} \int_D \left|D^\beta u(x)\right|^p \, dx + \eta^p \sum_{|\beta| = 1} \int_D \left|D^\beta u(x)\right|^p \, dx \right. \\ &\quad \left. + \eta^{2p} \int_D |u(x)|^p \, dx \right) \\ &= \pi \left(|u|_{2,p}^p + \eta^p |u|_{1,p}^p + \eta^{2p} \|u\|_p^p \right) . \end{aligned} \tag{10.12}$$

Therefore, carrying these inequalities (10.10), (10.11) and (10.12) into inequality (10.9) we have, with a constant $\widetilde{C}'(\vartheta) > 0$ independent of $\eta > 0$,

$$|u|_{2,p} + \eta \, |u|_{1,p} + \eta^2 \|u\|_p \leq \widetilde{C}'(\vartheta) \left(\left\|(A - \eta^2 e^{i\vartheta}) u\right\|_p + \eta \, \|u\|_p \right) .$$

If η is so large that

$$\eta \geq 2\widetilde{C}'(\vartheta) ,$$

then we can eliminate the last term on the right-hand side to obtain that

$$|u|_{2,p} + \eta \, |u|_{1,p} + \eta^2 \|u\|_p \leq 2\widetilde{C}'(\vartheta) \left\|(A - \eta^2 e^{i\vartheta}) u\right\|_p .$$

This proves inequality (10.2) if we take

$$\begin{aligned} \lambda &= \eta^2 e^{i\vartheta} , \\ R(\vartheta) &= 4\widetilde{C}'(\vartheta)^2 , \\ C(\vartheta) &= 2\widetilde{C}'(\vartheta) . \end{aligned}$$

The proof of Theorem 10.3 is now complete. □

11

Proof of Theorem 1.3

In this chapter we prove Theorem 1.3 (Theorems 11.1 and 11.3). Just as in Chaps. 8 and 10, we make use of Agmon's method to prove the surjectivity of the operator $A_p - \lambda I$ (Proposition 11.2).

11.1 Proof of Part (i) of Theorem 1.3

First, we prove part (i) of Theorem 1.3:

Theorem 11.1. *Let $1 < p < \infty$. Assume that conditions (H.1) and (H.2) are satisfied:*

(H.1) $\mu(x') \geq 0$ and $\gamma(x') \leq 0$ on ∂D.
(H.2) $\gamma(x') < 0$ on $M = \{x' \in \partial D : \mu(x') = 0\}$.

Then, for every $0 < \varepsilon < \pi/2$, there exists a constant $r_p(\varepsilon) > 0$ such that the resolvent set of A_p contains the set $\Sigma_p(\varepsilon) = \{\lambda = r^2 e^{i\vartheta} : r \geq r_p(\varepsilon), -\pi + \varepsilon \leq \vartheta \leq \pi - \varepsilon\}$, and that the resolvent $(A_p - \lambda I)^{-1}$ satisfies the estimate

$$\|(A_p - \lambda I)^{-1}\| \leq \frac{c_p(\varepsilon)}{|\lambda|}, \quad \lambda \in \Sigma_p(\varepsilon), \tag{1.2}$$

where $c_p(\varepsilon) > 0$ is a constant depending on ε.

Proof. The proof is divided into three steps.

Step 1: By estimate (10.2), it follows that if $\lambda = r^2 e^{i\vartheta}$, $-\pi < \vartheta < \pi$ and $|\lambda| = r^2 \geq R(\vartheta)$, then we have, for all $u \in \mathcal{D}(A_p)$,

$$|u|_{2,p} + |\lambda|^{1/2}|u|_{1,p} + |\lambda|\|u\|_p \leq C(\vartheta)\|(A_p - \lambda I)u\|_p .$$

However, we find from the proof of Theorem 10.3 that the constants $R(\vartheta)$ and $C(\vartheta)$ depend *continuously* on $\vartheta \in (-\pi, \pi)$, so that they may be chosen uniformly in $\vartheta \in [-\pi + \varepsilon, \pi + \varepsilon]$, for every $\varepsilon > 0$. This proves the existence of the constants $r_p(\varepsilon)$ and $c_p(\varepsilon)$, that is, we have, for all $\lambda = r^2 e^{i\vartheta}$ satisfying $r \geq r_p(\varepsilon)$ and $-\pi + \varepsilon \leq \vartheta \leq \pi + \varepsilon$,

$$|u|_{2,p} + |\lambda|^{1/2}|u|_{1,p} + |\lambda|\|u\|_p \leq c_p(\varepsilon)\|(A_p - \lambda I)u\|_p \,. \tag{11.1}$$

By estimate (11.1), it follows that the operator $A_p - \lambda I$ is injective and its range $\mathcal{R}(A_p - \lambda I)$ is closed in $L^p(D)$, for all $\lambda \in \Sigma_p(\varepsilon)$.

Step 2: We show that the operator $A_p - \lambda I$ is *surjective* for all $\lambda \in \Sigma_p(\varepsilon)$:

$$\mathcal{R}(A_p - \lambda I) = L^p(D), \quad \lambda \in \Sigma_p(\varepsilon) \,. \tag{11.2}$$

To do this, it suffices to show that the operator $A_p - \lambda I$ is a Fredholm operator with

$$\operatorname{ind}(A_p - \lambda I) = 0 \,, \quad \lambda \in \Sigma_p(\varepsilon) \,, \tag{11.3}$$

since $A_p - \lambda I$ is injective for all $\lambda \in \Sigma_p(\varepsilon)$.

(2-1) We reduce the study of the operator $A_p - \lambda I$ ($\lambda \in \Sigma_p(\varepsilon)$) to that of a pseudo-differential operator on the boundary, just as in the proof of Theorem 1.1.

Let $T(\lambda)$ be a first-order, classical pseudo-differential operator on the boundary ∂D defined by the formula

$$T(\lambda) = L_0 P(\lambda) = \mu(x')\,\Pi(\lambda) + \gamma(x') \,, \quad \lambda \in \Sigma_p(\varepsilon) \,,$$

where

$$\begin{aligned} \Pi(\lambda) : C^\infty(\partial D) &\longrightarrow C^\infty(\partial D) \\ \varphi &\longmapsto \frac{\partial}{\partial \mathbf{n}} (P(\lambda)\varphi) \,. \end{aligned}$$

Since the operator $T(\lambda) : C^\infty(\partial D) \to C^\infty(\partial D)$ extends to a continuous linear operator $T(\lambda) : B^{t,p}(\partial D) \to B^{t-1,p}(\partial D)$ for all $t \in \mathbf{R}$, we can introduce a densely defined, closed linear operator

$$\mathcal{T}_p(\lambda) : B^{2-1/p,p}(\partial D) \longrightarrow B^{2-1/p,p}(\partial D)$$

as follows.

(α) The domain $\mathcal{D}(\mathcal{T}_p(\lambda))$ of $\mathcal{T}_p(\lambda)$ is the space

$$\mathcal{D}(\mathcal{T}_p(\lambda)) = \left\{\varphi \in B^{2-1/p,p}(\partial D) : T(\lambda)\varphi \in B^{2-1/p,p}(\partial D)\right\}.$$

(β) $\mathcal{T}_p(\lambda)\varphi = T(\lambda)\varphi$, $\varphi \in \mathcal{D}(\mathcal{T}_p(\lambda))$.

Then we can obtain the following results (see Theorem 6.9):

(I) The null space $\mathcal{N}(A_p - \lambda I)$ of $A_p - \lambda I$ has finite dimension if and only if the null space $\mathcal{N}(\mathcal{T}_p(\lambda))$ of $\mathcal{T}_p(\lambda)$ has finite dimension, and we have

$$\dim \mathcal{N}(A_p - \lambda I) = \dim \mathcal{N}\left(\mathcal{T}_p(\lambda)\right).$$

(II) The range $\mathcal{R}(A_p - \lambda I)$ of $A_p - \lambda I$ is closed if and only if the range $\mathcal{R}\left(\mathcal{T}_p(\lambda)\right)$ of $\mathcal{T}_p(\lambda)$ is closed; and $\mathcal{R}(A_p - \lambda I)$ has finite codimension if and only if $\mathcal{R}\left(\mathcal{T}_p(\lambda)\right)$ has finite codimension, and we have

$$\operatorname{codim} \mathcal{R}(A_p - \lambda I) = \operatorname{codim} \mathcal{R}\left(\mathcal{T}_p(\lambda)\right).$$

(III) The operator $A_p - \lambda I$ is a Fredholm operator if and only if the operator $\mathcal{T}_p(\lambda)$ is a Fredholm operator, and we have

$$\operatorname{ind}(A_p - \lambda I) = \operatorname{ind} \mathcal{T}_p(\lambda) .$$

Therefore, assertion (11.3) is reduced to the following assertion:

$$\operatorname{ind} \mathcal{T}_p(\lambda) = 0 , \quad \lambda \in \Sigma_p(\varepsilon) . \tag{11.3'}$$

(2-2) To prove assertion (11.3′), we shall make use of the method of Agmon as in Chap. 10.

Let $\widetilde{T}(\vartheta)$ be the first-order, classical pseudo-differential operator on $\partial D \times S$ introduced in Chap. 10:

$$\widetilde{T}(\vartheta) = L_0 \widetilde{P}(\vartheta) = \mu(x')\widetilde{\Pi}(\vartheta) + \gamma(x') , \quad -\pi < \vartheta < \pi ,$$

where

$$\begin{aligned} \widetilde{\Pi}(\vartheta) : C^\infty(\partial D \times S) &\longrightarrow C^\infty(\partial D \times S) \\ \tilde{\varphi} &\longmapsto \frac{\partial}{\partial \mathbf{n}} \left(\widetilde{P}(\vartheta)\tilde{\varphi}\right). \end{aligned}$$

We define a densely defined, closed linear operator

$$\widetilde{\mathcal{T}}_p(\vartheta) : B^{2-1/p,p}(\partial D \times S) \longrightarrow B^{2-1/p,p}(\partial D \times S)$$

as follows.

($\tilde{\alpha}$) The domain $\mathcal{D}\left(\widetilde{\mathcal{T}}_p(\vartheta)\right)$ of $\widetilde{\mathcal{T}}_p(\vartheta)$ is the space

$$\mathcal{D}\left(\widetilde{\mathcal{T}}_p(\vartheta)\right) = \left\{\tilde{\varphi} \in B^{2-1/p,p}(\partial D \times S) : \widetilde{T}(\vartheta)\tilde{\varphi} \in B^{2-1/p,p}(\partial D \times S)\right\}.$$

($\tilde{\beta}$) $\widetilde{\mathcal{T}}_p(\vartheta)\tilde{\varphi} = \widetilde{T}(\vartheta)\tilde{\varphi}$, $\tilde{\varphi} \in \mathcal{D}\left(\widetilde{\mathcal{T}}_p(\vartheta)\right)$.

Then we have the most fundamental relationship between $\widetilde{\mathcal{T}}_p(\vartheta)$ and $\mathcal{T}_p(\lambda)$, analogous to Proposition 8.11:

Proposition 11.2. *If* $\operatorname{ind} \widetilde{\mathcal{T}}_p(\vartheta)$ *is finite, then there exists a finite subset* K *of* $\mathbf{Z}$ *such that the operator* $\mathcal{T}_p(\lambda')$ *is bijective for all* $\lambda' = \ell^2 e^{i\vartheta}$ *satisfying* $\ell \in \mathbf{Z} \setminus K$.

Step 3: End of proof of Theorem 11.1

(3-1) We show that if conditions (H.1) and (H.2) are satisfied, then we have

$$\operatorname{ind} \widetilde{T}_p(\vartheta) < \infty . \tag{11.4}$$

Now, estimate (10.8) with $s := 2 - 1/p$ gives that

$$|\tilde{\varphi}|_{2-1/p,p} \leq \widetilde{C}_t \left(|\widetilde{T}(\vartheta)\tilde{\varphi}|_{2-1/p,p} + |\tilde{\varphi}|_{t,p} \right), \quad \tilde{\varphi} \in \mathcal{D}(\widetilde{T}_p(\vartheta)) , \tag{11.5}$$

where $t < 2-1/p$. However, it follows from an application of Rellich's theorem that the injection $B^{2-1/p,p}(\partial D \times S) \to B^{t,p}(\partial D \times S)$ is *compact* for $t < 2-1/p$. Thus, applying Peetre's Lemma (Lemma 8.12) with

$$\begin{aligned} X = Y &:= B^{2-1/p,p}(\partial D \times S) , \\ Z &:= B^{t,p}(\partial D \times S) , \\ T &:= \widetilde{T}_p(\vartheta) , \end{aligned}$$

we obtain that the range $\mathcal{R}\left(\widetilde{T}_p(\vartheta)\right)$ is closed in $B^{2-1/p,p}(\partial D \times S)$ and

$$\dim \mathcal{N}\left(\widetilde{T}_p(\vartheta)\right) < \infty . \tag{11.6}$$

On the other hand, by formula (10.5) we find that the complete symbol of the adjoint $\widetilde{T}(\vartheta)^*$ is given by the following (cf. Theorem 5.10):

$$\begin{aligned} &\mu(x') \left(\tilde{p}_1(x',\xi',y,\eta;\vartheta) - \sqrt{-1}\tilde{q}_1(x',\xi',y,\eta;\vartheta)\right) \\ &+ \Bigg(\Bigg[\gamma(x') + \mu(x')\tilde{p}_0(x',\xi',y,\eta;\vartheta) - \sum_{j=1}^{n-1} \partial_{x_j} \left(\mu(x') \cdot \partial_{\xi_j} \tilde{q}_1(x',\xi',y,\eta;\vartheta)\right) \Bigg] \\ &- \sqrt{-1} \Bigg[\mu(x')\tilde{q}_0(x',\xi',y,\eta;\vartheta) + \sum_{j=1}^{n-1} \partial_{x_j} \left(\mu(x') \cdot \partial_{\xi_j} \tilde{p}_1(x',\xi',y,\eta;\vartheta)\right) \Bigg] \Bigg) \\ &+ \text{terms of order} \leq -1 . \end{aligned}$$

However, by virtue of Lemma 8.4 it follows that

$$\partial_{x_j} \mu(x') = 0 \text{ on } M = \{x' \in \partial D : \mu(x') = 0\} .$$

Thus we can easily verify that the pseudo-differential operator $\widetilde{T}(\vartheta)^*$ satisfies conditions (5.7a) and (5.7b) of Theorem 5.15 with $\mu := 0$, $\rho := 1$ and $\delta := 1/2$. This implies that estimate (11.5) remains valid for the adjoint operator $\widetilde{T}_p(\vartheta)^*$ of $\widetilde{T}_p(\vartheta)$:

$$|\tilde{\psi}|_{-2+1/p,p'} \leq \widetilde{C}_\tau \left(|\widetilde{T}(\vartheta)^*\tilde{\psi}|_{-2+1/p,p'} + |\tilde{\psi}|_{\tau,p'} \right), \quad \tilde{\psi} \in \mathcal{D}\left(\widetilde{T}_p(\vartheta)^*\right),$$

where $\tau < -2 + 1/p$ and $p' = p/(p-1)$, the exponent conjugate to p. Hence we have, by the closed range theorem (Theorem 8.13) and Peetre's Lemma (Lemma 8.12),

$$\operatorname{codim} \mathcal{R}\left(\widetilde{\mathcal{T}}_p(\vartheta)\right) = \dim \mathcal{N}\left(\widetilde{\mathcal{T}}_p(\vartheta)^*\right) < \infty\,, \tag{11.7}$$

since the injection $B^{-2+1/p,p'}(\partial D \times S) \to B^{\tau,p'}(\partial D \times S)$ is compact for $\tau < -2 + 1/p$.

Therefore, assertion (11.4) follows from assertions (11.6) and (11.7).

(3-2) By assertion (11.4), we can apply Proposition 11.2 to obtain that the operator $\mathcal{T}_p(\ell^2 e^{i\vartheta}) : B^{2-1/p,p}(\partial D) \to B^{2-1/p,p}(\partial D)$ is bijective if $\ell \in \mathbf{Z} \setminus K$ for some finite subset K of $\mathbf{Z}$. In particular, we have

$$\operatorname{ind} \mathcal{T}_p(\lambda_0) = 0 \quad \text{if} \quad \lambda_0 = \ell^2 e^{i\vartheta}\,, \quad \ell \in \mathbf{Z} \setminus K\,. \tag{11.8}$$

However, in view of formula (8.28) it follows that, for any given λ, $\lambda_0 \in \Sigma_p(\varepsilon)$, we can find a classical pseudo-differential operator $K(\lambda, \lambda_0)$ of order -1 on the boundary ∂D such that

$$T(\lambda) = T(\lambda_0) + K(\lambda, \lambda_0)\,, \quad \lambda\,, \ \lambda_0 \in \Sigma_p(\varepsilon)\,.$$

Furthermore, Rellich's theorem tells us that the operator

$$K(\lambda, \lambda_0) : B^{2-1/p,p}(\partial D) \longrightarrow B^{2-1/p,p}(\partial D)$$

is *compact*. Hence we have

$$\operatorname{ind} \mathcal{T}_p(\lambda) = \operatorname{ind} \mathcal{T}_p(\lambda_0)\,, \quad \lambda\,, \ \lambda_0 \in \Sigma_p(\varepsilon)\,. \tag{11.9}$$

Therefore, assertion (11.3′) (and hence assertion (11.2)) follows from assertions (11.8) and (11.9).

(3-3) Summing up, we have proved that the operator $A_p - \lambda I$ is bijective for all $\lambda \in \Sigma_p(\varepsilon)$ and its inverse $(A_p - \lambda I)^{-1}$ satisfies estimate (1.2).

Now the proof of Theorem 11.1 (and hence that of part (i) of Theorem 1.3) is complete. □

11.2 Proof of Part (ii) of Theorem 1.3

Part (ii) of Theorem 1.3 may be proved as follows. Theorem 11.1 tells us that, for $\mu_\varepsilon > 0$ large enough, the operator $A_p - \mu_\varepsilon I$ satisfies condition (2.20) (see Fig. 11.1).

Thus, applying Theorem 2.11 (and Remark 2.2) to the operator $A_p - \mu_\varepsilon I$ we obtain part (ii) of Theorem 1.3:

Theorem 11.3. *Assume that conditions (H.1) and (H.2) are satisfied. Then the operator A_p generates a semigroup e^{zA_p} on $L^p(D)$ which is* analytic *in the sector $\Delta_\varepsilon = \{z = t + is : z \neq 0, |\arg z| < \pi/2 - \varepsilon\}$ for any $0 < \varepsilon < \pi/2$, and enjoys the following properties:*

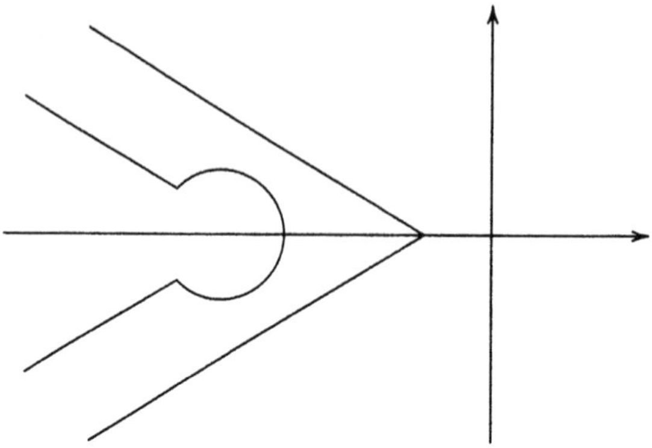

Fig. 11.1.

(a) *The operators* $A_p\, e^{zA_p}$ *and* $\frac{de^{zA_p}}{dz}$ *are bounded operators on* $L^p(D)$ *for each* $z \in \Delta_\varepsilon$, *and satisfy the relation*

$$\frac{de^{zA_p}}{dz} = A_p e^{zA_p}\,, \quad z \in \Delta_\varepsilon\,.$$

(b) *For each* $0 < \varepsilon < \pi/2$, *there exist constants* $M_0(\varepsilon) > 0$, $M_1(\varepsilon) > 0$ *and* $\mu_\varepsilon > 0$ *such that*

$$\|e^{zA_p}\| \leq M_0(\varepsilon)e^{\mu_\varepsilon \cdot \mathrm{Re}\, z}, \quad z \in \Delta_\varepsilon\,,$$
$$\|A_p\, e^{zA_p}\| \leq \frac{M_1(\varepsilon)}{|z|}e^{\mu_\varepsilon \cdot \mathrm{Re}\, z}\,, \quad z \in \Delta_\varepsilon\,.$$

(c) *For each* $u_0 \in L^p(D)$, *we have, as* $z \to 0, z \in \Delta_\varepsilon$,

$$e^{zA_p}u_0 \longrightarrow u_0 \quad \text{in } L^p(D)\,.$$

The proof of Theorem 1.3 is now complete. □

12

Proof of Theorem 1.4, Part (i)

This Chap. 12 and the next Chap. 13 are devoted to the proof of Theorem 1.4 and Theorem 1.5.

In this chapter we prove part (i) of Theorem 1.4. In the proof we make use of Sobolev's imbedding theorems (Theorems 12.1 and 12.2) and a λ-*dependent localization* argument essentially due to Masuda [Ma] (see Lemma 12.4) in order to adjust the estimate

$$\left\|(A_p - \lambda I)^{-1}\right\| \le \frac{c_p(\varepsilon)}{|\lambda|}, \quad \lambda \in \Sigma_p(\varepsilon), \tag{1.2}$$

to obtain the estimate

$$\|(\mathfrak{A} - \lambda I)^{-1}\| \le \frac{c(\varepsilon)}{|\lambda|}, \quad \lambda \in \Sigma(\varepsilon). \tag{1.4}$$

Here recall that

$$\mathcal{D}(A_p) = \left\{u \in H^{2,p}(D) \colon L_0 u = 0 \text{ on } \partial D\right\}.$$
$$\mathcal{D}(\mathfrak{A}) = \left\{u \in C_0(\overline{D} \setminus M) \colon Au \in C_0(\overline{D} \setminus M), \quad L_0 u = 0 \text{ on } \partial D\right\}.$$

12.1 Sobolev's Imbedding Theorems

It is the imbedding characteristics of Sobolev spaces that render these spaces so useful in the study of partial differential equations. We need the following imbedding properties of Sobolev spaces:

Theorem 12.1 (Sobolev). *Let D be a bounded domain in $\mathbf{R}^N$ with boundary ∂D of class C^2. Then we have the following two assertions:*

(i) If $1 \le p < N$, we have

$$H^{2,p}(D) \subset H^{1,q}(D), \quad \frac{1}{p} - \frac{1}{N} \le \frac{1}{q} \le \frac{1}{p},$$

with continuous injection.

(ii) If $N/2 < p < \infty$, $p \neq N$, we have

$$H^{2,p}(D) \subset C^{\nu}(\overline{D}) , \quad 0 < \nu \leq 2 - \frac{N}{p} ,$$

with continuous injection.

Theorem 12.2 (Gagliardo-Nirenberg). *Let D be a bounded domain in $\mathbf{R}^N$ with boundary of class C^2, and $1 \leq p, r \leq \infty$. If $p \neq N$ and if*

$$\frac{1}{q} = \theta \left(\frac{1}{p} - \frac{1}{N} \right) + (1 - \theta) \frac{1}{r} , \quad 0 \leq \theta \leq 1 ,$$

then we have, for all $u \in H^{1,p}(D) \cap L^r(D)$,

$$\|u\|_q \leq C |u|_{1,p}^{\theta} \|u\|_r^{1-\theta} , \tag{12.1}$$

with a constant $C = C(D, p, r, \theta) > 0$. Here and in the following

$$\|u\|_p = \|u\|_{L^p(D)} = \left(\int_D |u(x)|^p \, dx \right)^{1/p} ,$$

$$|u|_{1,p} = \|\nabla u\|_{L^p(D)} = \left(\int_D \sum_{|\alpha|=1} |D^{\alpha} u(x)|^p \, dx \right)^{1/p} .$$

For a proof of Theorems 12.1 and 12.2, see Adams [Ad, Theorem 5.4] and Friedman [Fr1, Part I, Theorem 10.1], and also Taira [Ta5, Chap. 2].

12.2 Proof of Part (i) of Theorem 1.4

The proof is carried out in a chain of auxiliary lemmas (Lemmas 12.3, 12.4 and 12.8), and is divided into three steps.

Step 1: We begin with a version of estimate (9.2):

Lemma 12.3. *Let $N < p < \infty$. Assume that conditions (H.1) and (H.2) are satisfied:*

(H.1) $\mu(x') \geq 0$ and $\gamma(x') \leq 0$ on ∂D.
(H.2) $\gamma(x') < 0$ on $M = \{x' \in \partial D \colon \mu(x') = 0\}$.

Then, for every $\varepsilon > 0$, there exists a constant $r_p(\varepsilon) > 0$ such that if $\lambda = r^2 e^{i\vartheta}$ with $r \geq r_p(\varepsilon)$ and $-\pi + \varepsilon \leq \vartheta \leq \pi - \varepsilon$, we have, for all $u \in \mathcal{D}(A_p)$,

$$|\lambda|^{1/2} \|u\|_{C^1(\overline{D})} + |\lambda| \, \|u\|_{C(\overline{D})} \leq C_p(\varepsilon) |\lambda|^{N/2p} \|(A - \lambda) u\|_p , \tag{12.2}$$

with a constant $C_p(\varepsilon) > 0$. Here recall that

$$\mathcal{D}(A_p) = \left\{ u \in H^{2,p}(D) \colon L_0 u = \mu(x') \frac{\partial u}{\partial \mathbf{n}} + \gamma(x') u = 0 \ \text{ on } \partial D \right\} .$$

Proof. First, applying Theorem 12.2 with

$$p = r > N\,, \quad q := \infty\,, \quad \theta := N/p\,,$$

we obtain from the Gagliardo–Nirenberg inequality (12.1) that

$$\|u\|_{C(\overline{D})} \leq C|u|_{1,p}^{N/p} \|u\|_p^{1-N/p}\,. \tag{12.3}$$

Here and in the following the letter C denotes a generic positive constant depending on p and ε, but independent of u and λ.

Combining inequality (10.2) with inequality (12.3), we find that

$$\begin{aligned}\|u\|_{C(\overline{D})} &\leq C\left(|\lambda|^{-1/2}\|(A-\lambda)u\|_p\right)^{N/p}\left(|\lambda|^{-1}\|(A-\lambda)u\|_p\right)^{1-N/p} \\ &= C|\lambda|^{-1+N/2p}\|(A-\lambda)u\|_p\,,\end{aligned}$$

so that

$$|\lambda|\,\|u\|_{C(\overline{D})} \leq C|\lambda|^{N/2p}\|(A-\lambda)u\|_p\,, \quad u \in \mathcal{D}(A_p)\,. \tag{12.4}$$

Similarly, applying inequality (12.3) to the functions $D_i u \in H^{1,p}(D)$, $1 \leq i \leq n$, we obtain that

$$\begin{aligned}\|D_i u\|_{C(\overline{D})} &\leq C|D_i u|_{1,p}^{N/p}\|D_i u\|_p^{1-N/p} \\ &\leq C|u|_{2,p}^{N/p}|u|_{1,p}^{1-N/p} \\ &\leq C\left(\|(A-\lambda)u\|_p\right)^{N/p}\left(|\lambda|^{-1/2}\|(A-\lambda)u\|_p\right)^{1-N/p} \\ &= C|\lambda|^{-1/2+N/2p}\|(A-\lambda)u\|_p\,.\end{aligned}$$

This proves that

$$|\lambda|^{1/2}\|u\|_{C^1(\overline{D})} \leq C|\lambda|^{N/2p}\|(A-\lambda)u\|_p\,, \quad u \in \mathcal{D}(A_p)\,. \tag{12.5}$$

Therefore, the desired inequality (12.2) follows from inequalities (12.4) and (12.5). □

Step 2: The next lemma proves estimate (1.4) for the resolvent $(\mathfrak{A}-\lambda I)^{-1}$:

Lemma 12.4. *If that conditions (H.1) and (H.2) are satisfied, then, for every $\varepsilon > 0$, there exists a constant $r(\varepsilon) > 0$ such that if $\lambda = r^2 e^{i\vartheta}$ with $r \geq r(\varepsilon)$ and $-\pi + \varepsilon \leq \vartheta \leq \pi - \varepsilon$, we have, for all $u \in \mathcal{D}(\mathfrak{A})$,*

$$|\lambda|^{1/2}\|u\|_{C^1(\overline{D})} + |\lambda|\,\|u\|_{C(\overline{D})} \leq c(\varepsilon)\|(A-\lambda)u\|_{C(\overline{D})}\,, \tag{12.6}$$

with a constant $c(\varepsilon) > 0$. Here recall that

$$\mathcal{D}(\mathfrak{A}) = \left\{u \in C_0(\overline{D} \setminus M) \colon Au \in C_0(\overline{D} \setminus M)\,,\ L_0 u = 0 \ \text{on } \partial D\right\}.$$

Proof. We shall make use of a λ-*dependent localization* argument due to Masuda [Ma] in order to adjust the term $\|(A-\lambda)u\|_p$ in inequality (12.2) to obtain inequality (12.6).

First, it should be noticed that

$$\mathfrak{A} \subset A_p \quad \text{for all} \quad 1 < p < \infty .$$

Indeed, since we have, for any $u \in \mathcal{D}(\mathfrak{A})$,

$$u \in C(\overline{D}) \subset L^p(D) , \quad Au \in C(\overline{D}) \subset L^p(D) \text{ and } L_0 u = 0 \text{ on } \partial D,$$

it follows from an application of Theorem 8.2 with $s := 2$ that

$$u \in H^{2,p}(D) .$$

(1) Let x_0 be an arbitrary point of $\overline{D} = D \cup \partial D$.

If $x_0 = x_0'$ is a *boundary point*, then we introduce a local coordinate system (x', x_N) near x_0' (see Fig. 12.1) such that $x' = (x_1, x_2, \ldots, x_{N-1})$ give local coordinates for the boundary ∂D and that

$$\begin{aligned} D &= \{(x', x_N) : x_N > 0\} , \\ \partial D &= \{(x', x_N) : x_N = 0\} , \\ x_0' &= (0, \ldots, 0, 0) , \\ \mathbf{n} &= (0, \ldots, 0, 1), \end{aligned}$$

and let

$$\begin{aligned} G_0 &= B(x_0', \eta_0) \cap D , \\ G' &= B(x_0', \eta) \cap D , \quad 0 < \eta < \eta_0 , \\ G'' &= B(x_0', \eta/2) \cap D , \quad 0 < \eta < \eta_0 . \end{aligned}$$

Here $B(x, \eta)$ denotes the ball of radius η about x.

Similarly, if x_0 is an *interior point*, then we let (see Fig. 12.2)

$$\begin{aligned} G_0 &= B(x_0, \eta_0) \subset D , \\ G' &= B(x_0, \eta) , \quad 0 < \eta < \eta_0 , \\ G'' &= B(x_0, \eta/2) , \quad 0 < \eta < \eta_0 . \end{aligned}$$

(2) Now take a function $\Theta(t) \in C_0^\infty(\mathbf{R})$ such that $\Theta(t)$ equals one near the origin, and define

$$\begin{aligned} \varphi(x) &= \Theta(|x'|^2)\Theta(x_N) , \\ x &= (x', x_N) = (x_1, \ldots, x_{N-1}, x_N) . \end{aligned}$$

Here we may assume that the function φ is chosen so that

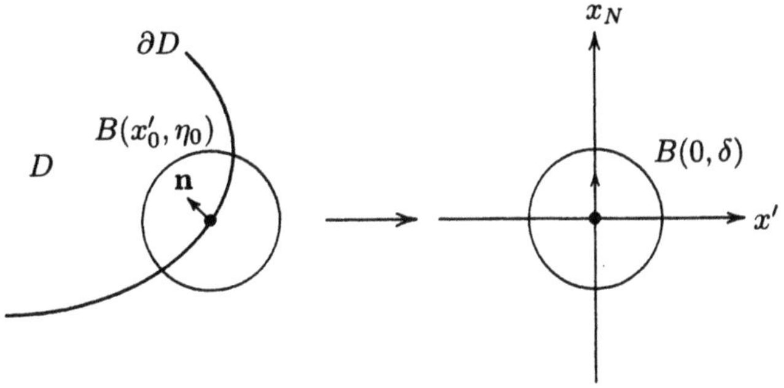

Fig. 12.1.

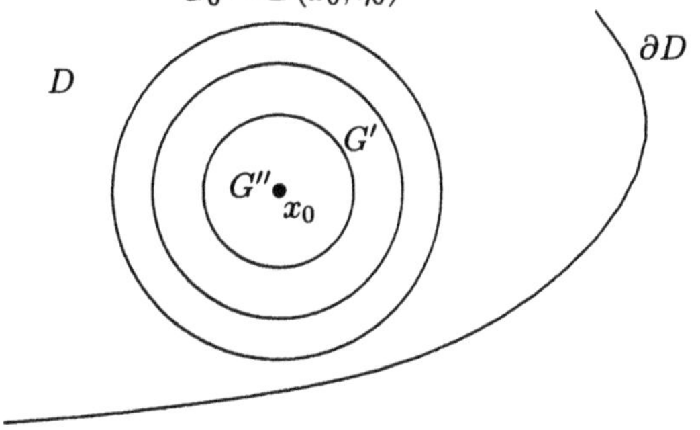

Fig. 12.2.

$$\begin{cases} \operatorname{supp} \varphi \subset B(0,1), \\ \varphi(x) = 1 \text{ on } B(0,1/2). \end{cases}$$

We introduce a localizing function

$$\varphi_0(x,\eta) := \varphi\left(\frac{x-x_0}{\eta}\right) = \Theta\left(\frac{|x'-x_0'|^2}{\eta^2}\right)\Theta\left(\frac{x_N-t_0}{\eta}\right),$$
$$x_0 = (x_0', t_0)\,. \tag{12.7}$$

It should be noticed that

$$\begin{cases} \operatorname{supp} \varphi_0 \subset B(x_0,\eta), \\ \varphi_0(x,\eta) = 1 \text{ on } B(x_0,\eta/2). \end{cases}$$

Then we have the following:

Claim 12.5. *If* $u \in \mathcal{D}(\mathfrak{A})$, *then it follows that* $\varphi_0 u \in \mathcal{D}(A_p)$ *for all* $1 < p < \infty$.

Proof. First, we recall that

$$u(x) \in H^{2,p}(D) \quad \text{for all} \quad 1 < p < \infty\,.$$

Hence we have

$$\varphi_0(x,\eta)u(x) = \varphi\left(\frac{x-x_0}{\eta}\right)u(x) \in H^{2,p}(D) .$$

Furthermore, it is easy to verify that the function $\varphi_0(x,\eta)u(x)$ satisfies the boundary condition

$$L_0(\varphi_0 u) = 0 \quad \text{on } \partial D .$$

Indeed, this is obvious if $x_0 \in D$, since we then have

$$\text{supp}(\varphi_0 u) \subset B(x_0,\eta) \subset D .$$

If $x_0 = x_0' \in \partial D$, then it follows from formula (12.7) that

$$\begin{aligned}\frac{\partial \varphi_0}{\partial \mathbf{n}} &= \frac{\partial}{\partial x_N}\left(\varphi_0(x,\eta)\right)\Big|_{x_N=0} \\ &= \frac{1}{\eta}\Theta'(0)\cdot\Theta\left(\frac{|x'-x_0'|^2}{\eta^2}\right) \\ &= 0 .\end{aligned}$$

This implies that

$$\begin{aligned}L_0(\varphi_0 u) &= \mu(x')\frac{\partial}{\partial \mathbf{n}}(\varphi_0 u) + \gamma(x')(\varphi_0 u) \\ &= \varphi_0\left(\mu(x')\frac{\partial u}{\partial \mathbf{n}} + \gamma(x')u\right) + \mu(x')\frac{\partial \varphi_0}{\partial \mathbf{n}}u \\ &= \varphi_0(L_0 u) \\ &= 0 \quad \text{on } \partial D .\end{aligned}$$

Summing up, we have proved that

$$\varphi_0 u \in \mathcal{D}(A_p) \quad \text{for all } 1 < p < \infty . \square$$

(3) Now take a number p such that

$$N < p < \infty .$$

Then, by Claim 12.5 we can apply inequality (12.2) to the function $\varphi_0 u$ ($u \in \mathcal{D}(\mathfrak{A})$) to obtain that

$$\begin{aligned}|\lambda|^{1/2}\|u\|_{C^1(\overline{G''})} + |\lambda|\,\|u\|_{C(\overline{G''})} &\le |\lambda|^{1/2}\,\|\varphi_0 u\|_{C^1(\overline{G'})} + |\lambda|\cdot\|\varphi_0 u\|_{C(\overline{G'})} \\ &= |\lambda|^{1/2}\,\|\varphi_0 u\|_{C^1(\overline{D})} + |\lambda|\cdot\|\varphi_0 u\|_{C(\overline{D})} \\ &\le C|\lambda|^{N/2p}\,\|(A-\lambda)(\varphi_0 u)\|_{L^p(D)} \\ &= C|\lambda|^{N/2p}\,\|(A-\lambda)(\varphi_0 u)\|_{L^p(G')} . \qquad (12.8)\end{aligned}$$

since we have

$$\begin{cases} \varphi_0(x,\eta) = 1 & \text{on } G'', \\ \operatorname{supp}(\varphi_0 u) \subset \overline{G'}. \end{cases}$$

Moreover, we can write the last term $(A-\lambda)(\varphi_0 u)$ of inequality (12.8) in the form

$$(A-\lambda)(\varphi_0 u) = \varphi_0 \left((A-\lambda)u\right) + [A, \varphi_0]u\,, \tag{12.9}$$

where $[A, \varphi_0]$ is the commutator of A and φ_0:

$$\begin{aligned} [A, \varphi_0]u &= A(\varphi_0 u) - \varphi_0 A u \\ &= 2\sum_{i,j=1}^{N} a^{ij} \frac{\partial \varphi_0}{\partial x_i}\frac{\partial u}{\partial x_j} + \left(\sum_{i,j=1}^{N} a^{ij}\frac{\partial^2 \varphi_0}{\partial x_i \partial x_j} + \sum_{i=1}^{N} b^i \frac{\partial \varphi_0}{\partial x_i}\right) u\,. \end{aligned}$$

In order to estimate the right-hand side of formula (12.9), we make use of the following elementary inequality:

Claim 12.6. *We have, for all $v \in C^j(\overline{G'})$, $j = 0, 1, 2$,*

$$\|v\|_{H^{j,p}(G')} \leq |G'|^{1/p} \|v\|_{C^j(\overline{G'})}\,,$$

where $|G'|$ is the measure of G'.

Proof. It suffices to note that

$$\left(\int_{G'} |w(x)|^p \, dx\right)^{1/p} \leq |G'|^{1/p} \, \|w\|_{C(\overline{G'}}\,, \quad w \in C(\overline{G'})\,. \qquad \square$$

We remark that there exists a constant $c > 0$, independent of η, such that

$$|G'| \leq |B(x_0, \eta)| \leq c\eta^N\,. \tag{12.10}$$

Thus, combining estimates (12.9) and (12.10) we obtain that

$$\|\varphi_0 \left((A-\lambda)u\right)\|_{L^p(G')} \leq c^{1/p}\eta^{N/p} \|(A-\lambda)u\|_{C(\overline{G'})}\,. \tag{12.11}$$

The next commutator estimate is essential in this step:

Lemma 12.7. *There exists a constant $C > 0$, independent of η, such that we have, as $\eta \downarrow 0$,*

$$\|[A, \varphi_0]u\|_{L^p(G')} \leq C\left(\eta^{-1+N/p}\|u\|_{C^1(\overline{D})} + \eta^{-2+N/p}\|u\|_{C(\overline{D})}\right). \tag{12.12}$$

Proof. First, we have, by formula (12.7),

$$|D_x^\alpha \varphi_0(x,\eta)| = O\left(\eta^{-|\alpha|}\right) \quad \text{as} \quad \eta \downarrow 0\,.$$

Hence it follows from an application of Claim 12.6 that

$$\left\| \frac{\partial \varphi_0}{\partial x_i} \frac{\partial u}{\partial x_j} \right\|_{L^p(G')} \leq C \frac{1}{\eta} |u|_{1,p,G'} \leq C \eta^{-1+N/p} \|u\|_{C^1(\overline{G'})},$$

$$\left\| \frac{\partial^2 \varphi_0}{\partial x_i \partial x_j} u \right\|_{L^p(G')} \leq C \frac{1}{\eta^2} |u|_{L^p(G')} \leq C \eta^{-2+N/p} \|u\|_{C(\overline{G'})},$$

$$\left\| \frac{\partial \varphi_0}{\partial x_i} u \right\|_{L^p(G')} \leq C \frac{1}{\eta} |u|_{L^p(G')} \leq C \eta^{-1+N/p} \|u\|_{C(\overline{G'})},$$

so that

$$\begin{aligned} \|[A, \varphi_0]u\|_{L^p(G')} &\leq C\eta^{-1+N/p}|u|_{C^1(\overline{G'})} + \eta^{-2+N/p}|u|_{C(\overline{G'})} \\ &\leq C\eta^{-1+N/p}|u|_{C^1(\overline{D})} + \eta^{-2+N/p}|u|_{C(\overline{D})} . \end{aligned}$$

This proves estimate (12.12). □

Therefore, it follows from estimates (12.11) and (12.12) that

$$\|(A-\lambda)u\|_{L^p(G')} \leq C\Big(\eta^{N/p}\|(A-\lambda)u\|_{C(\overline{D})} + \eta^{-1+N/p}|u|_{C^1(\overline{D})} + \eta^{-2+N/p}|u|_{C(\overline{D})} \Big) . \quad (12.13)$$

Finally, combining estimates (12.8) and (12.13) we have proved that

$$\begin{aligned} |\lambda|^{1/2}\|u\|_{C^1(\overline{G''})} + |\lambda|\,\|u\|_{C(\overline{G''})} &\leq C|\lambda|^{N/2p}\,\|(A-\lambda)(\varphi_0 u)\|_{L^p(G')} \\ &\leq C|\lambda|^{N/2p}\Big(\eta^{N/p}\|(A-\lambda)u\|_{C(\overline{D})} \\ &\quad + \eta^{-1+N/p}\|u\|_{C^1(\overline{D})} + \eta^{-2+N/p}\|u\|_{C(\overline{D})} \Big) . \end{aligned} \quad (12.14)$$

(4) Now it should be noticed (see Fig. 12.3) that the closure $\overline{D} = D \cup \partial D$ can be covered by a finite number of sets of the forms

$$\begin{aligned} &B(x_0', \eta/2) \cap \overline{D}, \quad x_0' \in \partial D, \\ &B(x_0, \eta/2), \quad x_0 \in D . \end{aligned}$$

Hence, taking the supremum of inequality (12.14) over $x \in \overline{D}$ we find that

$$\begin{aligned} &|\lambda|^{1/2}\|u\|_{C^1(\overline{D})} + |\lambda|\,\|u\|_{C(\overline{D})} \\ &\leq C|\lambda|^{N/2p}\eta^{N/p}\left(\|(A-\lambda)u\|_{C(\overline{D})} + \eta^{-1}\|u\|_{C^1(\overline{D})} + \eta^{-2}\|u\|_{C(\overline{D})} \right) . \end{aligned} \quad (12.15)$$

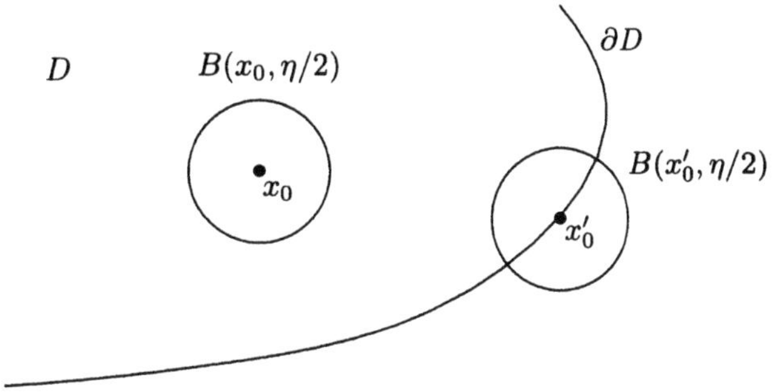

Fig. 12.3.

(5) We now choose the localization parameter η. We let

$$\eta = \frac{\eta_0}{|\lambda|^{1/2}} K ,$$

where K is a positive constant (to be chosen later on) satisfying the condition

$$0 < \eta = \frac{\eta_0}{|\lambda|^{1/2}} K < \eta_0 ,$$

that is,

$$0 < K < |\lambda|^{1/2} .$$

Then it follows from inequality (12.15) that

$$\begin{aligned} &|\lambda|^{1/2} \|u\|_{C^1(\overline{D})} + |\lambda| \|u\|_{C(\overline{D})} \\ &\leq C \eta_0^{N/p} K^{N/p} \|(A - \lambda) u\|_{C(\overline{D})} + \left(C \eta_0^{N/p-1} K^{-1+N/p} \right) |\lambda|^{1/2} \cdot \|u\|_{C^1(\overline{D})} \\ &\quad + \left(C \eta_0^{N/p-2} K^{-2+N/p} \right) |\lambda| \|u\|_{C(\overline{D})} . \end{aligned} \tag{12.16}$$

However, since the exponents $-1 + N/p$ and $-2 + N/p$ are negative, we can choose the constant K so large that

$$C \eta_0^{N/p-1} K^{-1+N/p} < 1 ,$$

and

$$C \eta_0^{N/p-2} K^{-2+N/p} < 1 .$$

Then the desired inequality (12.6) follows from inequality (12.16).
The proof of Lemma 12.4 is complete. □

Step 3: The next lemma, together with Lemma 12.4, proves that the resolvent set of the operator $\mathfrak{A}$ contains the set

$$\Sigma(\varepsilon) = \left\{ \lambda = r^2 e^{i\vartheta} : r \geq r(\varepsilon) , \ -\pi + \varepsilon \leq \vartheta \leq \pi - \varepsilon \right\} .$$

Lemma 12.8. *If $\lambda \in \Sigma(\varepsilon)$, then, for any $f \in C_0(\overline{D} \setminus M)$ there exists a unique function $u \in \mathcal{D}(\mathfrak{A})$ such that $(\mathfrak{A} - \lambda I)u = f$.*

Proof. Since we have

$$f \in C_0(\overline{D} \setminus M) \subset L^p(D) \quad \text{for all} \quad 1 < p < \infty ,$$

it follows from Theorem 1.3 that if $\lambda \in \Sigma(\varepsilon)$, there exists a unique function $u \in H^{2,p}(D)$ such that

$$(A - \lambda)u = f \quad \text{in } D , \tag{12.17}$$

and

$$L_0 u = \mu(x') \frac{\partial u}{\partial \mathbf{n}} + \gamma(x') u = 0 \quad \text{on } \partial D . \tag{12.18}$$

However, part (ii) of Theorem 12.1 tells us that

$$u \in H^{2,p}(D) \subset C^{2-N/p}(\overline{D}) \subset C^1(\overline{D}) \quad \text{if} \quad N < p < \infty .$$

Hence we have, by formula (12.18) and condition (H.2),

$$u = 0 \quad \text{on} \quad M = \{x' \in \partial D \colon \mu(x') = 0\} ,$$

so that

$$u \in C_0(\overline{D} \setminus M) .$$

Furthermore, in view of formula (12.17) it follows that

$$Au = f + \lambda u \in C_0(\overline{D} \setminus M) .$$

Summing up, we have proved that

$$\begin{cases} u \in \mathcal{D}(\mathfrak{A}), \\ (\mathfrak{A} - \lambda I)u = f. \end{cases} \qquad \square$$

Now the proof of part (i) of Theorem 1.4 is complete. □

13

Proofs of Theorem 1.5 and Theorem 1.4, Part (ii)

In this chapter we prove Theorem 1.5 and part (ii) of Theorem 1.4. First, we study Feller semigroups with reflecting barrier (Theorem 13.14) and then, using these Feller semigroups we construct Feller semigroups corresponding to such a physical phenomenon that either absorption or reflection phenomenon occurs at each point of the boundary (Theorem 13.18). Our proof is based on the generation theorems for Feller semigroups discussed in Sect. 3.3, just as in Chap. 7. Part (i) of Theorem 1.3, together with Theorem 1.5, proves part (ii) of Theorem 1.4.

13.1 Existence Theorem for Feller Semigroups

The purpose of this section is to give a general existence theorem for Feller semigroups in terms of boundary value problems, following Taira [Ta2, Sect. 9.6] (see also Sato–Ueno [SU] and Bony–Courrège–Priouret [BCP]).

Let A be a second-order, elliptic differential operator of the form (6.1) and let L be a general Ventcel' boundary condition of the form (7.2), respectively:

$$Au(x) = \sum_{i,j=1}^{n} a^{ij}(x)\frac{\partial^2 u}{\partial x_i \partial x_j}(x) + \sum_{i=1}^{n} b^i(x)\frac{\partial u}{\partial x_i}(x) + c(x)u(x)\,. \tag{6.1}$$

$$\begin{aligned} Lu(x') := {} & \sum_{i,j=1}^{N-1} \alpha^{ij}(x')\frac{\partial^2 u}{\partial x_i \partial x_j}(x') + \sum_{i=1}^{N-1} \beta^i(x')\frac{\partial u}{\partial x_i}(x') + \gamma(x')u(x') \\ & + \mu(x')\frac{\partial u}{\partial \mathbf{n}}(x') - \delta(x')Au(x') \\ & + \int_{\partial D} r(x',y')\left[u(y') - \tau(x',y')\left(u(x') + \sum_{j=1}^{N-1}(y_j - x_j)\frac{\partial u}{\partial x_j}(x')\right)\right] dy' \end{aligned}$$

$$+ \int_D t(x', y) \left[u(y) - \tau(x', y) \left(u(x') + \sum_{j=1}^{N-1} (y_j - x_j) \frac{\partial u}{\partial x_j}(x') \right) \right] dy \,. \tag{7.2}$$

Then we are interested in the following problem:

Problem. Given analytic data A and L, can we construct a Feller semigroup $\{T_t\}_{t \geq 0}$ on $\overline{D}$ whose infinitesimal generator $\mathfrak{A}$ is characterized by (A, L)?

First, we consider the following Dirichlet problem: Given functions f and φ defined in D and on ∂D, respectively, find a function u in D such that

$$\begin{cases} Au = f & \text{in } D, \\ u = \varphi & \text{on } \partial D. \end{cases} \tag{D}$$

The next theorem summarizes the basic facts about the Dirichlet problem in the framework of *Hölder spaces* (see Gilbarg–Trudinger [GT]):

Theorem 13.1. *(i)* (Existence and Uniqueness) *If $f \in C^\theta(D)$ $(0 < \theta < 1)$ and $\varphi \in C(\partial D)$, then problem (D) has a unique solution u in $C(\overline{D}) \cap C^{2+\theta}(D)$.*
(ii) (Interior Regularity) *If $u \in C^2(D)$ and $Au = f \in C^{k+\theta}(D)$ for some non-negative integer k, then we have $u \in C^{k+2+\theta}(D)$.*
(iii) (Global Regularity) *If $f \in C^{k+\theta}(\overline{D})$ and $\varphi \in C^{k+2+\theta}(\partial D)$ for some non-negative integer k, then a solution $u \in C(\overline{D}) \cap C^2(D)$ of problem (D) belongs to the space $C^{k+2+\theta}(\overline{D})$.*

Next we consider the following Dirichlet problem: For given functions f and φ defined in D and on ∂D, respectively, find a function u in D such that

$$\begin{cases} (\alpha - A)u = f & \text{in } D, \\ u = \varphi & \text{on } \partial D, \end{cases} \tag{D$'$}$$

where α is a positive parameter.

Theorem 13.1 with $k := 0$ tells us that problem (D$'$) has a unique solution u in the space $C^{2+\theta}(\overline{D})$ for any $f \in C^\theta(\overline{D})$ and any $\varphi \in C^{2+\theta}(\partial D)$ $(0 < \theta < 1)$. Therefore, we can introduce linear operators

$$G_\alpha^0 : C^\theta(\overline{D}) \longrightarrow C^{2+\theta}(\overline{D}) \,,$$

and

$$H_\alpha : C^{2+\theta}(\partial D) \longrightarrow C^{2+\theta}(\overline{D})$$

as follows.

(a) For any $f \in C^\theta(\overline{D})$, the function $G_\alpha^0 f \in C^{2+\theta}(\overline{D})$ is the unique solution of the problem

$$\begin{cases} (\alpha - A)G_\alpha^0 f = f & \text{in } D, \\ G_\alpha^0 f = 0 & \text{on } \partial D. \end{cases} \tag{13.1}$$

(b) For any $\varphi \in C^{2+\theta}(\partial D)$, the function $H_\alpha \varphi \in C^{2+\theta}(\overline{D})$ is the unique solution of the problem

$$\begin{cases} (\alpha - A)H_\alpha \varphi = 0 & \text{in } D, \\ H_\alpha \varphi = \varphi & \text{on } \partial D. \end{cases} \tag{13.2}$$

The operator G_α^0 is called the *Green operator* and the operator H_α is called the *harmonic operator*.

Then we have the following results:

Lemma 13.2. *The operator G_α^0, $\alpha > 0$, considered from $C(\overline{D})$ into itself, is non-negative and continuous with norm*

$$\|G_\alpha^0\| = \|G_\alpha^0 1\|_\infty = \sup_{x \in \overline{D}} G_\alpha^0 1(x) .$$

Proof. Let f be an arbitrary function in $C^\theta(\overline{D})$ such that $f \geq 0$ on $\overline{D}$. Then, applying the weak maximum principle (Theorem C.1 in Appendix C) with $A := A - \alpha$ to the function $-G_\alpha^0 f$ we obtain from formula (13.1) that

$$G_\alpha^0 f \geq 0 \quad \text{on } \overline{D} .$$

This proves the non-negativity of G_α^0.

Since G_α^0 is non-negative, we have, for all $f \in C^\theta(\overline{D})$,

$$-G_\alpha^0 \|f\|_\infty \leq G_\alpha^0 f \leq G_\alpha^0 \|f\|_\infty \quad \text{on } \overline{D} .$$

This implies the continuity of G_α^0 with norm

$$\|G_\alpha^0\| = \|G_\alpha^0 1\|_\infty .$$

The proof is complete. □

Lemma 13.3. *The operator H_α, $\alpha > 0$, considered from $C(\partial D)$ into $C(\overline{D})$, is non-negative and continuous with norm*

$$\|H_\alpha\| = \|H_\alpha 1\|_\infty = \sup_{x \in \overline{D}} H_\alpha 1(x) .$$

More precisely, we have the following:

Theorem 13.4. *(i) (a) The operator G_α^0, $\alpha > 0$, can be uniquely extended to a non-negative, bounded linear operator on $C(\overline{D})$ into itself, denoted again by G_α^0, with norm*

$$\|G_\alpha^0\| = \|G_\alpha^0 1\|_\infty \leq \frac{1}{\alpha} . \tag{13.3}$$

(b) For any $f \in C(\overline{D})$, we have

$$G_\alpha^0 f|_{\partial D} = 0 .$$

(c) For all α, $\beta > 0$, *the* resolvent equation *holds:*

$$G_\alpha^0 f - G_\beta^0 f + (\alpha - \beta) G_\alpha^0 G_\beta^0 f = 0 \,, \quad f \in C(\overline{D}) \,. \tag{13.4}$$

(d) For any $f \in C(\overline{D})$, *we have*

$$\lim_{\alpha \to +\infty} \alpha G_\alpha^0 f(x) = f(x) \,, \quad x \in D \,. \tag{13.5}$$

Furthermore, if $f|_{\partial D} = 0$, *then this convergence is uniform in* $x \in \overline{D}$, *that is,*

$$\lim_{\alpha \to +\infty} \alpha G_\alpha^0 f = f \ \text{in} \ C(\overline{D}) \,. \tag{13.5'}$$

(e) The operator G_α^0 *maps* $C^{k+\theta}(\overline{D})$ *continuously into* $C^{k+2+\theta}(\overline{D})$ *for any non-negative integer* k.

(ii) (a′) The operator H_α ($\alpha > 0$) *can be uniquely extended to a non-negative, bounded linear operator on* $C(\partial D)$ *into* $C(\overline{D})$, *denoted again by* H_α, *with norm* $\|H_\alpha\| = 1$.
(b′) For any $\varphi \in C(\partial D)$, *we have*

$$H_\alpha \varphi|_{\partial D} = \varphi \,.$$

(c′) For all α, $\beta > 0$, *we have*

$$H_\alpha \varphi - H_\beta \varphi + (\alpha - \beta) G_\alpha^0 H_\beta \varphi = 0 \,, \quad \varphi \in C(\partial D) \,. \tag{13.6}$$

(d′) The operator H_α *maps* $C^{k+2+\theta}(\partial D)$ *continuously into* $C^{k+2+\theta}(\overline{D})$ *for any non-negative integer* k.

Proof. (i) *Assertion (a)*: Making use of mollifiers (see Theorem 4.4), we find that the space $C^\theta(\overline{D})$ is dense in $C(\overline{D})$ and further that non-negative functions can be approximated by non-negative smooth functions. Hence, by Lemma 13.2 it follows that the operator $G_\alpha^0 : C^\theta(\overline{D}) \to C^{2+\theta}(\overline{D})$ can be uniquely extended to a non-negative, bounded linear operator $G_\alpha^0 : C(\overline{D}) \to C(\overline{D})$ with norm $\|G_\alpha^0\| = \|G_\alpha^0 1\|_\infty$.

Furthermore, since the function $G_\alpha^0 1$ satisfies the conditions

$$\begin{cases} (A - \alpha) G_\alpha^0 1 = -1 & \text{in } D, \\ G_\alpha^0 1 = 0 & \text{on } \partial D, \end{cases}$$

applying Theorem C.2 in Appendix C with $A := A - \alpha$ we obtain that

$$\|G_\alpha^0\| = \|G_\alpha^0 1\|_\infty \le \frac{1}{\alpha} \,.$$

Assertion (b): This follows from formula (13.1), since the space $C^\theta(\overline{D})$ is dense in $C(\overline{D})$ and the operator $G_\alpha^0 : C(\overline{D}) \to C(\overline{D})$ is bounded.

Assertion (c): We find from the uniqueness theorem for problem (D) (Theorem 13.1) that equation (13.4) holds for all $f \in C^\theta(\overline{D})$. Hence it holds for all $f \in C(\overline{D})$, since the space $C^\theta(\overline{D})$ is dense in $C(\overline{D})$ and the operators G^0_α are bounded.

Assertion (d): First, let f be an arbitrary function in $C^\theta(\overline{D})$ satisfying $f|_{\partial D} = 0$. Then it follows from the uniqueness theorem for problem (D) (Theorem 13.1) that we have, for all α, β,

$$f - \alpha G^0_\alpha f = G^0_\alpha\left((\beta - A)f\right) - \beta G^0_\alpha f \,.$$

Thus we have, by estimate (13.3),

$$\|f - \alpha G^0_\alpha f\|_\infty \leq \frac{1}{\alpha}\|(\beta - A)f\|_\infty + \frac{\beta}{\alpha}\|f\|_\infty \,,$$

so that

$$\lim_{\alpha \to +\infty} \|f - \alpha G^0_\alpha f\|_\infty = 0 \,.$$

Now let f be an arbitrary function in $C(\overline{D})$ satisfying $f|_{\partial D} = 0$. By means of mollifiers (see Theorem 4.4), we can find a sequence $\{f_j\}$ in $C^\theta(\overline{D})$ such that

$$\begin{cases} f_j \longrightarrow f & \text{in } C(\overline{D}) \text{ as } j \to \infty, \\ f_j = 0 & \text{on } \partial D. \end{cases}$$

Then we have, by estimate (13.3),

$$\begin{aligned} \|f - \alpha G^0_\alpha f\|_\infty &\leq \|f - f_j\|_\infty + \|f_j - \alpha G^0_\alpha f_j\|_\infty + \|\alpha G^0_\alpha f_j - \alpha G^0_\alpha f\|_\infty \\ &\leq 2\|f - f_j\|_\infty + \|f_j - \alpha G^0_\alpha f_j\|_\infty \,, \end{aligned}$$

and hence

$$\limsup_{\alpha \to +\infty} \|f - \alpha G^0_\alpha f\|_\infty \leq 2\|f - f_j\|_\infty \,.$$

This proves assertion (13.5′), since $\|f - f_j\|_\infty \to 0$ as $j \to \infty$.

To prove assertion (13.5), let f be an arbitrary function in $C(\overline{D})$ and x an arbitrary point of D. Take a function $\psi(x) \in C(\overline{D})$ such that

$$\begin{cases} 0 \leq \psi \leq 1 & \text{on } \overline{D}, \\ \psi = 0 & \text{in a neighborhood of } x, \\ \psi = 1 & \text{near the set } \partial D. \end{cases}$$

Then it follows from the non-negativity of G^0_α and estimate (13.3) that

$$0 \leq \alpha G^0_\alpha \psi(x) + \alpha G^0_\alpha (1 - \psi)(x) = \alpha G^0_\alpha 1(x) \leq 1 \,. \tag{13.7}$$

However, applying assertion (13.5′) to the function $1 - \psi$ we have

$$\lim_{\alpha \to +\infty} \alpha G^0_\alpha (1 - \psi)(x) = (1 - \psi)(x) = 1 \,.$$

In view of inequalities (13.7), this implies that

$$\lim_{\alpha \to +\infty} \alpha G_\alpha^0 \psi(x) = 0 \,.$$

Thus, since $-\|f\|_\infty \psi \leq f\psi \leq \|f\|_\infty \psi$ on $\overline{D}$, it follows that

$$|\alpha G_\alpha^0 (f\psi)(x)| \leq \|f\|_\infty \; \alpha G_\alpha^0 \psi(x) \longrightarrow 0 \quad \text{as } \alpha \to +\infty \,.$$

Therefore, applying assertion (13.5′) to the function $(1-\psi)f$ we obtain that

$$f(x) = ((1-\psi)f)(x) = \lim_{\alpha \to +\infty} \alpha G_\alpha^0 \left((1-\psi)f\right)(x) = \lim_{\alpha \to +\infty} \alpha G_\alpha^0 f(x) \,.$$

Assertion (e): This is an immediate consequence of part (iii) of Theorem 13.1.

(ii) *Assertion* (a′): Since the space $C^{2+\theta}(\partial D)$ is dense in $C(\partial D)$, by Lemma 13.3 it follows that the operator $H_\alpha : C^{2+\theta}(\partial D) \to C^{2+\theta}(\overline{D})$ can be uniquely extended to a non-negative, bounded linear operator $H_\alpha : C(\partial D) \to C(\overline{D})$. Moreover, applying Theorem C.1 in Appendix C with $A := A - \alpha$ we have

$$\|H_\alpha\| = \|H_\alpha 1\|_\infty = 1 \,.$$

Assertion (b′): This follows from formula (13.2), since the space $C^{2+\theta}(\partial D)$ is dense in $C(\partial D)$ and the operator $H_\alpha : C(\partial D) \to C(\overline{D})$ is bounded.

Assertion (c′): We find from the uniqueness theorem for problem (D) that formula (13.6) holds for all $\varphi \in C^{2+\theta}(\partial D)$. Hence it holds for all $\varphi \in C(\partial D)$, since the space $C^{2+\theta}(\partial D)$ is dense in $C(\partial D)$ and the operators G_α^0 and H_α are bounded.

Assertion (d′): This is an immediate consequence of part (iii) of Theorem 13.1.

The proof of Theorem 13.4 is now complete. □

Now we consider the following general boundary value problem (∗) in the framework of the spaces of *continuous functions*:

$$\begin{cases} (\alpha - A)u = f & \text{in } D, \\ Lu = 0 & \text{on } \partial D. \end{cases} \tag{$*$}$$

To do this, we introduce three linear operators associated with problem (∗).

(I) First, we introduce a linear operator

$$A : C(\overline{D}) \longrightarrow C(\overline{D})$$

as follows.

(a) The domain $\mathcal{D}(A)$ of A is the space $C^2(\overline{D})$.

(b) $Au = \sum_{i,j=1}^N a^{ij}(x) \dfrac{\partial^2 u}{\partial x_i \partial x_j} + \sum_{i=1}^N b^i(x) \dfrac{\partial u}{\partial x_i} + c(x)u, \; u \in \mathcal{D}(A).$

Then we have the following:

Lemma 13.5. *The operator A has its minimal closed extension $\overline{A}$ in the space $C(\overline{D})$.*

Proof. We apply part (i) of Theorem 3.5 to the operator A.

Assume that a function $u \in C^2(\overline{D})$ takes a positive maximum at a point x_0 of D:

$$u(x_0) = \max_{x \in \overline{D}} u(x) > 0 .$$

Then it follows that

$$\frac{\partial u}{\partial x_i}(x_0) = 0, \quad 1 \leq i \leq N ,$$
$$\sum_{i,j=1}^{N} a^{ij}(x_0) \frac{\partial^2 u}{\partial x_i \partial x_j}(x_0) \leq 0 ,$$

since the matrix $(a^{ij}(x))$ is positive definite. Hence we have

$$Au(x_0) = \sum_{i,j=1}^{N} a^{ij}(x_0) \frac{\partial^2 u}{\partial x_i \partial x_j}(x_0) + c(x_0)u(x_0) \leq 0 ,$$

This implies that the operator A satisfies condition (β) of Theorem 3.5 with $K_0 := D$ and $K := \overline{D}$. Therefore, Lemma 13.5 follows from an application of the same theorem. □

Remark 13.1. Since the injection: $C(\overline{D}) \to \mathcal{D}'(D)$ is continuous, we have the formula

$$\overline{A}u = \sum_{i,j=1}^{N} a^{ij}(x) \frac{\partial^2 u}{\partial x_i \partial x_j} + \sum_{i=1}^{N} b^i(x) \frac{\partial u}{\partial x_i} + c(x)u ,$$

where the right-hand side is taken in the sense of *distributions*.

The extended operators $G_\alpha^0 : C(\overline{D}) \to C(\overline{D})$ and $H_\alpha : C(\partial D) \to C(\overline{D})$ $(\alpha > 0)$ still satisfy formulas (13.1) and (13.2) respectively in the following sense:

Lemma 13.6. *(i) For any $f \in C(\overline{D})$, we have*

$$\begin{cases} G_\alpha^0 f \in \mathcal{D}(\overline{A}), \\ (\alpha I - \overline{A}) G_\alpha^0 f = f \text{ in } D. \end{cases}$$

(ii) For any $\varphi \in C(\partial D)$, we have

$$\begin{cases} H_\alpha \varphi \in \mathcal{D}(\overline{A}), \\ (\alpha I - \overline{A}) H_\alpha \varphi = 0 \text{ in } D. \end{cases}$$

Here $\mathcal{D}(\overline{A})$ is the domain of the closed extension $\overline{A}$.

Proof. *Assertion (i)*: Choose a sequence $\{f_j\}$ in $C^\theta(\overline{D})$ such that $f_j \to f$ in $C(\overline{D})$ as $j \to \infty$. Then it follows from the boundedness of G_α^0 that

$$G_\alpha^0 f_j \longrightarrow G_\alpha^0 f \quad \text{in } C(\overline{D}) ,$$

and also

$$(\alpha - A)G_\alpha^0 f_j = f_j \longrightarrow f \quad \text{in } C(\overline{D}) .$$

Hence we have

$$\begin{cases} G_\alpha^0 f \in \mathcal{D}(\overline{A}), \\ (\alpha I - \overline{A})G_\alpha^0 f = f \quad \text{in } D . \end{cases}$$

since the operator $\overline{A} : C(\overline{D}) \to C(\overline{D})$ is closed.

Assertion (ii): Similarly, part (ii) is proved, since the space $C^{2+\theta}(\partial D)$ is dense in $C(\partial D)$ and the operator $H_\alpha : C(\partial D) \to C(\overline{D})$ is bounded. □

Corollary 13.7. *Every function u in $\mathcal{D}(\overline{A})$ can be written in the following form:*

$$u = G_\alpha^0 \left((\alpha I - \overline{A})u\right) + H_\alpha(u|_{\partial D}) , \quad \alpha > 0 . \tag{13.8}$$

Proof. We let

$$w = u - G_\alpha^0 \left((\alpha I - \overline{A})u\right) - H_\alpha(u|_{\partial D}) .$$

Then it follows from Lemma 13.6 that the function w is in $\mathcal{D}(\overline{A})$ and satisfies the conditions

$$\begin{cases} (\alpha I - \overline{A})w = 0 & \text{in } D, \\ w = 0 & \text{on } \partial D. \end{cases}$$

Thus, in view of Remark 13.1 we can apply Theorem 13.1 to obtain that

$$w = 0 .$$

This proves formula (13.8). □

(II) Secondly, we introduce a linear operator

$$LG_\alpha^0 : C(\overline{D}) \longrightarrow C(\partial D)$$

as follows.

(a) The domain $\mathcal{D}\left(LG_\alpha^0\right)$ of LG_α^0 is the space $C^\theta(\overline{D})$.
(b) $LG_\alpha^0 f = L\left(G_\alpha^0 f\right)$, $f \in \mathcal{D}\left(LG_\alpha^0\right)$.

Then we have the following:

Lemma 13.8. *The operator LG_α^0, $\alpha > 0$, can be uniquely extended to a non-negative, bounded linear operator $\overline{LG_\alpha^0} : C(\overline{D}) \to C(\partial D)$.*

Proof. Let f be an arbitrary function in $\mathcal{D}(LG_\alpha^0)$ such that $f \geq 0$ on $\overline{D}$. Then we have

$$\begin{cases} G_\alpha^0 f \in C^2(\overline{D}), & \\ G_\alpha^0 f \geq 0 & \text{on } \overline{D}, \\ G_\alpha^0 f = 0 & \text{on } \partial D, \end{cases}$$

and so

$$\begin{aligned} LG_\alpha^0 f(x') &= \mu(x') \frac{\partial}{\partial \mathbf{n}} (G_\alpha^0 f)(x') + \delta(x') f(x') + \int_D t(x', y)\, G_\alpha^0 f(y)\, dy \\ &\geq 0 \quad \text{on } \partial D\,. \end{aligned}$$

This proves that the operator LG_α^0 is non-negative.

By the non-negativity of LG_α^0, we have, for all $f \in \mathcal{D}(LG_\alpha^0)$,

$$-LG_\alpha^0 \|f\|_\infty \leq LG_\alpha^0 f \leq LG_\alpha^0 \|f\|_\infty \quad \text{on } \partial D\,.$$

This implies the boundedness of LG_α^0 with norm

$$\|LG_\alpha^0\| = \|LG_\alpha^0 1\|_\infty\,.$$

Recall that the space $C^\theta(\overline{D})$ is dense in $C(\overline{D})$ and that non-negative functions can be approximated by non-negative smooth functions (see Theorem 4.4). Hence we find that the operator LG_α^0 can be uniquely extended to a non-negative, bounded linear operator $\overline{LG_\alpha^0} : C(\overline{D}) \to C(\partial D)$. □

The next lemma states a fundamental relationship between the operators $\overline{LG_\alpha^0}$ and $\overline{LG_\beta^0}$ for α, $\beta > 0$:

Lemma 13.9. *For any $f \in C(\overline{D})$, we have*

$$\overline{LG_\alpha^0} f - \overline{LG_\beta^0} f + (\alpha - \beta) \overline{LG_\alpha^0}\, G_\beta^0 f = 0\,, \quad \alpha, \beta > 0\,. \tag{13.9}$$

Proof. Choose a sequence $\{f_j\}$ in $C^\theta(\overline{D})$ such that $f_j \to f$ in $C(\overline{D})$ as $j \to \infty$. Then, using the resolvent equation (13.4) with $f := f_j$ we have

$$LG_\alpha^0 f_j - LG_\beta^0 f_j + (\alpha - \beta) LG_\alpha^0 G_\beta^0 f_j = 0\,.$$

Hence formula (13.9) follows by letting $j \to \infty$, since the operators $\overline{LG_\alpha^0}$, $\overline{LG_\beta^0}$ and G_β^0 are all bounded. □

(III) Finally, we introduce a linear operator

$$LH_\alpha : C(\partial D) \longrightarrow C(\partial D)$$

as follows.

(a) The domain $\mathcal{D}(LH_\alpha)$ of LH_α is the space $C^{2+\theta}(\partial D)$.
(b) $LH_\alpha \psi = L(H_\alpha \psi)$, $\psi \in \mathcal{D}(LH_\alpha)$.

Then we have the following:

Lemma 13.10. *The operator LH_α, $\alpha > 0$, has its minimal closed extension $\overline{LH_\alpha}$ in the space $C(\partial D)$.*

Proof. We apply part (i) of Theorem 3.5 to the operator LH_α. To do this, it suffices to show that the operator LH_α satisfies condition (β') with $K := \partial D$ (or condition (β) with $K := K_0 = \partial D$) of the same theorem.

Assume that a function ψ in $\mathcal{D}(LH_\alpha) = C^{2+\theta}(\partial D)$ takes its positive maximum at some point x' of ∂D. Since the function $H_\alpha\psi$ is in $C^{2+\theta}(\overline{D})$ and satisfies the conditions

$$\begin{cases} (A-\alpha)H_\alpha\psi = 0 & \text{in } D, \\ H_\alpha\psi = \psi & \text{on } \partial D, \end{cases}$$

applying the weak maximum principle (Theorem C.1 in Appendix C) with $A := A - \alpha$ to the function $H_\alpha\psi$, we find that the function $H_\alpha\psi$ takes its positive maximum at a boundary point $x_0' \in \partial D$. Thus we can apply the boundary point lemma (Lemma C.5 in Appendix C) to obtain that

$$\frac{\partial}{\partial \mathbf{n}}(H_\alpha\psi)(x_0') < 0 .$$

Hence we have

$$\begin{aligned}
&LH_\alpha\psi(x_0') \\
&= \sum_{i,j=1}^{N-1} \alpha^{ij}(x_0')\frac{\partial^2\psi}{\partial x_i \partial x_j}(x_0') + \mu(x_0')\frac{\partial}{\partial \mathbf{n}}(H_\alpha\psi)(x_0') + \gamma(x_0')\psi(x_0') \\
&\quad - \alpha\delta(x_0')\psi(x_0') + \int_D t(x_0', y)\,[H_\alpha\psi(y) - \psi(x_0')\tau(x_0', y)]\,dy \\
&\quad + \int_{\partial D} r(x_0', y')\,[\psi(y') - \psi(x_0')\tau(x_0', y')]\,dy' \\
&= \sum_{i,j=1}^{N-1} \alpha^{ij}(x_0')\frac{\partial^2\psi}{\partial x_i \partial x_j}(x_0') + \mu(x_0')\frac{\partial}{\partial \mathbf{n}}(H_\alpha\psi)(x_0') + \gamma(x_0')\psi(x_0') \\
&\quad - \alpha\delta(x_0')\psi(x_0') + \int_D t(x_0', y)\,[H_\alpha\psi(y) - \psi(x_0')]\,dy \\
&\quad + \int_{\partial D} r(x_0', y')\,[\psi(y') - \psi(x_0')]\,dy' \\
&\quad + \left(\int_{\partial D} r(x_0', y')\,[1 - \tau(x_0', y')]\,dy' + \int_D t(x_0', y)\,[1 - \tau(x_0', y)]\,dy\right)\psi(x_0') \\
&\leq 0.
\end{aligned}$$

This verifies condition (β') of Theorem 3.5. Therefore, Lemma 13.10 follows from an application of the same theorem. □

Remark 13.2. The operator $\overline{LH_\alpha}$ enjoys the following property:

> If a function ψ in the domain $\mathcal{D}\left(\overline{LH_\alpha}\right)$ takes its *positive* maximum at some point x' of ∂D, then we have
> $$\overline{LH_\alpha}\psi(x') \leq 0\,. \tag{13.10}$$

The next lemma states a fundamental relationship between the operators $\overline{LH_\alpha}$ and $\overline{LH_\beta}$ for α, $\beta > 0$:

Lemma 13.11. *The domain $\mathcal{D}\left(\overline{LH_\alpha}\right)$ of $\overline{LH_\alpha}$ does not depend on $\alpha > 0$; so we denote by $\mathcal{D}$ the common domain. Then we have*

$$\overline{LH_\alpha}\psi - \overline{LH_\beta}\psi + (\alpha - \beta)\overline{LG_\alpha^0}H_\beta\psi = 0\,, \quad \alpha, \beta > 0\,, \quad \psi \in \mathcal{D}\,. \tag{13.11}$$

Proof. Let ψ be an arbitrary function in $\mathcal{D}(\overline{LH_\beta})$, and choose a sequence $\{\psi_j\}$ in $\mathcal{D}(LH_\beta) = C^{2+\theta}(\partial D)$ such that

$$\begin{cases} \psi_j \longrightarrow \psi & \text{in } C(\partial D), \\ LH_\beta\psi_j \longrightarrow \overline{LH_\beta}\psi & \text{in } C(\partial D). \end{cases}$$

Then it follows from the boundedness of H_β and $\overline{LG_\alpha^0}$ that

$$LG_\alpha^0(H_\beta\psi_j) = \overline{LG_\alpha^0}(H_\beta\psi_j) \longrightarrow \overline{LG_\alpha^0}(H_\beta\psi) \quad \text{in } C(\partial D)\,.$$

Therefore, using formula (13.6) with $\varphi := \psi_j$ we obtain that

$$\begin{aligned} LH_\alpha\psi_j &= LH_\beta\psi_j - (\alpha - \beta)LG_\alpha^0(H_\beta\psi_j) \\ &\longrightarrow \overline{LH_\beta}\psi - (\alpha - \beta)\overline{LG_\alpha^0}(H_\beta\psi) \quad \text{in } C(\partial D)\,. \end{aligned}$$

This implies that

$$\begin{cases} \psi \in \mathcal{D}(\overline{LH_\alpha}), \\ \overline{LH_\alpha}\psi = \overline{LH_\beta}\psi - (\alpha - \beta)\overline{LG_\alpha^0}(H_\beta\psi). \end{cases}$$

Conversely, interchanging α and β we have

$$\mathcal{D}(\overline{LH_\alpha}) \subset \mathcal{D}(\overline{LH_\beta})\,,$$

and so

$$\mathcal{D}(\overline{LH_\alpha}) = \mathcal{D}(\overline{LH_\beta})\,.$$

This proves the lemma. $\square$

Now we can give a general existence theorem for Feller semigroups on ∂D in terms of boundary value problem $(*)$. The next theorem tells us that the operator $\overline{LH_\alpha}$ is the infinitesimal generator of some Feller semigroup on ∂D if and only if problem $(*)$ is solvable for sufficiently *many* functions φ in the space $C(\partial D)$:

Theorem 13.12. *(i) If the operator* $\overline{LH_\alpha}$, $\alpha > 0$, *is the infinitesimal generator of a Feller semigroup on* ∂D, *then, for each constant* $\lambda > 0$, *the boundary value problem*

$$\begin{cases} (\alpha - A)u = 0 & \text{in } D, \\ (\lambda - L)u = \varphi & \text{on } \partial D \end{cases} \tag{$*'$}$$

has a solution $u \in C^{2+\theta}(\overline{D})$ *for any* φ *in some* dense *subset of* $C(\partial D)$.

(ii) Conversely, if, for some constant $\lambda \geq 0$, *problem* $(*')$ *has a solution* $u \in C^{2+\theta}(\overline{D})$ *for any* φ *in some* dense *subset of* $C(\partial D)$, *then the operator* $\overline{LH_\alpha}$ *is the infinitesimal generator of some Feller semigroup on* ∂D.

Proof. Assertion (i): If the operator $\overline{LH_\alpha}$ generates a Feller semigroup on ∂D, applying part (i) of Theorem 3.5 with $K := \partial D$ to the operator $\overline{LH_\alpha}$ we obtain that

$$\mathcal{R}\left(\lambda I - \overline{LH_\alpha}\right) = C(\partial D) \quad \text{for each } \lambda > 0\,.$$

This implies that the range $\mathcal{R}\left(\lambda I - LH_\alpha\right)$ is a dense subset of $C(\partial D)$ for each $\lambda > 0$. However, if $\varphi \in C(\partial D)$ is in the range $\mathcal{R}\left(\lambda I - LH_\alpha\right)$, and if $\varphi = (\lambda I - LH_\alpha)\psi$ with $\psi \in C^{2+\theta}(\partial D)$, then the function $u = H_\alpha\psi \in C^{2+\theta}(\overline{D})$ is a solution of problem $(*')$. This proves assertion (i).

Assertion (ii): We apply part (ii) of Theorem 3.5 with $K := \partial D$ to the operator LH_α. To do this, it suffices to show that the operator LH_α satisfies condition (γ) of the same theorem, since it satisfies condition (β'), as is shown in the proof of Lemma 13.10.

By the uniqueness theorem for problem (D), it follows that any function $u \in C^{2+\theta}(\overline{D})$ which satisfies the equation

$$(\alpha - A)u = 0 \quad \text{in } D$$

can be written in the form

$$u = H_\alpha\left(u|_{\partial D}\right), \quad u|_{\partial D} \in C^{2+\theta}(\partial D) = \mathcal{D}\left(LH_\alpha\right).$$

Thus we find that if there exists a solution $u \in C^{2+\theta}(\overline{D})$ of problem $(*')$ for a function $\varphi \in C(\partial D)$, then we have

$$(\lambda I - LH_\alpha)\left(u|_{\partial D}\right) = \varphi\,,$$

and so

$$\varphi \in \mathcal{R}\left(\lambda I - LH_\alpha\right).$$

Therefore, if there exists a constant $\lambda \geq 0$ such that problem $(*')$ has a solution u in $C^{2+\theta}(\overline{D})$ for any φ in some dense subset of $C(\partial D)$, then the range $\mathcal{R}\left(\lambda I - LH_\alpha\right)$ is dense in $C(\partial D)$. This verifies condition (γ) (with $\alpha_0 := \lambda$) of Theorem 3.5. Hence assertion (ii) follows from an application of the same theorem.

The proof of Theorem 13.12 is complete. □

We conclude this section by giving a precise meaning to the boundary conditions Lu for functions u in $\mathcal{D}(\overline{A})$.

We let

$$\mathcal{D}(L) = \left\{u \in \mathcal{D}(\overline{A}) : u|_{\partial D} \in \mathcal{D}\right\},$$

where $\mathcal{D}$ is the common domain of the operators $\overline{LH_\alpha}$, $\alpha > 0$. It should be noticed that the domain $\mathcal{D}(L)$ contains $C^{2+\theta}(\overline{D})$, since $C^{2+\theta}(\partial D) = \mathcal{D}\left(LH_\alpha\right) \subset \mathcal{D}$. Moreover, Corollary 13.7 tells us that every function u in $\mathcal{D}(L) \subset \mathcal{D}(\overline{A})$ can be written in the form

$$u = G_\alpha^0\left((\alpha I - \overline{A})u\right) + H_\alpha\left(u|_{\partial D}\right), \quad \alpha > 0. \tag{13.8}$$

Then we define

$$Lu = \overline{LG_\alpha^0}\left((\alpha I - \overline{A})u\right) + \overline{LH_\alpha}\left(u|_{\partial D}\right). \tag{13.12}$$

The next lemma justifies definition (13.12) of Lu for $u \in \mathcal{D}(L)$:

Lemma 13.13. *The right-hand side of formula (13.12) depends only on* u, not *on the choice of expression (13.8).*

Proof. Assume that

$$\begin{aligned} u &= G_\alpha^0\left(\left(\alpha I - \overline{A}\right)u\right) + H_\alpha\left(u|_{\partial D}\right) \\ &= G_\beta^0\left(\left(\beta I - \overline{A}\right)u\right) + H_\beta\left(u|_{\partial D}\right), \end{aligned}$$

where $\alpha > 0$, $\beta > 0$. Then it follows from formula (13.9) with $f := \left(\alpha I - \overline{A}\right)u$ and formula (13.12) with $\psi := u|_{\partial D}$ that

$$\begin{aligned} &\overline{LG_\alpha^0}\left(\left(\alpha I - \overline{A}\right)u\right) + \overline{LH_\alpha}\left(u|_{\partial D}\right) \\ &= \overline{LG_\beta^0}\left(\left(\alpha I - \overline{A}\right)u\right) - (\alpha - \beta)\overline{LG_\alpha^0}G_\beta^0\left(\left(\alpha I - \overline{A}\right)u\right) \\ &\quad + \overline{LH_\beta}\left(u|_{\partial D}\right) - (\alpha - \beta)\overline{LG_\alpha^0}H_\beta\left(u|_{\partial D}\right) \\ &= \overline{LG_\beta^0}\left((\beta I - A)u\right) + \overline{LH_\beta}\left(u|_{\partial D}\right) \\ &\quad + (\alpha - \beta)\left\{\overline{LG_\beta^0}u - \overline{LG_\alpha^0}G_\beta^0\left(\alpha I - \overline{A}\right)u - \overline{LG_\alpha^0}H_\beta\left(u|_{\partial D}\right)\right\}. \end{aligned} \tag{13.13}$$

However, the last term of formula (13.13) vanishes. Indeed, it follows from formula (13.9) with $f := u$ that

$$\begin{aligned} &\overline{LG_\beta^0}u - \overline{LG_\alpha^0}\left(G_\beta^0\left(\alpha I - \overline{A}\right)u\right) - \overline{LG_\alpha^0}H_\beta\left(u|_{\partial D}\right) \\ &= \overline{LG_\beta^0}u - \overline{LG_\alpha^0}\left(G_\beta^0\left(\beta I - \overline{A}\right)u + H_\beta\left(u|_{\partial D}\right) + (\alpha - \beta)G_\beta^0 u\right) \\ &= \overline{LG_\beta^0}u - \overline{LG_\alpha^0}u - (\alpha - \beta)\overline{LG_\alpha^0}G_\beta^0 u \\ &= 0. \end{aligned}$$

Therefore, we obtain from formula (13.13) that

$$\overline{LG_\alpha^0}\left(\left(\alpha I - \overline{A}\right)u\right) + \overline{LH_\alpha}\left(u|_{\partial D}\right) = \overline{LG_\beta^0}\left(\left(\beta I - \overline{A}\right)u\right) + \overline{LH_\beta}\left(u|_{\partial D}\right).$$

This proves Lemma 13.13. □

13.2 Feller Semigroups with Reflecting Barrier

Now we consider the Neumann condition

$$L_N u = \frac{\partial u}{\partial \mathbf{n}} .$$

We recall that the boundary condition L_N is supposed to correspond to the *reflection phenomenon.*

The next theorem (formula (13.15)) asserts that we can "piece together" a Markov process on the boundary ∂D with A-diffusion in D to construct a Markov process on the whole $\overline{D} = D \cup \partial D$ with reflecting barrier (see Bony–Courrège–Priouret [BCP, Théorème XIX]):

Theorem 13.14. *We define a linear operator*

$$\mathfrak{A}_N : C(\overline{D}) \longrightarrow C(\overline{D})$$

as follows.

(a) The domain $\mathcal{D}(\mathfrak{A}_N)$ of $\mathfrak{A}_N$ is the space

$$\mathcal{D}(\mathfrak{A}_N) = \left\{ u \in \mathcal{D}(\overline{A}) : u|_{\partial D} \in \mathcal{D}_N \, , \ \ L_N u = 0 \ \ on \ \partial D \right\} , \tag{13.14}$$

where $\mathcal{D}_N$ is the common domain of the operators $\overline{L_N H_\alpha}$, $\alpha > 0$.
(b) $\mathfrak{A}_N u = \overline{A} u$, $u \in \mathcal{D}(\mathfrak{A}_N)$.

Then the operator $\mathfrak{A}_N$ is the infinitesimal generator of some Feller semigroup $\{e^{t\mathfrak{A}_N}\}_{t\geq 0}$ on $\overline{D}$, and the Green operator $G_\alpha^N = (\alpha I - \mathfrak{A}_N)^{-1}$, $\alpha > 0$, is given by the following formula:

$$G_\alpha^N f = G_\alpha^0 f - H_\alpha \left(\overline{L_N H_\alpha}^{-1} \left(\overline{L_N G_\alpha^0 f} \right) \right) , \quad f \in C(\overline{D}) . \tag{13.15}$$

Proof. We apply part (ii) of Theorem 3.3 to the operator $\mathfrak{A}_N$ defined by formula (13.14). The proof is divided into several steps.

Step 1: First, we prove that

The operator $\overline{L_N H_\alpha}$ is the generator of some Feller semigroup on ∂D, for any $\alpha > 0$ sufficiently large.

We recall that the linear operator

$$L_N H_\alpha : C(\partial D) \longrightarrow C(\partial D)$$

is defined as follows.

(a) The domain $\mathcal{D}(L_N H_\alpha)$ of $L_N H_\alpha$ is the space $C^{2+\theta}(\partial D)$.
(b) $L_N H_\alpha \varphi = L_N (H_\alpha \varphi)$, $\varphi \in \mathcal{D}(L_N H_\alpha)$.

It is worth while pointing out here that the operators H_α and $L_N H_\alpha$ are respectively the operators $P(\alpha)$ and $\Pi(\alpha)$ introduced in Sect. 8.3.

Then the proof of Theorem 11.1 with $\mu(x') \equiv 1$ and $\gamma(x') \equiv 0$ tells us that the operator $L_N H_\alpha$ maps the space $C^\infty(\partial D)$ onto itself for any sufficiently large $\alpha > 0$. This implies that the range $\mathcal{R}(L_N H_\alpha)$ is a *dense* subset of $C(\partial D)$. Hence, applying part (ii) of Theorem 13.12 we obtain that the operator $\overline{L_N H_\alpha}$ generates a Feller semigroup on ∂D, for any $\alpha > 0$ sufficiently large.

Step 2: Next we prove that

The operator $\overline{L_N H_\beta}$ generates a Feller semigroup on ∂D, for any $\beta > 0$.

Take a constant $\alpha > 0$ so large that the operator $\overline{L_N H_\alpha}$ generates a Feller semigroup on ∂D. We apply Corollary 3.6 with $K := \partial D$ to the operator $\overline{L_N H_\beta}$, $\beta > 0$. By formula (13.11), it follows that the operator $\overline{L_N H_\beta}$ can be written as

$$\overline{L_N H_\beta} = \overline{L_N H_\alpha} + N_{\alpha,\beta} \,,$$

where $N_{\alpha,\beta} = (\alpha - \beta)\overline{L_N G_\alpha^0}\, H_\beta$ is a bounded linear operator on $C(\partial D)$ into itself. Furthermore, assertion (13.10) implies that the operator $\overline{L_N H_\beta}$ satisfies condition (β') of Theorem 3.5. Therefore, it follows from an application of Corollary 3.6 that the operator $\overline{L_N H_\beta}$ also generates a Feller semigroup on ∂D.

Step 3: Now we prove that

The equation

$$\overline{L_N H_\alpha}\, \psi = \varphi$$

has a unique solution ψ in $\mathcal{D}(\overline{L_N H_\alpha})$ for any $\varphi \in C(\partial D)$; hence the inverse $\overline{L_N H_\alpha}^{-1}$ of $\overline{L_N H_\alpha}$ can be defined on the whole space $C(\partial D)$.

Furthermore, the operator $-\overline{L_N H_\alpha}^{-1}$ is non-negative and bounded on $C(\partial D)$. (13.16)

Since the function $H_\alpha 1$ takes its positive maximum 1 only on the boundary ∂D, we can apply the boundary point lemma (see Lemma C.5 in Appendix C) to obtain that

$$\frac{\partial}{\partial \mathbf{n}}(H_\alpha 1) < 0 \quad \text{on } \partial D \,. \tag{13.17}$$

Hence the Neumann condition implies that

$$L_N H_\alpha 1 = \frac{\partial}{\partial \mathbf{n}}(H_\alpha 1) < 0 \quad \text{on } \partial D \,,$$

and so

$$\ell_\alpha = - \sup_{x' \in \partial D} L_N H_\alpha 1(x') > 0 \,.$$

Furthermore, using Corollary 3.4 with

$$K := \partial D\,, \quad A := \overline{L_N H_\alpha}\,, \quad c := \ell_\alpha\,,$$

we obtain that the operator $\overline{L_N H_\alpha} + \ell_\alpha I$ is the infinitesimal generator of some Feller semigroup on ∂D. Therefore, since $\ell_\alpha > 0$, it follows from an application of part (i) of Theorem 3.3 with $A := \overline{L_N H_\alpha} + \ell_\alpha I$ that the equation

$$-\overline{L_N H_\alpha}\psi = \left(\ell_\alpha I - \left(\overline{L_N H_\alpha} + \ell_\alpha I\right)\right)\psi = \varphi$$

has a unique solution $\psi \in \mathcal{D}(\overline{L_N H_\alpha})$ for any $\varphi \in C(\partial D)$, and further that the operator $-\overline{L_N H_\alpha}^{-1} = \left(\ell_\alpha I - \left(\overline{L_N H_\alpha} + \ell_\alpha I\right)\right)^{-1}$ is non-negative and bounded on the space $C(\partial D)$ with norm

$$\left\|-\overline{L_N H_\alpha}^{-1}\right\|_\infty = \left\|\left(\ell_\alpha I - \left(\overline{L_N H_\alpha} + \ell_\alpha I\right)\right)^{-1}\right\| \leq \frac{1}{\ell_\alpha}\,.$$

Step 4: By assertion (13.16), we can define the right-hand side of formula (13.15) for all $\alpha > 0$. We prove that

$$G_\alpha^N = (\alpha I - \mathfrak{A}_N)^{-1}\,, \quad \alpha > 0\,. \tag{13.18}$$

In view of Lemmas 13.6 and 13.11, it follows that we have, for any $f \in C(\overline{D})$,

$$\begin{cases} G_\alpha^N f = G_\alpha^0 f - H_\alpha\left(\overline{L_N H_\alpha}^{-1}\left(\overline{L_N G_\alpha^0 f}\right)\right) \in \mathcal{D}(\overline{A}), \\ G_\alpha^N f|_{\partial D} = -\overline{L_N H_\alpha}^{-1}\left(\overline{L_N G_\alpha^0 f}\right) \in \mathcal{D}\left(\overline{L_N H_\alpha}\right) = \mathcal{D}_N, \\ L_N G_\alpha^N f = \overline{L_N G_\alpha^0 f} - \overline{L_N H_\alpha}\left(\overline{L_N H_\alpha}^{-1}\left(\overline{L_N G_\alpha^0 f}\right)\right) = 0, \end{cases}$$

and that

$$(\alpha I - \overline{A})G_\alpha^N f = f\,.$$

This proves that

$$\begin{cases} G_\alpha^N f \in \mathcal{D}(\mathfrak{A}_N)\,, \\ (\alpha I - \mathfrak{A}_N)\, G_\alpha^N f = f, \end{cases}$$

that is,

$$(\alpha I - \mathfrak{A}_N)\, G_\alpha^N = I \quad \text{on } C(\overline{D})\,.$$

Therefore, to prove formula (13.18) it suffices to show the injectivity of the operator $\alpha I - \mathfrak{A}_N$ for $\alpha > 0$.

Assume that

$$u \in \mathcal{D}(\mathfrak{A}_N) \quad \textit{and} \quad (\alpha I - \mathfrak{A}_N)\, u = 0\,.$$

Then, by Corollary 13.7 the function u can be written as

$$u = H_\alpha\left(u|_{\partial D}\right), \quad u|_{\partial D} \in \mathcal{D}_N = \mathcal{D}\left(\overline{L_N H_\alpha}\right).$$

Thus we have

$$\overline{L_N H_\alpha}\,(u|_{\partial D}) = L_N u = 0\,.$$

In view of assertion (13.16), this implies that

$$u|_{\partial D} = 0\,,$$

so that

$$u = H_\alpha\,(u|_{\partial D}) = 0 \quad \text{in } D\,.$$

Step 5: The non-negativity of G_α^N ($\alpha > 0$) follows immediately from formula (13.15), since the operators G_α^0, H_α, $-\overline{L_N H_\alpha}^{-1}$ and $\overline{L_N G_\alpha^0}$ are all non-negative.

Step 6: We prove that the operator G_α^N is bounded on the space $C(\overline{D})$ with norm

$$\left\|G_\alpha^N\right\| \leq \frac{1}{\alpha}\,, \quad \alpha > 0\,. \tag{13.19}$$

To do this, it suffices to show that

$$G_\alpha^N 1 \leq \frac{1}{\alpha} \quad \text{on } \overline{D}\,. \tag{13.19$'$}$$

since G_α^N is non-negative on $C(\overline{D})$.

First, it follows from the uniqueness property of solutions of problem (D′) that

$$\alpha G_\alpha^0 1 + H_\alpha 1 = 1 + G_\alpha^0 c \quad \text{on } \overline{D}\,, \tag{13.20}$$

where $c(x)$ is the termination coefficient of A. Indeed, the both sides have the same boundary value 1 and satisfy the same equation: $(\alpha - A)u = \alpha$ in D.

Applying the operator L_N to the both hand sides of equality (13.20), we obtain that

$$\begin{aligned} -L_N H_\alpha 1 &= -L_N 1 - L_N G_\alpha^0 c + \alpha L_N G_\alpha^0 1 \\ &= -\frac{\partial}{\partial \mathbf{n}}\left(G_\alpha^0 c\right) + \alpha L_N G_\alpha^0 1 \\ &\geq \alpha L_N G_\alpha^0 1 \quad \text{on } \partial D\,, \end{aligned}$$

since $G_\alpha^0 c|_{\partial D} = 0$ and $G_\alpha^0 c \leq 0$ on $\overline{D}$. Hence we have, by the non-negativity of $-\overline{L_N H_\alpha}^{-1}$,

$$-\overline{L_N H_\alpha}^{-1}\left(L_N G_\alpha^0 1\right) \leq \frac{1}{\alpha} \quad \text{on } \partial D\,. \tag{13.21}$$

Using formula (13.15) with $f := 1$, inequality (13.21) and equality (13.20), we obtain that

$$\begin{aligned} G_\alpha^N 1 &= G_\alpha^0 1 + H_\alpha\left(-\overline{L_N H_\alpha}^{-1}\left(L_N G_\alpha^0 1\right)\right) \\ &\leq G_\alpha^0 1 + \frac{1}{\alpha} H_\alpha 1 \\ &= \frac{1}{\alpha} + \frac{1}{\alpha} G_\alpha^0 c \\ &\leq \frac{1}{\alpha} \quad \text{on } \overline{D}\,, \end{aligned}$$

since the operators H_α and G_α^0 are both non-negative.

Step 7: Finally, we prove that

$$\textit{The domain } \mathcal{D}(\mathfrak{A}_N) \textit{ is } \text{dense} \textit{ in the space } C(\overline{D}) . \tag{13.22}$$

(7-1) Before the proof, we need some lemmas on the behavior of G_α^0, H_α and $\overline{-L_N H_\alpha}^{-1}$ as $\alpha \to +\infty$:

Lemma 13.15. *For all $f \in C(\overline{D})$, we have*

$$\lim_{\alpha \to +\infty} \left[\alpha G_\alpha^0 f + H_\alpha \left(f|_{\partial D}\right)\right] = f \quad \textit{in } C(\overline{D}) . \tag{13.23}$$

Proof. Choose a constant $\beta > 0$ and let

$$g = f - H_\beta(f|_{\partial D}) .$$

Then, using formula (13.6) with $\varphi := f|_{\partial D}$ we obtain that

$$\alpha G_\alpha^0 g - g = \left[\alpha G_\alpha^0 f + H_\alpha(f|_{\partial D}) - f\right] - \beta G_\alpha^0 H_\beta(f|_{\partial D}) . \tag{13.24}$$

However, we have, by estimate (13.3),

$$\lim_{\alpha \to +\infty} G_\alpha^0 H_\beta(f|_{\partial D}) = 0 \quad \text{in } C(\overline{D}) ,$$

and by assertion (13.5′)

$$\lim_{\alpha \to +\infty} \alpha G_\alpha^0 g = g \quad \text{in } C(\overline{D}) ,$$

since $g|_{\partial D} = 0$. Therefore, formula (13.23) follows by letting $\alpha \to +\infty$ in formula (13.24). □

Lemma 13.16. *The function*

$$\frac{\partial}{\partial \mathbf{n}} (H_\alpha 1)$$

diverges to $-\infty$ uniformly and monotonically as $\alpha \to +\infty$.

Proof. First, formula (13.6) with $\varphi := 1$ gives that

$$H_\alpha 1 = H_\beta 1 - (\alpha - \beta) G_\alpha^0 H_\beta 1 .$$

Thus, in view of the non-negativity of G_α^0 and H_α it follows that

$$\alpha \geq \beta \implies H_\alpha 1 \leq H_\beta 1 \quad \text{on } \overline{D} .$$

Since $H_\alpha 1|_{\partial D} = H_\beta 1|_{\partial D} = 1$, this implies that the functions

$$\frac{\partial}{\partial \mathbf{n}} (H_\alpha 1)$$

are monotonically non-increasing in α. Furthermore, using formula (13.5) with $f := H_\beta 1$ we find that the function

$$H_\alpha 1(x) = H_\beta 1(x) - \left(1 - \frac{\beta}{\alpha}\right) \alpha G_\alpha^0 H_\beta 1(x)$$

converges to zero monotonically as $\alpha \to +\infty$, for each interior point x of D.

Now, for any given constant $K > 0$ we can construct a function $u \in C^2(\overline{D})$ such that

$$u = 1 \quad \text{on } \partial D \,, \tag{13.25a}$$

$$\frac{\partial u}{\partial \mathbf{n}} \leq -K \quad \text{on } \partial D \,. \tag{13.25b}$$

Indeed it follows from Theorem 13.1 that, for any integer $m > 0$, the function

$$u = (H_{\alpha_0} 1)^m \,, \quad \alpha_0 > 0 \,,$$

belongs to $C^{2+\theta}(\overline{D})$ and satisfies condition (13.25a). Further we have

$$\begin{aligned} \frac{\partial u}{\partial \mathbf{n}} &= m \frac{\partial}{\partial \mathbf{n}} (H_{\alpha_0} 1) \\ &\leq m \sup_{x' \in \partial D} \frac{\partial}{\partial \mathbf{n}} (H_{\alpha_0} 1)(x') \,. \end{aligned}$$

In view of inequality (13.17), this implies that the function $u = (H_{\alpha_0} 1)^m$ satisfies condition (13.25b) for m sufficiently large.

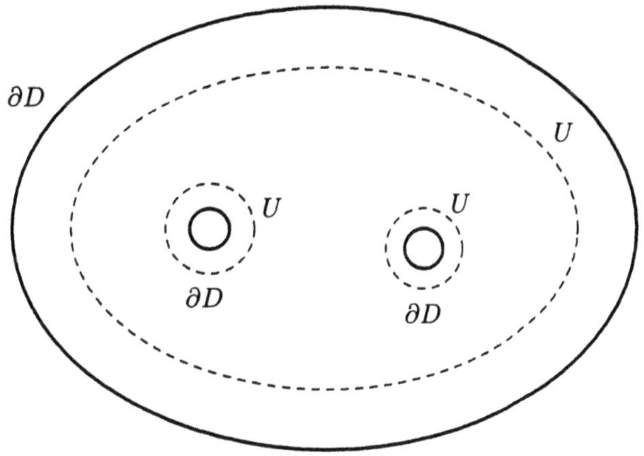

Fig. 13.1.

Take a function u in $C^2(\overline{D})$ satisfying conditions (13.25a) and (13.25b), and choose a neighborhood U of ∂D, relative to $\overline{D}$, with smooth boundary ∂U such that (see Fig. 13.1)

$$u \geq \frac{1}{2} \quad \text{on } U \,. \tag{13.26}$$

Recall that the function $H_\alpha 1$ converges to zero in D monotonically as $\alpha \to +\infty$. Since $u|_{\partial D} = H_\alpha 1|_{\partial D} = 1$, by using Dini's theorem we can find a constant $\alpha > 0$ (depending on u and hence on K) such that

$$H_\alpha 1 \leq u \quad \text{on } \partial U \setminus \partial D \,, \tag{13.27a}$$

$$\alpha > 2\|Au\|_\infty \,. \tag{13.27b}$$

It follows from inequalities (13.26) and (13.27b) that

$$\begin{aligned}(A-\alpha)(H_\alpha 1 - u) &= \alpha u - Au \\ &\geq \frac{\alpha}{2} - \|Au\|_\infty \\ &> 0 \quad \text{in } U \,.\end{aligned}$$

Thus, applying the weak maximum principle (Theorem C.1 in Appendix C) with $A := A - \alpha$ to the function $H_\alpha 1 - u$ we obtain that the function $H_\alpha 1 - u$ may take its positive maximum only on the boundary ∂U. However, conditions (13.25a) and (13.27a) imply that

$$H_\alpha 1 - u \leq 0 \quad \text{on } \partial U = (\partial U \setminus \partial D) \cup \partial D \,.$$

Therefore, we have

$$H_\alpha 1 \leq u \quad \text{on } \overline{U} = U \cup \partial U \,,$$

and hence

$$\frac{\partial}{\partial \mathbf{n}}(H_\alpha 1) \leq \frac{\partial u}{\partial \mathbf{n}} \leq -K \quad \text{on } \partial D \,,$$

since $u|_{\partial D} = H_\alpha 1|_{\partial D} = 1$.

The proof of Lemma 13.16 is complete. □

Corollary 13.17. $\lim_{\alpha \to +\infty} \| -\overline{L_N H_\alpha}^{-1} \| = 0$.

Proof. By Lemma 13.16, it follows that the function

$$L_N H_\alpha 1(x') = \frac{\partial}{\partial \mathbf{n}} (H_\alpha 1)(x') \,, \quad x' \in \partial D \,,$$

diverges to $-\infty$ monotonically as $\alpha \to +\infty$. By Dini's theorem, this convergence is uniform in $x' \in \partial D$. Hence the function

$$\frac{1}{L_N H_\alpha 1(x')}$$

converges to zero uniformly in $x' \in \partial D$ as $\alpha \to +\infty$. This gives that

$$\begin{aligned}\left\| -\overline{L_N H_\alpha}^{-1} \right\| &= \left\| -\overline{L_N H_\alpha}^{-1} 1 \right\|_\infty \\ &\leq \left\| \frac{1}{L_N H_\alpha 1} \right\|_\infty \longrightarrow 0 \quad \text{as} \quad \alpha \to +\infty \,,\end{aligned}$$

since we have

$$1 = \frac{-L_N H_\alpha 1(x')}{|L_N H_\alpha 1(x')|} \leq \left\| \frac{1}{L_N H_\alpha 1} \right\|_\infty (-L_N H_\alpha 1(x')), \quad x' \in \partial D. \quad \square$$

(7-2) **Proof of assertion (13.22)**
In view of formula (13.18) and inequality (13.19), it suffices to prove that

$$\lim_{\alpha \to +\infty} \|\alpha G_\alpha^N f - f\|_\infty = 0, \quad f \in C^\infty(\overline{D}), \tag{13.28}$$

since the space $C^\infty(\overline{D})$ is dense in $C(\overline{D})$.

First, it should be noticed that

$$\begin{aligned}
\|\alpha G_\alpha^N f - f\|_\infty &= \left\| \alpha G_\alpha^0 f - \alpha H_\alpha \left(\overline{L_N H_\alpha}^{-1} (L_N G_\alpha^0 f) \right) - f \right\|_\infty \\
&\leq \left\| \alpha G_\alpha^0 f + H_\alpha (f|_{\partial D}) - f \right\|_\infty \\
&\quad + \left\| -\alpha H_\alpha \left(\overline{L_N H_\alpha}^{-1} (L_N G_\alpha^0 f) \right) - H_\alpha (f|_{\partial D}) \right\|_\infty \\
&\leq \left\| \alpha G_\alpha^0 f + H_\alpha (f|_{\partial D}) - f \right\|_\infty \\
&\quad + \left\| -\alpha \overline{L_N H_\alpha}^{-1} (L_N G_\alpha^0 f) - f|_{\partial D} \right\|_\infty .
\end{aligned}$$

Thus, in view of formula (13.23) it suffices to show that

$$\lim_{\alpha \to +\infty} \left[-\alpha \overline{L_N H_\alpha}^{-1} (L_N G_\alpha^0 f) - f|_{\partial D} \right] = 0 \quad \text{in } C(\partial D). \tag{13.29}$$

Take a constant β such that $0 < \beta < \alpha$, and write

$$f = G_\beta^0 g + H_\beta \varphi ,$$

where (cf. formula (13.8)):

$$\begin{cases} g = (\beta - A) f \in C^\theta(\overline{D}), \\ \varphi = f|_{\partial D} \in C^{2+\theta}(\partial D). \end{cases}$$

Then, using equations (13.4) (with $f := g$) and (13.6), we obtain that

$$\begin{aligned}
G_\alpha^0 f &= G_\alpha^0 G_\beta^0 g + G_\alpha^0 H_\beta \varphi \\
&= \frac{1}{\alpha - \beta} \left(G_\beta^0 g - G_\alpha^0 g + H_\beta \varphi - H_\alpha \varphi \right).
\end{aligned}$$

Hence we have

$$\begin{aligned}
&\left\| -\alpha \overline{L_N H_\alpha}^{-1} (L_N G_\alpha^0 f) - f|_{\partial D} \right\|_\infty \\
&= \left\| \frac{\alpha}{\alpha - \beta} \left(-\overline{L_N H_\alpha}^{-1} \right) \left(L_N G_\beta^0 g - L_N G_\alpha^0 g + L_N H_\beta \varphi \right) + \frac{\alpha}{\alpha - \beta} \varphi - \varphi \right\|_\infty \\
&\leq \frac{\alpha}{\alpha - \beta} \left\| -\overline{L_N H_\alpha}^{-1} \right\| \cdot \left\| L_N G_\beta^0 g + L_N H_\beta \varphi \right\|_\infty \\
&\quad + \frac{\alpha}{\alpha - \beta} \left\| -\overline{L_N H_\alpha}^{-1} \right\| \cdot \left\| L_N G_\alpha^0 \right\| \cdot \|g\|_\infty + \frac{\beta}{\alpha - \beta} \|\varphi\|_\infty .
\end{aligned}$$

By Corollary 13.17, it follows that the first term on the last inequality converges to zero as $\alpha \to +\infty$. For the second term, using formula (13.4) with $f := 1$ and the non-negativity of G_β^0 and $L_N G_\alpha^0$ we find that

$$\begin{aligned}\|L_N G_\alpha^0\| &= \|L_N G_\alpha^0 1\|_\infty \\ &= \|L_N G_\beta^0 1 - (\alpha - \beta) L_N G_\alpha^0 G_\beta^0 1\|_\infty \\ &\leq \|L_N G_\beta^0 1\|_\infty .\end{aligned}$$

Hence the second term also converges to zero as $\alpha \to +\infty$. It is clear that the third term converges to zero as $\alpha \to +\infty$. This completes the proof of assertion (13.29) and hence of assertion (13.28).

Step 8: Summing up, we have proved that the operator $\mathfrak{A}_N$, defined by formula (13.14), satisfies conditions (a) through (d) in Theorem 3.3. Hence it follows from an application of the same theorem that the operator $\mathfrak{A}_N$ is the infinitesimal generator of some Feller semigroup $\{e^{t\mathfrak{A}_N}\}_{t\geq 0}$ on $\overline{D}$.

The proof of Theorem 13.14 is now complete. □

13.3 Proof of Theorem 1.5

Our proof of Theorem 1.5 is essentially the same as that of Theorem 7.21 in Sect. 7.3. We apply part (ii) of Theorem 3.3 to the operator $\mathfrak{A}$ defined by formula (1.2).

First, we simplify the boundary condition

$$L_0 u = \mu(x') \frac{\partial u}{\partial \mathbf{n}} + \gamma(x') u = 0 \quad \text{on } \partial D .$$

Assume that conditions (H.1) and (H.2) are satisfied:

(H.1) $\mu(x') \geq 0$ and $\gamma(x') \leq 0$ on ∂D.
(H.2) $\gamma(x') < 0$ on $M = \{x' \in \partial D : \mu(x') = 0\}$.

Then we find that the boundary condition

$$\mu(x') \frac{\partial u}{\partial \mathbf{n}} + \gamma(x') u = 0 \quad \text{on } \partial D$$

is equivalent to the condition

$$\left(\frac{\mu(x')}{\mu(x') - \gamma(x')} \right) \frac{\partial u}{\partial \mathbf{n}} + \left(\frac{\gamma(x')}{\mu(x') - \gamma(x')} \right) u = 0 \quad \text{on } \partial D .$$

Hence, if we let

$$\tilde{\mu}(x') = \frac{\mu(x')}{\mu(x') - \gamma(x')}, \quad \tilde{\gamma}(x') = \frac{\gamma(x')}{\mu(x') - \gamma(x')} ,$$

then we have

$$\tilde{\mu}(x')\frac{\partial u}{\partial \mathbf{n}} + \tilde{\gamma}(x')u = 0 \quad \text{on } \partial D\,,$$

with

$$0 \leq \tilde{\mu}(x') \leq 1 \quad \text{on } \partial D\,,$$
$$\tilde{\gamma}(x') = \tilde{\mu}(x') - 1 \quad \text{on } \partial D\,.$$

Namely, we may assume that the boundary condition L_0 is of the form

$$L_0 u = \mu(x')\frac{\partial u}{\partial \mathbf{n}} + (\mu(x') - 1)u = 0 \quad \text{on } \partial D\,,$$

with

$$0 \leq \mu(x') \leq 1 \quad \text{on } \partial D\,.$$

Next, we express the boundary condition L_0 in terms of the Dirichlet and Neumann conditions.

It should be noticed that

$$\overline{L_0 G_\alpha^0} = \mu(x')\overline{L_N G_\alpha^0}\,,$$

and that

$$\overline{L_0 H_\alpha} = \mu(x')\overline{L_N H_\alpha} + \mu(x') - 1\,.$$

Hence, in view of definition (13.12) it follows that

$$\begin{aligned} L_0 u &= \overline{L_0 G_\alpha^0}\left((\alpha I - \overline{A})u\right) + \overline{L_0 H_\alpha}\,(u|_{\partial D}) \\ &= \mu(x')\overline{L_N G_\alpha^0}\left((\alpha I - \overline{A})u\right) + \mu(x')\overline{L_N H_\alpha}\,(u|_{\partial D}) + (\mu(x') - 1)\,(u|_{\partial D}) \\ &= \mu(x')\left(\overline{L_N G_\alpha^0}\left((\alpha I - \overline{A})u\right) + \overline{L_N H_\alpha}\,(u|_{\partial D})\right) + (\mu(x') - 1)\,(u|_{\partial D}) \\ &= \mu(x')L_N u + (\mu(x') - 1)\,(u|_{\partial D})\,, \end{aligned}$$

so that

$$L_0 = \mu(x')L_N + \mu(x') - 1\,. \tag{13.30}$$

Therefore, the next theorem proves Theorem 1.5:

Theorem 13.18. *We define a linear operator*

$$\mathfrak{A} : C_0(\overline{D} \setminus M) \longrightarrow C_0(\overline{D} \setminus M)$$

as follows (cf. formula (1.3)).

(a) The domain $\mathcal{D}(\mathfrak{A})$ of $\mathfrak{A}$ is the space

$$\begin{aligned} \mathcal{D}(\mathfrak{A}) = \{u \in C_0(\overline{D} \setminus M) : \overline{A}u \in C_0(\overline{D} \setminus M)\,, \\ L_0 u = \mu(x')L_N u + (\mu(x') - 1)u = 0 \ \text{on } \partial D\}\,. \end{aligned}$$

(b) $\mathfrak{A}u = \overline{A}u$, $u \in \mathcal{D}(\mathfrak{A})$.

Here recall that

$$C_0(\overline{D} \setminus M) = \{u \in C(\overline{D}) : u = 0 \ \text{ on } M\} .$$

Assume that the following condition (H′) is satisfied:

(H′) $0 \leq \mu(x') \leq 1$ *on* ∂D.

Then the operator $\mathfrak{A}$ *is the infinitesimal generator of some Feller semigroup* $\{e^{t\mathfrak{A}}\}_{t \geq 0}$ *on* $\overline{D} \setminus M$*, and the Green operator* $G_\alpha = (\alpha I - \mathfrak{A})^{-1}$*,* $\alpha > 0$*, is given by the following formula:*

$$G_\alpha f = G_\alpha^N f - H_\alpha \left(\overline{L_0 H_\alpha}^{-1} \left(L_0 G_\alpha^N f \right) \right), \quad f \in C_0(\overline{D} \setminus M) . \tag{13.31}$$

Here G_α^N *is the Green operator for the Neumann condition* L_N *given by formula (13.15).*

Proof. We apply part (ii) of Theorem 3.3 to the operator $\mathfrak{A}$. The proof is divided into several steps.

Step 1: First, we prove that

> *If condition (H′) is satisfied, then the operator* $\overline{L_0 H_\alpha}$ *is the generator of some Feller semigroup on* ∂D*, for any sufficiently large* $\alpha > 0$.

We recall that the linear operator

$$L_0 H_\alpha : C(\partial D) \longrightarrow C(\partial D)$$

is defined as follows.

(a) The domain $\mathcal{D}(L_0 H_\alpha)$ of $L_0 H_\alpha$ is the space $C^{2+\theta}(\partial D)$.
(b) $L_0 H_\alpha \varphi = L_0 (H_\alpha \varphi)$, $\varphi \in \mathcal{D}(L_0 H_\alpha)$.

Then the proof of Theorem 11.1 with $\gamma(x') := \mu(x') - 1$ tells us that the operator $L_0 H_\alpha$ maps the space $C^\infty(\partial D)$ onto itself for any sufficiently large $\alpha > 0$. This implies that the range $\mathcal{R}(L_0 H_\alpha)$ is a *dense* subset of $C(\partial D)$. Hence, applying part (ii) of Theorem 13.14, we obtain that the operator $\overline{L_0 H_\alpha}$ generates a Feller semigroup on ∂D, for any $\alpha > 0$ sufficiently large.

Step 2: Next we prove that

> *The operator* $\overline{L_0 H_\beta}$ *generates a Feller semigroup on* ∂D*,*
> *for any* $\beta > 0$.

Take a constant $\alpha > 0$ so large that the operator $\overline{L_0 H_\alpha}$ generates a Feller semigroup on ∂D. We apply Corollary 3.6 with $K := \partial D$ to the operator $\overline{L_0 H_\beta}$ for $\beta > 0$. By formula (13.11), it follows that the operator $\overline{L_0 H_\beta}$ can be written as

$$\overline{L_0 H_\beta} = \overline{L_0 H_\alpha} + M_{\alpha\beta} ,$$

where $M_{\alpha\beta} = (\alpha - \beta)\overline{L_0 G_\alpha^0}\, H_\beta$ is a bounded linear operator on $C(\partial D)$ into itself. Furthermore, assertion (13.10) implies that the operator $\overline{L_0 H_\beta}$ satisfies condition (β') of Theorem 3.5. Therefore, it follows from an application of Corollary 3.6 that the operator $\overline{L_0 H_\beta}$ also generates a Feller semigroup on ∂D.

Step 3: Now we prove that

$$\begin{aligned}&\textit{If condition (H') is satisfied, then the equation}\\ &\quad \overline{L_0 H_\alpha}\psi = \varphi\\ &\textit{has a unique solution } \psi \textit{ in } \mathcal{D}\left(\overline{L_0 H_\alpha}\right) \textit{ for any } \varphi \in C(\partial D);\\ &\textit{hence the inverse } \overline{L_0 H_\alpha}^{-1} \textit{ of } \overline{L_0 H_\alpha} \textit{ can be defined on}\\ &\textit{the whole space } C(\partial D)\,.\\ &\textit{Furthermore, the operator } -\overline{L_0 H_\alpha}^{-1} \textit{ is non-negative}\\ &\textit{and bounded on } C(\partial D)\,.\end{aligned} \tag{13.32}$$

Since we have, by inequality (13.17),

$$L_0 H_\alpha 1 = \mu(x') \frac{\partial}{\partial \mathbf{n}} (H_\alpha 1) + \mu(x') - 1 < 0 \quad \text{on } \partial D\,,$$

it follows that

$$k_\alpha = -\sup_{x' \in \partial D} L_0 H_\alpha 1(x') > 0\,.$$

Here it should be noticed that the constants k_α are increasing in $\alpha > 0$:

$$\alpha \geq \beta > 0 \implies k_\alpha \geq k_\beta\,.$$

Furthermore, using Corollary 3.4 with

$$K := \partial D\,, \quad A := \overline{L_0 H_\alpha}\,, \quad c := k_\alpha\,,$$

we obtain that the operator $\overline{L_0 H_\alpha} + k_\alpha I$ is the infinitesimal generator of some Feller semigroup on ∂D. Therefore, since $k_\alpha > 0$, it follows from an application of part (i) of Theorem 3.3 with $A := \overline{L_0 H_\alpha} + k_\alpha I$ that the equation

$$-\overline{L_0 H_\alpha}\psi = \left(k_\alpha I - (\overline{L_0 H_\alpha} + k_\alpha I)\right)\psi = \varphi$$

has a unique solution $\psi \in \mathcal{D}\left(\overline{L_0 H_\alpha}\right)$ for any $\varphi \in C(\partial D)$, and further that the operator $-\overline{L_0 H_\alpha}^{-1} = \left(k_\alpha I - (\overline{L_0 H_\alpha} + k_\alpha I)\right)^{-1}$ is non-negative and bounded on the space $C(\partial D)$ with norm

$$\left\| -\overline{L_0 H_\alpha}^{-1} \right\| = \left\| \left(k_\alpha I - (\overline{L_0 H_\alpha} + k_\alpha I)\right)^{-1} \right\| \leq \frac{1}{k_\alpha}\,.$$

Step 4: By assertion (13.32), we can define the operator G_α by formula (13.31) for all $\alpha > 0$. We prove that

$$G_\alpha = (\alpha I - \mathfrak{A})^{-1}\,, \quad \alpha > 0\,. \tag{13.33}$$

By Lemma 13.6 and Theorem 13.14, it follows that, for any $f \in C_0(\overline{D}\backslash M)$,

$$G_\alpha f \in \mathcal{D}(\overline{A}) ,$$

and

$$\overline{A}G_\alpha f = \alpha G_\alpha f - f .$$

Furthermore, we have

$$L_0 G_\alpha f = L_0 G_\alpha^N f - \overline{L_0 H_\alpha}\left(\overline{L_0 H_\alpha}^{-1}\left(L_0 G_\alpha^N f\right)\right) = 0 \quad \text{on } \partial D . \tag{13.34}$$

However, we recall that

$$L_0 u = \mu(x')L_N u + (\mu(x') - 1)u , \quad u \in \mathcal{D}(L_0) . \tag{13.30'}$$

Hence formula (13.34) is equivalent to the following:

$$\mu(x')L_N(G_\alpha f) + (\mu(x') - 1)G_\alpha f = 0 \quad \text{on } \partial D . \tag{13.34'}$$

This implies that

$$G_\alpha f = 0 \quad \text{on} \quad M = \{x' \in \partial D : \mu(x') = 0\} ,$$

and so

$$\overline{A}G_\alpha f = \alpha G_\alpha f - f = 0 \quad \text{on } M .$$

Summing up, we have proved that

$$G_\alpha f \in \mathcal{D}(\mathfrak{A}) = \left\{u \in C_0(\overline{D} \setminus M) : \overline{A}u \in C_0(\overline{D} \setminus M) , \ L_0 u = 0 \text{ on } \partial D\right\} ,$$

and

$$(\alpha I - \mathfrak{A})G_\alpha f = f , \quad f \in C_0(\overline{D} \setminus M) ,$$

that is,

$$(\alpha I - \mathfrak{A})G_\alpha = I \quad \text{on} \quad C_0(\overline{D} \setminus M) .$$

Therefore, to prove formula (13.33) it suffices to show the injectivity of the operator $\alpha I - \mathfrak{A}$ for $\alpha > 0$.

Assume that

$$u \in \mathcal{D}(\mathfrak{A}) \quad \text{and} \quad (\alpha I - \mathfrak{A})u = 0 .$$

Then, by Corollary 13.7 we can write the function u in the form

$$u = H_\alpha\left(u|_{\partial D}\right) , \quad u|_{\partial D} \in \mathcal{D} = \mathcal{D}\left(\overline{L_0 H_\alpha}\right) .$$

Thus we have

$$\overline{L_0 H_\alpha}\left(u|_{\partial D}\right) = L_0 u = 0 .$$

In view of assertion (13.32), this implies that

$$u|_{\partial D} = 0 ,$$

so that

$$u = H_\alpha\left(u|_{\partial D}\right) = 0 \quad \text{in } D .$$

Step 5: Now we prove the following three assertions:

(i) The operator G_α is non-negative on the space $C_0(\overline{D} \setminus M)$:

$$f \in C_0(\overline{D} \setminus M)\,, \quad f \geq 0 \quad \text{on } \overline{D} \setminus M \implies G_\alpha f \geq 0 \quad \text{on } \overline{D} \setminus M\,.$$

(ii) The operator G_α is bounded on the space $C_0(\overline{D} \setminus M)$ with norm

$$\|G_\alpha\| \leq \frac{1}{\alpha}\,, \quad \alpha > 0\,.$$

(iii) The domain $\mathcal{D}(\mathfrak{A})$ is dense in the space $C_0(\overline{D} \setminus M)$.

(i) First, we show the *non-negativity* of the operator G_α on the space $C(\overline{D})$:

$$f \in C(\overline{D})\,, \quad f \geq 0 \quad \text{on } \overline{D} \implies G_\alpha f \geq 0 \quad \text{on } \overline{D}\,.$$

Recall that the Dirichlet problem

$$\begin{cases} (\alpha - A)u = f & \text{in } D, \\ u = \varphi & \text{on } \partial D \end{cases} \tag{D'}$$

is uniquely solvable. Hence it follows that

$$G_\alpha^N f = H_\alpha \left(G_\alpha^N f|_{\partial D}\right) + G_\alpha^0 f \quad \text{on } \overline{D}\,. \tag{13.35}$$

Indeed, the both sides have the same boundary values $G_\alpha^N f|_{\partial D}$ and satisfy the same equation: $(\alpha - A)u = f$ in D.

Thus, applying the operator L_0 to the both sides of formula (13.35) we obtain that

$$L_0 G_\alpha^N f = \overline{L_0 H_\alpha}\left(G_\alpha^N f|_{\partial D}\right) + \overline{L_0 G_\alpha^0} f\,.$$

Since the operators $-\overline{L_0 H_\alpha}^{-1}$ and $\overline{L_0 G_\alpha^0}$ are non-negative, it follows that

$$\begin{aligned} \left(-\overline{L_0 H_\alpha}^{-1}\right)\left(L_0 G_\alpha^N f\right) &= -G_\alpha^N f + \left(-\overline{L_0 H_\alpha}^{-1}\right)\left(\overline{L_0 G_\alpha^0} f\right) \\ &\geq -G_\alpha^N f \quad \text{on } \partial D\,. \end{aligned}$$

Therefore, by the non-negativity of H_α and G_α^0 we find that

$$\begin{aligned} G_\alpha f &= G_\alpha^N f + H_\alpha\left(-\overline{L_0 H_\alpha}^{-1}\left(L_0 G_\alpha^N f\right)\right) \\ &\geq G_\alpha^N f - H_\alpha\left(G_\alpha^N f|_{\partial D}\right) \\ &= G_\alpha^0 f \geq 0 \quad \text{on } \overline{D}\,. \end{aligned}$$

(ii) Next we prove the *boundedness* of the operator G_α on the space $C(\overline{D})$: To do this, it suffices to show that

$$G_\alpha 1 \leq \frac{1}{\alpha} \quad \text{on } \overline{D}\,,$$

since G_α is non-negative on $C(\overline{D})$.

We remark (cf. formula (13.30′)) that

$$\begin{aligned} L_0 G_\alpha^N f &= \mu(x')\, L_N G_\alpha^N f + (\mu(x') - 1) G_\alpha^N f \\ &= (\mu(x') - 1) G_\alpha^N f \quad \text{on } \partial D \,, \end{aligned}$$

so that

$$\begin{aligned} G_\alpha f &= G_\alpha^N f - H_\alpha \left(\overline{L_0 H_\alpha}^{-1} \left(L_0 G_\alpha^N f \right) \right) \\ &= G_\alpha^N f + H_\alpha \left(-\overline{L_0 H_\alpha}^{-1} \left((\mu(x') - 1) G_\alpha^N f|_{\partial D} \right) \right) . \end{aligned}$$

Hence it follows that

$$G_\alpha 1 = G_\alpha^N 1 - H_\alpha \left(-\overline{L_0 H_\alpha}^{-1} \left((1 - \mu(x')) G_\alpha^N 1|_{\partial D} \right) \right) .$$

However, we have, by inequality (13.19′),

$$0 \leq G_\alpha^N 1 \leq \frac{1}{\alpha} \quad \text{on } \overline{D} \,,$$

and also

$$H_\alpha \left(-\overline{L_0 H_\alpha}^{-1} \left((1 - \mu(x')) G_\alpha^N 1|_{\partial D} \right) \right) \geq 0 \quad \text{on } \overline{D} \,,$$

since the operators H_α, $-\overline{L_0 H_\alpha}^{-1}$ and G_α^N are all non-negative and $1 - \mu(x') \geq 0$ on ∂D.

Therefore, we find that

$$G_\alpha 1 \leq G_\alpha^N 1 \leq \frac{1}{\alpha} \quad \text{on } \overline{D} \,.$$

(iii) Finally we prove the *density* of the domain $\mathcal{D}(\mathfrak{A})$ in the space $C_0(\overline{D} \setminus M)$: In view of formula (13.33), it suffices to show that

$$\lim_{\alpha \to +\infty} \| \alpha G_\alpha f - f \|_\infty = 0 \,, \quad f \in C_0(\overline{D} \setminus M) \cap C^\infty(\overline{D}) \,, \tag{13.36}$$

since the space $C_0(\overline{D} \setminus M) \cap C^\infty(\overline{D})$ is dense in $C_0(\overline{D} \setminus M)$.

It should be noticed that

$$\begin{aligned} \alpha G_\alpha f - f &= \alpha G_\alpha^N f - f - \alpha H_\alpha \left(\overline{L_0 H_\alpha}^{-1} \left(L_0 G_\alpha^N f \right) \right) \\ &= \left(\alpha G_\alpha^N f - f \right) + H_\alpha \left(\overline{L_0 H_\alpha}^{-1} \left(\alpha (1 - \mu(x')) G_\alpha^N f|_{\partial D} \right) \right) . \end{aligned} \tag{13.37}$$

We estimate the two terms of the second equality in formula (13.37) as follows:

(iii-1) By assertion (13.28), it follows that the first term tends to zero:

$$\lim_{\alpha \to +\infty} \left\| \alpha G_\alpha^N f - f \right\|_\infty = 0 \,.$$

(iii-2) To estimate the second term, it should be noticed that

$$\begin{aligned}
&H_\alpha\left(\overline{L_0 H_\alpha}^{-1}\left(\alpha(1-\mu(x'))G_\alpha^N f|_{\partial D}\right)\right)\\
&= H_\alpha\left(\overline{L_0 H_\alpha}^{-1}\left((1-\mu(x'))f|_{\partial D}\right)\right)\\
&\quad + H_\alpha\left(\overline{L_0 H_\alpha}^{-1}\left((1-\mu(x'))(\alpha G_\alpha^N f - f)|_{\partial D}\right)\right).
\end{aligned}$$

However, we have, as $\alpha \to +\infty$,

$$\begin{aligned}
&\left\|H_\alpha\left(\overline{L_0 H_\alpha}^{-1}\left((1-\mu(x'))\left(\alpha G_\alpha^N f - f\right)|_{\partial D}\right)\right)\right\|_\infty\\
&\leq \left\|-\overline{L_0 H_\alpha}^{-1}\right\| \cdot \left\|(1-\mu(x'))\left(\alpha G_\alpha^N f - f\right)|_{\partial D}\right\|_\infty\\
&\leq \frac{1}{k_\alpha}\left\|(1-\mu(x'))\left(\alpha G_\alpha^N f - f\right)|_{\partial D}\right\|_\infty\\
&\leq \frac{1}{k_1}\left\|\alpha G_\alpha^N f - f\right\|_\infty \longrightarrow 0\,.
\end{aligned} \tag{13.38}$$

Here we have used the following facts (cf. the proof of assertion (13.32)):

$$\begin{aligned}
&\left\|-\overline{L_0 H_\alpha}^{-1}\right\| \leq \frac{1}{k_\alpha}\,, \quad \alpha > 0\,.\\
&k_1 = -\sup_{x'\in\partial D} L_0 H_1 1(x') \leq k_\alpha = -\sup_{x'\in\partial D} L_0 H_\alpha 1(x')\,, \quad \alpha \geq 1\,.
\end{aligned}$$

Thus we are reduced to the study of the term

$$H_\alpha\left(\overline{L_0 H_\alpha}^{-1}\left((1-\mu(x'))f|_{\partial D}\right)\right).$$

Now, for any given $\varepsilon > 0$, we can find a function $h(x') \in C^\infty(\partial D)$ such that

$$\begin{cases} h = 0 \quad \text{near } M = \{x' \in \partial D : \mu(x') = 0\},\\ \|(1-\mu(x'))f|_{\partial D} - h\|_\infty < \varepsilon. \end{cases}$$

Then we have, for all $\alpha \geq 1$,

$$\begin{aligned}
&\left\|H_\alpha\left(\overline{L_0 H_\alpha}^{-1}\left((1-\mu(x'))f|_{\partial D}\right)\right) - H_\alpha\left(\overline{L_0 H_\alpha}^{-1} h\right)\right\|_\infty\\
&\leq \left\|-\overline{L_0 H_\alpha}^{-1}\right\| \cdot \|(1-\mu(x'))f|_{\partial D} - h\|_\infty\\
&\leq \frac{\varepsilon}{k_\alpha}\\
&\leq \frac{\varepsilon}{k_1}\,.
\end{aligned} \tag{13.39}$$

Furthermore, we can find a function $\Theta(x') \in C_0^\infty(\partial D)$ such that

$$\begin{cases} \Theta = 1 & \text{near } M,\\ (1-\Theta)h = h & \text{on } \partial D. \end{cases}$$

Then we have

$$\begin{aligned} h(x') &= (1 - \Theta(x'))\, h(x') \\ &= (-L_0 H_\alpha 1(x')) \left(\frac{1 - \Theta(x')}{-L_0 H_\alpha 1(x')} \right) h(x') \\ &\le \left[\sup_{x' \in \partial D} \left(\frac{1 - \Theta(x')}{-L_0 H_\alpha 1(x')} \right) \right] \|h\|_\infty \, (-L_0 H_\alpha 1(x')) . \end{aligned}$$

Since the operator $-\overline{L_0 H_\alpha}^{-1}$ is non-negative on the space $C(\partial D)$, it follows that

$$-\overline{L_0 H_\alpha}^{-1} h \le \sup_{x' \in \partial D} \left(\frac{1 - \Theta(x')}{-L_0 H_\alpha 1(x')} \right) \|h\|_\infty \quad \text{on } \partial D ,$$

and hence

$$\left\| H_\alpha \left(\overline{L_0 H_\alpha}^{-1} h \right) \right\| \le \left\| -\overline{L_0 H_\alpha}^{-1} h \right\|_\infty \le \sup_{x' \in \partial D} \left(\frac{1 - \Theta(x')}{-L_0 H_\alpha 1(x')} \right) \|h\|_\infty . \tag{13.40}$$

However, there exists a constant $\delta_0 > 0$ such that

$$0 \le \frac{1 - \Theta(x')}{\mu(x')} \le \delta_0 , \quad x' \in \partial D .$$

Thus it follows that

$$\begin{aligned} \frac{1 - \Theta(x')}{-L_0 H_\alpha 1(x')} &= \frac{1 - \Theta(x')}{\mu(x') \left(-\frac{\partial}{\partial \mathbf{n}} \left(H_\alpha 1(x') \right) \right) + (1 - \mu(x'))} \\ &\le \left(\frac{1 - \Theta(x')}{\mu(x')} \right) \frac{1}{\left(-\frac{\partial}{\partial \mathbf{n}} \left(H_\alpha 1(x') \right) \right)} \\ &\le \delta_0 \frac{1}{\inf_{x' \in \partial D} \left(-\frac{\partial}{\partial \mathbf{n}} \left(H_\alpha 1(x') \right) \right)} , \end{aligned}$$

so that, by Lemma 13.16,

$$\lim_{\alpha \to +\infty} \left[\sup_{x' \in \partial D} \left(\frac{1 - \Theta(x')}{-L_0 H_\alpha 1(x')} \right) \right] = 0 .$$

Summing up, we obtain from inequalities (13.39) and (13.40) that

$$\begin{aligned} &\limsup_{\alpha \to +\infty} \left\| H_\alpha \left(\overline{L_0 H_\alpha}^{-1} ((1 - \mu(x')) f|_{\partial D}) \right) \right\|_\infty \\ &\le \limsup_{\alpha \to +\infty} \Bigg[\left\| H_\alpha \left(\overline{L_0 H_\alpha}^{-1} h \right) \right\|_\infty \\ &\qquad + \left\| H_\alpha \left(\overline{L_0 H_\alpha}^{-1} ((1 - \mu(x')) f|_{\partial D}) \right) - H_\alpha \left(\overline{L_0 H_\alpha}^{-1} h \right) \right\|_\infty \Bigg] \\ &\le \lim_{\alpha \to +\infty} \left[\sup_{x' \in \partial D} \left(\frac{1 - \Theta(x')}{-L_0 H_\alpha 1(x')} \right) \right] \|h\|_\infty + \frac{\varepsilon}{k_1} \\ &\le \frac{\varepsilon}{k_1} . \end{aligned}$$

Since ε is arbitrary, this proves that

$$\lim_{\alpha\to+\infty} \left\| H_\alpha \left(\overline{L_0 H_\alpha}^{-1} ((1-\mu(x'))f|_{\partial D}) \right) \right\|_\infty = 0 . \tag{13.41}$$

Therefore, combining assertions (13.38) and (13.41) we find that the second term in formula (13.37) also tends to zero:

$$\lim_{\alpha\to+\infty} \left\| H_\alpha \left(\overline{L_0 H_\alpha}^{-1} (\alpha(1-\mu(x'))G_\alpha^N f|_{\partial D}) \right) \right\|_\infty = 0.$$

This completes assertion (13.36) and hence assertion (iii).

Step 6: Summing up, we have proved that the operator $\mathfrak{A}$, defined by formula (13.31), satisfies conditions (a) through (d) in Theorem 3.3. Hence, in view of assertion (7.31) it follows from an application of part (ii) of the same theorem that the operator $\mathfrak{A}$ is the infinitesimal generator of some Feller semigroup $\{e^{t\mathfrak{A}}\}_{t\geq 0}$ on $\overline{D}\setminus M$.

The proof of Theorem 13.18 and hence that of Theorem 1.5 is now complete. □

13.4 Proof of Part (ii) of Theorem 1.4

We apply Theorem 2.2 to the operator $\mathfrak{A}$.

In the proof of Theorem 13.18, we have proved that the domain $\mathcal{D}(\mathfrak{A})$ is *dense* in the space $C_0(\overline{D}\setminus M)$. Furthermore, part (i) of Theorem 1.4 verifies condition (2.1). Therefore, it follows from an application of Theorem 2.11 that

> *The semigroup* $\{e^{t\mathfrak{A}}\}_{t\geq 0}$ *can be extended to a semigroup* $e^{z\mathfrak{A}}$ *which is* analytic *in the sector* $\Delta_\varepsilon = \{z = t + is : z \neq 0\ ,\ |\arg z| < \pi/2 - \varepsilon\}$ *for any* $0 < \varepsilon < \pi/2$.

This (together with Theorem 1.5) proves part (ii) of Theorem 1.4. □

14

Boundary Value Problems for Waldenfels Operators

In this final chapter we study a class of degenerate boundary value problems for second-order elliptic *integro-differential operators*, called Waldenfels operators, and generalize Theorems 1.2, 1.3, and 1.4 (Theorems 1.6, 1.8 and 1.9). As an application, we construct a Feller semigroup corresponding to such a physical phenomenon that a Markovian particle moves both by jumps and continuously in the state space until it "dies" at the time when it reaches the set where the particle is definitely absorbed, generalizing Theorem 1.5 (Theorem 1.7). The results discussed here are adapted from Taira [Ta6] (cf. Garroni-Menaldi [GM] and Galakhov-Skubachevskiĭ [GB]).

14.1 Formulation of a Boundary Value Problem

In this chapter we assume that the domain D is *convex.*

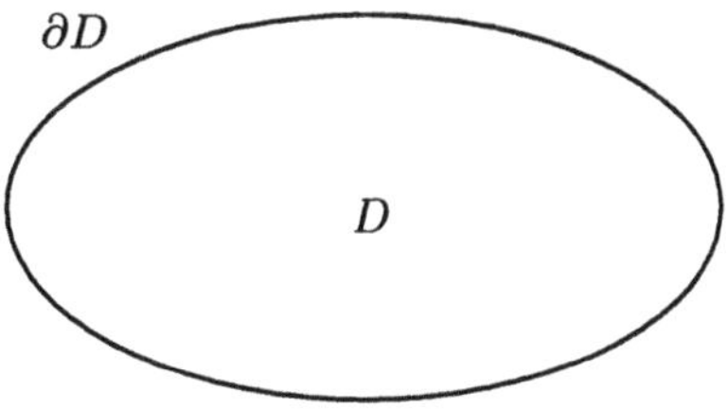

Fig. 14.1.

Let W be a second-order, elliptic integro-differential operator with real coefficients such that

$$Wu(x) = Au(x) + Su(x)$$

$$:= \sum_{i,j=1}^{N} a^{ij}(x) \frac{\partial^2 u}{\partial x_i \partial x_j}(x) + \sum_{i=1}^{N} b^i(x) \frac{\partial u}{\partial x_i}(x) + c(x)u(x)$$
$$+ \int_{\mathbf{R}^N \setminus \{0\}} \left(u(x+z) - u(x) - \sum_{j=1}^{N} z_j \frac{\partial u}{\partial x_j}(x) \right) s(x,z)\, m(dz) .$$

Here:

(1) $a^{ij}(x) \in C^\infty(\overline{D})$, $a^{ij}(x) = a^{ji}(x)$ and there exists a constant $a_0 > 0$ such that

$$\sum_{i,j=1}^{N} a^{ij}(x)\xi_i\xi_j \geq a_0|\xi|^2 , \quad x \in \overline{D},\ \xi \in \mathbf{R}^N .$$

(2) $b^i(x) \in C^\infty(\overline{D})$.
(3) $c(x) \in C^\infty(\overline{D})$, and $c(x) \leq 0$ in D, but $c(x) \not\equiv 0$ in D.
(4) $s(x,z) \in C(\overline{D} \times \mathbf{R}^N)$ and $0 \leq s(x,z) \leq 1$ on $\overline{D} \times \mathbf{R}^N$, and there exist constants $C_0 > 0$ and $0 < \theta < 1$ such that

$$|s(x,z) - s(y,z)| \leq C_0|x-y|^\theta , \quad x,\ y \in \overline{D},\ z \in \mathbf{R}^N , \tag{1.5}$$

and

$$s(x,z) = 0 \quad \text{if } x + z \notin \overline{D} . \tag{1.6}$$

Condition (1.6) implies that the integral operator S may be considered as an operator acting on functions u defined on the closure $\overline{D}$ (see Garroni-Menaldi [GM, Chap. II, Remark 1.19]).
(5) The measure $m(dz)$ is a Radon measure on the space $\mathbf{R}^N \setminus \{0\}$ which satisfies the *moment condition*

$$\int_{\{|z|\leq 1\}} |z|^2\, m(dz) + \int_{\{|z|>1\}} |z|\, m(dz) < \infty . \tag{1.7}$$

Condition (1.7) is a standard condition for the measure $m(dz)$, and it implies that a Markovian particle does not move by jumps so far.

The operator W is called a second-order *Waldenfels operator* (cf. [BCP], [Wa]). The differential operator A is called a diffusion operator which describes analytically a strong Markov process with continuous paths in the interior D such as Brownian motion. The integral operator S is called a second-order *Lévy operator* which is supposed to correspond to the jump phenomenon in the closure $\overline{D}$ (see Fig. 14.2). In this context, condition (1.6) implies that any Markovian particle does not move by jumps from $x \in D$ to the outside of $\overline{D}$.

Let L_0 be a special case of general Ventcel' boundary conditions such that

$$L_0 u(x') = \mu(x') \frac{\partial u}{\partial \mathbf{n}}(x') + \gamma(x')u(x') .$$

Here:

(1) $\mu(x') \in C^\infty(\partial D)$ and $\mu(x') \geq 0$ on ∂D.
(2) $\gamma(x') \in C^\infty(\partial D)$ and $\gamma(x') \leq 0$ on ∂D.
(3) $\mathbf{n} = (n_1, n_2, \ldots, n_N)$ is the unit interior normal to the boundary ∂D (see Fig. 1.1).

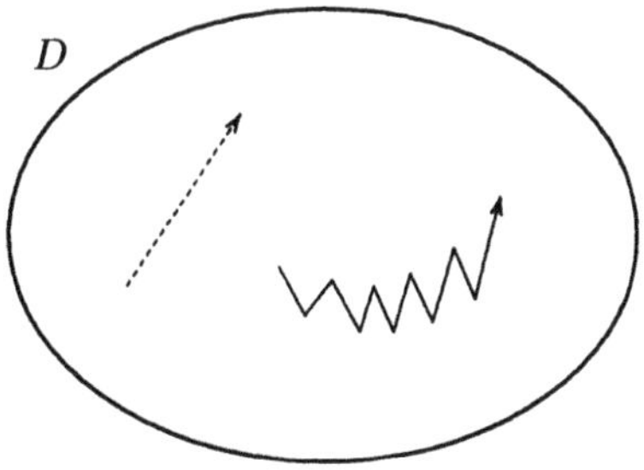

Fig. 14.2.

Now we can formulate our boundary value problem for (W, L_0) as follows: Given functions f and φ defined in D and on ∂D, respectively, find a function u in D such that

$$\begin{cases} Wu = f & \text{in } D\,, \\ L_0 u = \varphi & \text{on } \partial D\,. \end{cases} \tag{1.8}$$

The first purpose of this chapter is to prove an existence and uniqueness theorem for the boundary value problem (1.8) in the framework of Hölder spaces. The second purpose of this chapter is to study problem (1.8) from the point of view of analytic semigroup theory in functional analysis.

14.2 Proof of Theorem 1.6

In this section we study the boundary value problem (1.8) in the framework of Hölder spaces and prove Theorem 1.6, generalizing Theorem 1.2 to the integro-differential operator case. Namely, we prove that the operator

$$(W, L_0) = (A + S, L_0)\colon C^{2+\theta}(\overline{D}) \longrightarrow C^\theta(\overline{D}) \bigoplus C^{1+\theta}_{L_0}(\partial D)$$

is an algebraic and topological *isomorphism*. The essential point in the proof is to estimate the integral operator S in terms of Hölder norms (Lemmas 14.2 and 14.3). We show that the operator (W, L_0) may be considered as a perturbation of a compact operator to the operator (A, L_0) in the framework of Hölder spaces. Thus the proof of Theorem 1.6 is reduced to the differential operator case which is studied in detail in Chap. 9.

Step 1: First, we study the integral operator S in the framework of Hölder spaces. To do this, we need the following elementary estimates for the measure $m(dz)$:

Claim 14.1. *For each $\varepsilon > 0$, we let*

$$\sigma(\varepsilon) = \int_{\{|z| \leq \varepsilon\}} |z|^2 \, m(dz) \, ,$$
$$\delta(\varepsilon) = \int_{\{|z| > \varepsilon\}} |z| \, m(dz),$$
$$\tau(\varepsilon) = \int_{\{|z| > \varepsilon\}} m(dz).$$

Then we have, as $\varepsilon \downarrow 0$,

$$\sigma(\varepsilon) \longrightarrow 0, \tag{14.1a}$$
$$\delta(\varepsilon) \leq \frac{C_1}{\varepsilon} + C_2, \tag{14.1b}$$
$$\tau(\varepsilon) \leq \frac{C_1}{\varepsilon^2} + C_2, \tag{14.1c}$$

where

$$C_1 = \int_{\{|z| \leq 1\}} |z|^2 \, m(dz), \quad C_2 = \int_{\{|z| > 1\}} |z| \, m(dz) \, .$$

Proof. Assertion (14.1a) follows immediately from the moment condition (1.7).

The term $\delta(\varepsilon)$ can be estimated as follows:

$$\begin{aligned} \delta(\varepsilon) &= \int_{\{|z|>1\}} |z| \, m(dz) + \int_{\{\varepsilon < |z| \leq 1\}} |z| \, m(dz) \\ &\leq \int_{\{|z|>1\}} |z| \, m(dz) + \frac{1}{\varepsilon} \int_{\{\varepsilon < |z| \leq 1\}} |z|^2 \, m(dz) \\ &\leq \int_{\{|z|>1\}} |z| \, m(dz) + \frac{1}{\varepsilon} \int_{\{|z| \leq 1\}} |z|^2 \, m(dz) \\ &= C_2 + \frac{C_1}{\varepsilon} \, . \end{aligned}$$

Similarly we have, for the term $\tau(\varepsilon)$,

$$\begin{aligned} \tau(\varepsilon) &= \int_{\{|z|>1\}} m(dz) + \int_{\{\varepsilon < |z| \leq 1\}} m(dz) \\ &\leq \int_{\{|z|>1\}} |z| \, m(dz) + \frac{1}{\varepsilon^2} \int_{\{\varepsilon < |z| \leq 1\}} |z|^2 \, m(dz) \\ &\leq \int_{\{|z|>1\}} |z| \, m(dz) + \frac{1}{\varepsilon^2} \int_{\{|z| \leq 1\}} |z|^2 \, m(dz) \\ &= C_2 + \frac{C_1}{\varepsilon^2} \, . \quad \square \end{aligned}$$

By virtue of Claim 14.1, we can estimate the term Su in terms of Hölder norms, just as in [GM, Chap. II, Lemmas 1.2 and 1.5]:

Lemma 14.2. *For every $\eta > 0$, there exists a constant $C_\eta > 0$ such that we have, for all $u \in C^2(\overline{D})$,*

$$\|Su\|_\infty \leq \eta \|\nabla^2 u\|_\infty + C_\eta \left(\|u\|_\infty + \|\nabla u\|_\infty\right) . \tag{14.2}$$

Here

$$\|u\|_\infty = \sup_{x \in D} |u(x)| .$$

Proof. We decompose the term Su into the following two terms:

$$\begin{aligned} Su(x) &= \int_0^1 (1-t)\, dt \int_{\{|z| \leq \varepsilon\}} z \cdot \nabla^2 u(x+tz) z \, s(x,z)\, m(dz) \\ &\quad + \int_{\{|z|>\varepsilon\}} (u(x+z) - u(x) - z \cdot \nabla u(x))\, s(x,z)\, m(dz) \\ &:= S_\varepsilon^{(1)} u(x) + S_\varepsilon^{(2)} u(x) . \end{aligned} \tag{14.3}$$

By Claim 14.1, we can estimate the terms $S_\varepsilon^{(1)} u$ and $S_\varepsilon^{(2)} u$ as follows:

$$\|S_\varepsilon^{(1)} u\|_\infty \leq \frac{1}{2} \left(\int_{\{|z| \leq \varepsilon\}} |z|^2\, m(dz) \right) \|\nabla^2 u\|_\infty = \frac{1}{2} \sigma(\varepsilon) \|\nabla^2 u\|_\infty ,$$

$$\begin{aligned} \|S_\varepsilon^{(2)} u\|_\infty &\leq 2 \left(\int_{\{|z|>\varepsilon\}} m(dz) \right) \|u\|_\infty + \left(\int_{\{|z|>\varepsilon\}} |z|\, m(dz) \right) \|\nabla u\|_\infty \\ &\leq 2 \left(\frac{C_1}{\varepsilon^2} + C_2 \right) \|u\|_\infty + \left(\frac{C_1}{\varepsilon} + C_2 \right) \|\nabla u\|_\infty . \end{aligned}$$

Therefore, we have proved that

$$\begin{aligned} \|Su\|_\infty &\leq \|S_\varepsilon^{(1)} u\|_\infty + \|S_\varepsilon^{(2)} u\|_\infty \\ &\leq \frac{1}{2} \sigma(\varepsilon) \|\nabla^2 u\|_\infty + 2 \left(\frac{C_1}{\varepsilon^2} + C_2 \right) \|u\|_\infty + \left(\frac{C_1}{\varepsilon} + C_2 \right) \|\nabla u\|_\infty . \end{aligned}$$

In view of assertion (14.1a), this proves estimate (14.2) if we choose ε sufficiently small. □

Lemma 14.3. *For every $\eta > 0$, there exists a constant $C_\eta > 0$ such that we have, for all $u \in C^{2+\theta}(\overline{D})$,*

$$\|Su\|_{C^\theta(\overline{D})} \leq \eta \|\nabla^2 u\|_{C^\theta(\overline{D})} + C_\eta \left(\|u\|_{C^\theta(\overline{D})} + \|\nabla u\|_{C^\theta(\overline{D})} \right) . \tag{14.4}$$

Here

$$\|u\|_{C^\theta(\overline{D})} = \|u\|_\infty + [u]_\theta, \quad [u]_\theta = \sup_{\substack{x,y \in D \\ x \neq y}} \frac{|u(x) - u(y)|}{|x-y|^\theta} .$$

Proof. We estimate the two functions $S_\varepsilon^{(1)}u$ and $S_\varepsilon^{(2)}u$ of formula (14.3) in terms of Hölder norms.

To estimate the function

$$S_\varepsilon^{(2)}u(x) = \int_{\{|z|>\varepsilon\}} (u(x+z) - u(x) - z \cdot \nabla u(x))\, s(x,z)\, m(dz) \, ,$$

we write the difference $S_\varepsilon^{(2)}u(x) - S_\varepsilon^{(2)}u(y)$ in the following form:

$$\begin{aligned}
& S_\varepsilon^{(2)}u(x) - S_\varepsilon^{(2)}u(y) \\
&= \int_{\{|z|>\varepsilon\}} (u(x+z) - u(y+z))\, s(x,z)\, m(dz) \\
&\quad + \int_{\{|z|>\varepsilon\}} u(y+z)\, (s(x,z) - s(y,z))\, m(dz) \\
&\quad + \int_{\{|z|>\varepsilon\}} (u(y) - u(x))\, s(x,z)\, m(dz) \\
&\quad - \int_{\{|z|>\varepsilon\}} u(y)\, (s(x,z) - s(y,z))\, m(dz) \\
&\quad - \int_{\{|z|>\varepsilon\}} (\nabla u(x) - \nabla u(y)) \cdot z\, s(x,z)\, m(dz) \\
&\quad - \int_{\{|z|>\varepsilon\}} \nabla u(y) \cdot z\, (s(x,z) - s(y,z))\, m(dz) \\
&:= A(x,y) + B(x,y) + C(x,y) - D(x,y) - E(x,y) - F(x,y) \, .
\end{aligned}$$

Then, by using estimates (14.1c), (14.1b) and condition (1.5) we can estimate the terms $A(x,y)$ through $F(x,y)$ as follows:

$$\begin{aligned}
|A(x,y)|,\ |C(x,y)| &\leq [u]_\theta\, \tau(\varepsilon)\, |x-y|^\theta \\
&\leq \left(\frac{C_1}{\varepsilon^2} + C_2\right) [u]_\theta\, |x-y|^\theta \, . \\
|E(x,y)| &\leq [\nabla u]_\theta\, \delta(\varepsilon)\, |x-y|^\theta \\
&\leq \left(\frac{C_1}{\varepsilon} + C_2\right) [\nabla u]_\theta\, |x-y|^\theta \, . \\
|B(x,y)|,\ |D(x,y)| &\leq \|u\|_\infty\, C_0\, \tau(\varepsilon)\, |x-y|^\theta \\
&\leq C_0 \left(\frac{C_1}{\varepsilon^2} + C_2\right) \|u\|_\infty\, |x-y|^\theta \, . \\
|F(x,y)| &\leq \|\nabla u\|_\infty\, C_0\, \delta(\varepsilon)\, |x-y|^\theta \\
&\leq C_0 \left(\frac{C_1}{\varepsilon} + C_2\right) \|\nabla u\|_\infty\, |x-y|^\theta
\end{aligned}$$

Summing up, we have proved that

$$\begin{aligned}
[S_\varepsilon^{(2)}u]_\theta &\leq 2\left(\frac{C_1}{\varepsilon^2}+C_2\right)[u]_\theta + 2C_0\left(\frac{C_1}{\varepsilon^2}+C_2\right)\|u\|_\infty \\
&\quad + \left(\frac{C_1}{\varepsilon}+C_2\right)[\nabla u]_\theta + C_0\left(\frac{C_1}{\varepsilon}+C_2\right)\|\nabla u\|_\infty \\
&\leq 2(1+C_0)\left(\frac{C_1}{\varepsilon^2}+C_2\right)\|u\|_{C^\theta(\overline{D})} \\
&\quad + (1+C_0)\left(\frac{C_1}{\varepsilon}+C_2\right)\|\nabla u\|_{C^\theta(\overline{D})} \,. \qquad (14.5)
\end{aligned}$$

The function

$$S_\varepsilon^{(1)}u(x) = \int_0^1 (1-t)\,dt \int_{\{|z|\leq\varepsilon\}} z\cdot\nabla^2 u(x+tz)z\,s(x,z)\,m(dz)$$

can be estimated in a similar way:

$$\begin{aligned}
[S_\varepsilon^{(1)}u]_\theta &\leq \frac{1}{2}\sigma(\varepsilon)\,[\nabla^2 u]_\theta + \frac{1}{2}\sigma(\varepsilon)\,C_0\,\|\nabla^2 u\|_\infty \\
&\leq \left(\frac{1+C_0}{2}\right)\sigma(\varepsilon)\,\|\nabla^2 u\|_{C^\theta(\overline{D})} \,. \qquad (14.6)
\end{aligned}$$

Thus, combining estimates (14.5) and (14.6) we obtain that

$$\begin{aligned}
[Su]_\theta &\leq [S_\varepsilon^{(1)}u]_\theta + [S_\varepsilon^{(2)}u]_\theta \\
&\leq \eta\|\nabla^2 u\|_{C^\theta(\overline{D})} + C_\eta\left(\|u\|_{C^\theta(\overline{D})} + \|\nabla u\|_{C^\theta(\overline{D})}\right), \qquad (14.7)
\end{aligned}$$

if we choose ε sufficiently small.

Therefore, the desired estimate (14.4) follows by combining estimates (14.2) and (14.7). □

Remark 14.1. By the proof of Lemmas 14.2 and 14.3, it follows that the operator $S_\varepsilon^{(2)}$ may be considered as a *drift vector filed* for each $\varepsilon > 0$.

Step 2: End of Proof of Theorem 1.6. First, Theorem 1.2 (or Theorem 9.1) implies that

$$\text{ind}(A, L_0) = 0\,.$$

On the other hand, we find from the proof of Lemmas 14.2 and 14.3 that the operator $(W, L_0) = (A+S, L_0)$ is a perturbation of a *compact* operator to the operator (A, L_0). Hence we have

$$\text{ind}(W, L_0) = \text{ind}(A, L_0) = 0\,.$$

Indeed, since the operator $S_\varepsilon^{(2)}$ maps $C^{1+\theta}(\overline{D})$ continuously into $C^\theta(\overline{D})$, it follows from an application of the Ascoli-Arzelà theorem that the operator

$$S_\varepsilon^{(2)}: C^{2+\theta}(\overline{D}) \longrightarrow C^\theta(\overline{D})$$

is compact. Moreover, the operator norm of the operator

$$S_\varepsilon^{(1)} = S - S_\varepsilon^{(2)} : C^{2+\theta}(\overline{D}) \longrightarrow C^\theta(\overline{D})$$

tends to zero as $\varepsilon \downarrow 0$. This proves that the operator

$$S : C^{2+\theta}(\overline{D}) \longrightarrow C^\theta(\overline{D})$$

is compact.

Therefore, to show the bijectivity of (W, L_0) it suffices to prove its *injectivity*:

$$\begin{cases} u \in C^{2+\theta}(\overline{D}), \ Wu = 0 \text{ in } D, \ L_0 u = 0 \text{ on } \partial D \\ \Longrightarrow \ u = 0 \text{ in } D\,. \end{cases}$$

However, this is an immediate consequence of the following *maximum principle*:

Proposition 14.4. *If condition (H) is satisfied, then we have*

$$\begin{cases} u \in C^2(\overline{D}), \ Wu \geq 0 \text{ in } D, \ L_0 u \geq 0 \text{ on } \partial D \\ \Longrightarrow u \leq 0 \text{ on } \overline{D}\,. \end{cases}$$

Proof. If $u(x)$ is a constant m, then we have $0 \leq Wu(x) = m\,c(x)$ in D. This implies that $u(x) \equiv m$ is non-positive, since $c(x) \leq 0$ and $c(x) \not\equiv 0$ in D.

Now we consider the case where $u(x)$ is not a constant. Assume, to the contrary, that

$$m = \max_{\overline{D}} u > 0\,.$$

Then, applying the weak maximum principle (see Theorem C.1 in Appendix C) to the operator W we obtain that there exists a point x_0' of ∂D such that

$$\begin{cases} u(x_0') = m\,, \\ u(x) < u(x_0') \quad \text{for all } x \in D\,. \end{cases}$$

Furthermore, it follows from an application of the boundary point lemma (see Lemma C.5 in Appendix C) that

$$\frac{\partial u}{\partial \mathbf{n}}(x_0') < 0\,.$$

Hence we have

$$\mu(x_0') = 0, \quad \gamma(x_0') = 0\,,$$

since $L_0 u(x_0') \geq 0$. This contradicts hypothesis (H) that $\mu(x') + |\gamma(x')| > 0$ on ∂D. □

The proof of Theorem 1.6 is now complete. □

14.3 Proof of Theorem 1.7

This section is devoted to the proof of Theorem 1.7.

Step 1: Our proof is based on the following version of the Hille-Yosida theorem in terms of the maximum principle (cf. [BCP, Théorème de Hille-Yosida-Ray]):

Theorem 14.5 (Hille-Yosida-Ray). *Let A be a linear operator from the space $C_0(\overline{D} \setminus M)$ into itself, and assume that*

(a) The domain $\mathcal{D}(A)$ is dense in the space $C_0(\overline{D} \setminus M)$.
(b) For any $u \in \mathcal{D}(A)$ such that $\max_{\overline{D}} u > 0$, there exists a point $x \in \overline{D} \setminus M$ such that

$$\begin{cases} u(x) = \max_{\overline{D}} u \,, \\ Au(x) \leq 0 \,. \end{cases}$$

(c) For all $\alpha > 0$, the range $\mathcal{R}(\alpha I - A)$ is dense in the space $C_0(\overline{D} \setminus M)$.

Then the operator A is closable *in the space $C_0(\overline{D} \setminus M)$, and its minimal closed extension $\overline{A}$ generates a Feller semigroup $\{T_t\}_{t \geq 0}$ on $\overline{D} \setminus M$.*

Proof. We apply a version of the Hille-Yosida theorem (Theorem 3.3). The proof is divided into six steps.

(1) First, we prove that, for all $\alpha > 0$

$$\begin{aligned} &u \in \mathcal{D}(A), \ (\alpha - A)u = f \geq 0 \quad \text{on } \overline{D} \\ &\Longrightarrow 0 \leq u \leq \frac{1}{\alpha} \max_{\overline{D}} f \quad \text{on } \overline{D} \,. \end{aligned} \tag{14.8}$$

Indeed, if we assume, to the contrary, that

$$\min_{\overline{D}} u < 0 \,,$$

then it follows from condition (b) with $u := -u$ that there exists a point $x_0 \in \overline{D} \setminus M$ such that

$$\begin{aligned} &-u(x_0) = \max_{\overline{D}}(-u) = -\min_{\overline{D}} u > 0 \,, \\ &-Au(x_0) \leq 0 \,. \end{aligned}$$

Hence we have

$$0 \leq f(x_0) = \alpha u(x_0) - Au(x_0) < 0 \,.$$

This contradiction proves that

$$u \geq 0 \quad \text{on } \overline{D} \,.$$

Similarly, if we assume, to the contrary, that

$$\max_{\overline{D}} u > \frac{1}{\alpha} \max_{\overline{D}} f ,$$

then we can find a point $x_1 \in \overline{D} \setminus M$ such that

$$u(x_1) = \max_{\overline{D}} u > \frac{1}{\alpha} \max_{\overline{D}} f ,$$
$$Au(x_1) \leq 0 .$$

Hence it follows that

$$f(x_1) = \alpha u(x_1) - Au(x_1) \geq \alpha u(x_1) > \max_{\overline{D}} f .$$

This contradiction proves that

$$\max_{\overline{D}} u \leq \frac{1}{\alpha} \max_{\overline{D}} f .$$

(2) Secondly, by virtue of assertion (14.8) we can define an inverse

$$G_\alpha = (\alpha I - A)^{-1} : \mathcal{R}(\alpha I - A) \longrightarrow \mathcal{D}(A)$$

which is non-negative and bounded on the range $\mathcal{R}(\alpha I - A)$:

$$f \in \mathcal{R}(\alpha I - A),\ f(x) \geq 0 \quad \text{on } \overline{D} \Longrightarrow 0 \leq G_\alpha f(x) \leq \frac{1}{\alpha} \max_{\overline{D}} f \quad \text{on } \overline{D} .$$

Moreover, since the range $\mathcal{R}(\alpha I - A)$ is dense in $C_0(\overline{D} \setminus M)$, we obtain that the operator G_α can be uniquely extended to a non-negative, bounded linear operator on the whole space $C_0(\overline{D} \setminus M)$, denoted again by G_α.

$$\begin{array}{ccc} C_0(\overline{D} \setminus M) & \xrightarrow{G_\alpha} & C_0(\overline{D} \setminus M) \\ \uparrow & & \downarrow \\ \mathcal{R}(\alpha I - A) & \xrightarrow[(\alpha I - A)^{-1}]{} & \mathcal{D}(A) \end{array}$$

More precisely, it should be noticed that

$$f \in C_0(\overline{D} \setminus M),\ f(x) \geq 0 \quad \text{on } \overline{D} \Longrightarrow 0 \leq G_\alpha f(x) \leq \frac{1}{\alpha} \max_{\overline{D}} f \quad \text{on } \overline{D} , \tag{14.9}$$

and that

$$\|G_\alpha\| \leq \frac{1}{\alpha}, \quad \alpha > 0 . \tag{14.10}$$

(3) Thirdly, we prove that, for each $u \in C_0(\overline{D} \setminus M)$,

$$\lim_{\alpha \to \infty} \|\alpha G_\alpha u - u\| = 0 . \tag{14.11}$$

For any given $\varepsilon > 0$, we can find a function $v \in \mathcal{D}(A)$ such that

$$\|u - v\| < \varepsilon .$$

Then we have, by assertion (14.10),

$$\begin{aligned} \|\alpha G_\alpha u - u\| &\leq \alpha \|G_\alpha (u - v)\| + \|\alpha G_\alpha v - v\| + \|v - u\| \\ &\leq 2\|u - v\| + \|\alpha G_\alpha v - v\| \\ &\leq 2\varepsilon + \|\alpha G_\alpha v - v\| . \end{aligned} \tag{14.12}$$

However, it should be noticed that

$$v = G_\alpha(\alpha I - A)v = \alpha G_\alpha v - G_\alpha(Av) .$$

Hence we have, by assertion (14.10),

$$\|\alpha G_\alpha v - v\| = \|G_\alpha(Av)\| \leq \frac{1}{\alpha} \|Av\| . \tag{14.13}$$

Therefore, combining inequalities (14.12) and (14.13) we obtain that

$$\|\alpha G_\alpha u - u\| \leq 2\varepsilon + \frac{1}{\alpha} \|Av\| ,$$

so that

$$\limsup_{\alpha \to \infty} \|\alpha G_\alpha u - u\| \leq 2\varepsilon .$$

This proves assertion (14.11), since ε is arbitrary.

(4) Now we can prove that the operator A is *closable* in $C_0(\overline{D} \setminus M)$. To do this, we assume that

$$\{u_n\} \subset \mathcal{D}(A), \; u_n \to 0 \text{ and } Au_n \to v \quad \text{in } C_0(\overline{D} \setminus M) .$$

Then we have, for all $\alpha > 0$,

$$(\alpha I - A)u_n = \alpha u_n - Au_n \longrightarrow -v \quad \text{in } C_0(\overline{D} \setminus M) ,$$

and so

$$u_n = G_\alpha(\alpha I - A)u_n \longrightarrow -G_\alpha v \quad \text{in } C_0(\overline{D} \setminus M) .$$

Hence it follows that

$$G_\alpha v = 0 .$$

Therefore, we obtain from assertion (14.11) that

$$v = \lim_{\alpha \to \infty} \alpha G_\alpha v = 0 .$$

This proves that A is closable in $C_0(\overline{D} \setminus M)$.

(5) Let $\overline{A}$ be the minimal closed extension of A in the space $C_0(\overline{D} \setminus M)$. Finally, it remains to prove the formula

$$G_\alpha = (\alpha I - \overline{A})^{-1}, \quad \alpha > 0 . \tag{14.14}$$

(a) First, we prove the formula

$$G_\alpha(\alpha I - \overline{A}) = I \quad \text{on } \mathcal{D}(\overline{A}) . \tag{14.14a}$$

For any given function $u \in \mathcal{D}(\overline{A})$, we can find a sequence $\{u_n\}$ in $\mathcal{D}(A)$ such that

$$u_n \to u \text{ and } Au_n \to \overline{A}u \quad \text{in } C_0(\overline{D} \setminus M) .$$

Then it follows that

$$\begin{aligned} u_n &= G_\alpha(\alpha I - A)u_n = \alpha G_\alpha u_n - G_\alpha(Au_n) \\ &\longrightarrow \alpha G_\alpha u - G_\alpha(\overline{A}u) \quad \text{in } C_0(\overline{D} \setminus M) , \end{aligned}$$

so that

$$G_\alpha(\alpha I - \overline{A})u = u, \quad u \in \mathcal{D}(\overline{A}) .$$

This proves formula (14.14a).

(b) Next, we prove the formula

$$(\alpha I - \overline{A})G_\alpha = I \quad \text{on } C_0(\overline{D} \setminus M). \tag{14.14b}$$

Since the range $\mathcal{R}(\alpha I - A)$ is dense, for any given function $f \in C_0(\overline{D} \setminus M)$, we can find a sequence $\{u_n\}$ in $\mathcal{D}(A)$ such that

$$(\alpha - A)u_n \longrightarrow f \quad \text{in } C_0(\overline{D} \setminus M) .$$

Then we have

$$u_n = G_\alpha(\alpha - A)u_n \longrightarrow G_\alpha f \quad \text{in } C_0(\overline{D} \setminus M) ,$$

and also

$$Au_n = (A - \alpha)u_n + u_n \longrightarrow -f + \alpha G_\alpha f \quad \text{in } C_0(\overline{D} \setminus M) .$$

Hence it follows that

$$\begin{aligned} &G_\alpha f \in \mathcal{D}(\overline{A}) , \\ &\overline{A}(G_\alpha f) = -f + \alpha G_\alpha f , \end{aligned}$$

or equivalently,

$$(\alpha I - \overline{A})G_\alpha f = f, \quad f \in C_0(\overline{D} \setminus M) .$$

This proves formula (14.14b).

(6) By combining assertions (14.14), (14.9) and (14.10), we obtain from part (ii) of Theorem 3.3 that the minimal closed extension $\overline{A}$ is the infinitesimal generator of some Feller semigroup $\{T_t\}_{t \geq 0}$ on $\overline{D} \setminus M$. □

Step 2: End of Proof of Theorem 1.7. Recall that $\mathcal{W}$ is a linear operator from $C_0(\overline{D} \setminus M)$ into itself defined as follows:

(a) The domain of definition $\mathcal{D}(\mathcal{W})$ is the set

$$\begin{aligned} &\mathcal{D}(\mathcal{W}) \\ &= \left\{u \in C^2(\overline{D}) \cap C_0(\overline{D} \setminus M) : Wu \in C_0(\overline{D} \setminus M),\ L_0 u = 0 \text{ on } \partial D\right\} . \end{aligned}$$

(b) $\mathcal{W}u = Wu$, $u \in \mathcal{D}(\mathcal{W})$.

We verify conditions (a), (b) and (c) in Theorem 14.5 for the operator $\mathcal{W}$ to prove that the minimal closed extension $\overline{\mathcal{W}}$ is the infinitesimal generator of some Feller semigroup $\{e^{t\mathcal{W}}\}_{t\geq 0}$ on $\overline{D} \setminus M$.

Condition (c): We obtain from Theorem 1.6 (and its proof) that the operator

$$(W - \alpha, L_0) \colon C^{2+\theta}(\overline{D}) \longrightarrow C^{\theta}(\overline{D}) \bigoplus C^{1+\theta}_{L_0}(\partial D)$$

is an algebraic and topological isomorphism for all $\alpha > 0$. This verifies condition (c), since the range $\mathcal{R}(\mathcal{W} - \alpha I)$ contains the space $C^{\theta}(\overline{D}) \cap C_0(\overline{D} \setminus M)$ which is dense in $C_0(\overline{D} \setminus M)$.

Condition (b): First, let x_0 be a point of D such that $u(x_0) = \max_{\overline{D}} u > 0$. Then we have

$$\begin{aligned} Wu(x_0) &= Au(x_0) + Su(x_0) \\ &:= \sum_{i,j=1}^{N} a^{ij}(x_0) \frac{\partial^2 u}{\partial x_i \partial x_j}(x_0) + c(x_0)u(x_0) \\ &\quad + \int_{\mathbf{R}^N \setminus \{0\}} (u(x_0 + z) - u(x_0))\, s(x_0, z)\, m(dz) \\ &\leq 0 . \end{aligned}$$

since the matrix $(a^{ij}(x))$ is positive definite, $c(x) \leq 0$ in D and $0 \leq s(x,z) \leq 1$ on $\overline{D} \times \mathbf{R}^N$.

Next, let x_0' be a point of $\partial D \setminus M$ such that $u(x_0') = \max u > 0$. Assume, to the contrary, that

$$\mathcal{W}u(x_0') = Wu(x_0') > 0 .$$

We have only to consider the case where u is not a constant. Then it follows from an application of the boundary point lemma (Lemma C.5 in Appendix C) that

$$\frac{\partial u}{\partial \mathbf{n}}(x_0') < 0 .$$

Since we have

$$0 = L_0 u(x_0') = \mu(x_0') \frac{\partial u}{\partial \mathbf{n}}(x_0') + \gamma(x_0') u(x_0') ,$$

we obtain that

$$\mu(x_0') = 0, \quad \gamma(x_0') = 0 .$$

This contradicts the hypothesis: $x_0' \in \partial D \setminus M$, that is, $\mu(x_0') > 0$.

Condition (a): The density of the domain $\mathcal{D}(\mathcal{W})$ can be proved just as in the proof of Theorem 13.18. Indeed, it suffices to note that Theorem 13.4 and Lemma 13.16 remain valid for elliptic Waldenfels operators W (see [BCP, Proposition III.1.6]).

The proof of Theorem 1.7 is complete. □

14.4 Proof of Theorem 1.8

In this section we prove Theorem 1.8, generalizing Theorem 1.3 to the integro-differential operator case.

Recall that W_p is a linear operator from $L^p(D)$ into itself defined as follows:

(a) The domain of definition $\mathcal{D}(W_p)$ is the set

$$\mathcal{D}(W_p) = \{u \in H^{2,p}(D) : L_0 u = 0 \text{ on } \partial D\} .$$

(b) $W_p u = Wu = (A + S)u$, $u \in \mathcal{D}(W_p)$.

We estimate the integral operator S in terms of L^p norms, and show that S is an A_p-*completely continuous* operator in the sense of Gohberg-Kreĭn [GK] (Lemma 14.8).

Step 1: First, we prove part (i) of Theorem 1.8:

Theorem 14.6. *Let* $1 < p < \infty$. *Assume that condition (H) is satisfied:*
(H) $\mu(x') + |\gamma(x')| > 0$ *on* ∂D.
Then, for every $0 < \varepsilon < \pi/2$, *there exists a constant* $r_p(\varepsilon) > 0$ *such that the resolvent set of* W_p *contains the set*

$$\Sigma_p(\varepsilon) = \{\lambda = r^2 e^{i\vartheta} : r \geq r_p(\varepsilon), -\pi + \varepsilon \leq \vartheta \leq \pi - \varepsilon\} ,$$

and that the resolvent $(W_p - \lambda I)^{-1}$ *satisfies the estimate*

$$\left\|(W_p - \lambda I)^{-1}\right\| \leq \frac{c_p(\varepsilon)}{|\lambda|}, \quad \lambda \in \Sigma_p(\varepsilon) . \tag{1.9}$$

Proof. The proof is divided into three steps.

(1) We show that there exist constants $r_p(\varepsilon)$ and $c_p(\varepsilon)$ such that we have, for all $\lambda = r^2 e^{i\vartheta}$ satisfying $r \geq r_p(\varepsilon)$ and $-\pi + \varepsilon \leq \vartheta \leq \pi + \varepsilon$,

$$|u|_{2,p} + |\lambda|^{1/2}|u|_{1,p} + |\lambda|\|u\|_p \leq c_p(\varepsilon)\|(W_p - \lambda I)u\|_p . \tag{14.15}$$

Here

$$\|u\|_p = \|u\|_{L^p(D)}, \quad |u|_{1,p} = \|\nabla u\|_{L^p(D)}, \quad |u|_{2,p} = \|\nabla^2 u\|_{L^p(D)} .$$

First, we recall (see formula (9.2)) that estimate (14.15) is proved for the differential operator A:

$$|u|_{2,p} + |\lambda|^{1/2}|u|_{1,p} + |\lambda|\,\|u\|_p \leq c_p'(\varepsilon)\|(A_p - \lambda I)u\|_p \,. \tag{14.16}$$

Here recall that A_p is a linear operator from $L^p(D)$ into itself defined as follows:

(a) The domain of definition $\mathcal{D}(A_p)$ is the set

$$\mathcal{D}(A_p) = \{u \in H^{2,p}(D) : L_0 u = 0 \text{ on } \partial D\} \,.$$

(b) $A_p u = Au$, $u \in \mathcal{D}(A_p)$.

To replace the last term $\|(A_p - \lambda I)u\|_p$ by the term $\|(W_p - \lambda I)u\|_p$, we need the following L^p-estimate for the integral operator S:

Lemma 14.7. *For every $\eta > 0$, there exists a constant $C_\eta > 0$ such that we have, for all $u \in H^{2,p}(D)$,*

$$\|Su\|_p \leq \eta |u|_{2,p} + C_\eta \left(\|u\|_p + |u|_{1,p}\right) \,. \tag{14.17}$$

Proof. We decompose the term Su into the following three terms:

$$\begin{aligned}
Su(x) &= \int_0^1 (1-t)\,dt \int_{\{|z|\leq\varepsilon\}} z\cdot\nabla^2 u(x+tz)z\, s(x,z)\, m(dz) \\
&\quad + \int_{\{|z|>\varepsilon\}} (u(x+z) - u(x)) s(x,z)\, m(dz) \\
&\quad - \int_{\{|z|>\varepsilon\}} z\cdot\nabla u(x)\, s(x,z)\, m(dz) \\
&:= S_1 u(x) + S_2 u(x) - S_3 u(x) \,.
\end{aligned}$$

First, we estimate the L^p norm of the term $S_3 u$. By using estimate (14.1b), we obtain that

$$\left|\int_{\{|z|>\varepsilon\}} z\cdot\nabla u(x)\, s(x,z)\, m(dz)\right| \leq \delta(\varepsilon)\,|\nabla u(x)| \leq \left(\frac{C_1}{\varepsilon} + C_2\right)|\nabla u(x)| \,.$$

Hence we have the L^p estimate of the term $S_3 u$:

$$\|S_3 u\|_p \leq \left(\frac{C_1}{\varepsilon} + C_2\right)\|\nabla u\|_p \,.$$

Secondly, we have

$$\left\|\int_{\{|z|>\varepsilon\}} u(\cdot)\, s(\cdot,z)\, m(dz)\right\|_p \leq \left(\frac{C_1}{\varepsilon^2} + C_2\right)\|u\|_p \,.$$

Furthermore, by using Hölder's inequality and Fubini's theorem we obtain from condition (1.6) that

$$\begin{aligned}
&\int_{\mathbf{R}^N}\left|\int_{\{|z|>\varepsilon\}} u(x+z)\, s(x,z)\, m(dz)\right|^p dx \\
&\leq \int_{\mathbf{R}^N}\left(\int_{\{|z|>\varepsilon\}} |u(x+z)|\, s(x,z)\, m(dz)\right)^p dx \\
&\leq \int_{\mathbf{R}^N}\left(\int_{\{|z|>\varepsilon\}} |u(x+z)|^p s(x,z)^p m(dz)\right)\left(\int_{\{|z|>\varepsilon\}} m(dz)\right)^{p/q} dx \\
&= \tau(\varepsilon)^{p/q}\int_{\mathbf{R}^N}\int_{\{|z|>\varepsilon\}} |u(x+z)|^p s(x,z)^p\, m(dz)\, dx \\
&= \tau(\varepsilon)^{p/q}\int_{\{|z|>\varepsilon\}}\left(\int_{\mathbf{R}^N} |u(x+z)|^p s(x,z)^p\, dx\right) m(dz) \\
&\leq \tau(\varepsilon)^{p/q}\left(\int_D |u(y)|^p\, dy\right)\left(\int_{\{|z|>\varepsilon\}} m(dz)\right) \\
&= \tau(\varepsilon)^p \|u\|_p^p.
\end{aligned}$$

By estimate (14.1c), we have the L^p estimate of the term $S_2 u$:

$$\|S_2 u\|_p \leq \left(\frac{C_1}{\varepsilon^2} + C_2\right) \|u\|_p\, .$$

Similarly, by using Hölder's inequality and Fubini's theorem we find that

$$\begin{aligned}
&\int_{\mathbf{R}^N}\left|\int_0^1 (1-t)\, dt \int_{\{|z|\leq\varepsilon\}} z\cdot\nabla^2 u(x+tz) z\, s(x,z)\, m(dz)\right|^p dx \\
&\leq \int_{\mathbf{R}^N}\left(\int_0^1 dt \int_{\{|z|\leq\varepsilon\}} |z|^2 |\nabla^2 u(x+tz)|\, s(x,z)\, m(dz)\right)^p dx \\
&\leq \int_{\mathbf{R}^N}\int_0^1 dt\left(\int_{\{|z|\leq\varepsilon\}} |z|^2 |\nabla^2 u(x+tz)|^p\, s(x,z)^p\, m(dz)\right) \\
&\quad\times\left(\int_{\{|z|\leq\varepsilon\}} |z|^2\, m(dz)\right)^{p/q} dx \\
&= \sigma(\varepsilon)^{p/q}\int_{\mathbf{R}^N}\int_0^1 dt\left(\int_{\{|z|\leq\varepsilon\}} |z|^2 |\nabla^2 u(x+tz)|^p\, s(x,z)^p\, m(dz)\right) dx \\
&= \sigma(\varepsilon)^{p/q}\int_0^1 dt \int_{\{|z|\leq\varepsilon\}} |z|^2 \left(\int_{\mathbf{R}^N} |\nabla^2 u(x+tz)|^p\, s(x,z)^p\, dx\right) m(dz) \\
&\leq \sigma(\varepsilon)^{p/q}\left(\int_D |\nabla^2 u(y)|^p\, dy\right)\left(\int_{\{|z|\leq\varepsilon\}} |z|^2\, m(dz)\right)
\end{aligned}$$

$$\leq \sigma(\varepsilon)^p \left(\int_D |\nabla^2 u(y)|^p \, dy \right) .$$

Hence we have the L^p estimate of the term $S_1 u$:

$$\|S_1 u\|_p \leq \sigma(\varepsilon) \, \|\nabla^2 u\|_p .$$

Summing up, we have proved that

$$\begin{aligned}\|Su\|_p &\leq \|S_1 u\|_p + \|S_2 u\|_p + \|S_3 u\|_p \\ &\leq \sigma(\varepsilon) \, |u|_{2,p} + \left(\frac{C_1}{\varepsilon} + C_2 \right) |u|_{1,p} + \left(\frac{C_1}{\varepsilon^2} + C_2 \right) \|u\|_p .\end{aligned}$$

In view of assertion (14.1a), this proves estimate (14.17) if we choose ε sufficiently small. □

Since we have

$$(A - \lambda)u = (W - \lambda)u - Su ,$$

it follows from estimate (14.17) that

$$\|(A_p - \lambda I)u\|_p \leq \|(W_p - \lambda I)u\|_p + \eta |u|_{2,p} + C_\eta \left(|u|_{1,p} + \|u\|_p \right) .$$

Thus, carrying this estimate into estimate (14.16) we obtain that

$$\begin{aligned}&|u|_{2,p} + |\lambda|^{1/2} |u|_{1,p} + |\lambda| \|u\|_p \\ &\leq c'_p(\varepsilon) \|(W_p - \lambda I)u\|_p + \eta c'_p(\varepsilon) |u|_{2,p} + C_\eta c'_p(\varepsilon) \left(|u|_{1,p} + \|u\|_p \right) . \quad (14.18)\end{aligned}$$

Therefore, the desired estimate (14.15) follows from estimate (14.18) if we take the constant η so small that

$$\eta c'_p(\varepsilon) < 1$$

and the parameter λ so large that

$$|\lambda|^{1/2} > C_\eta c'_p(\varepsilon) .$$

(2) By estimate (14.15), we find that the operator $W_p - \lambda I$ is injective and its range $\mathcal{R}(W_p - \lambda I)$ is closed in $L^p(D)$, for all $\lambda \in \Sigma_p(\varepsilon)$.

We show that the operator $W_p - \lambda I$ is surjective for all $\lambda \in \Sigma_p(\varepsilon)$:

$$\mathcal{R}(W_p - \lambda I) = L^p(D), \quad \lambda \in \Sigma_p(\varepsilon) .$$

To do this, it suffices to show that the operator $W_p - \lambda I$ is a Fredholm operator with

$$\operatorname{ind}(W_p - \lambda I) = 0, \quad \lambda \in \Sigma_p(\varepsilon) , \tag{14.19}$$

since $W_p - \lambda I$ is injective for all $\lambda \in \Sigma_p(\varepsilon)$.

To prove assertion (14.19), we need the following:

Lemma 14.8. *The operator S is A_p-*completely continuous, *that is, the operator $S: \mathcal{D}(A_p) \to L^p(D)$ is completely continuous where the domain $\mathcal{D}(A_p)$ is endowed with the graph norm of A_p.*

Proof. Let $\{u_j\}$ be an arbitrary bounded sequence in $\mathcal{D}(A_p)$; hence there exists a constant $K > 0$ such that

$$\|u_j\|_p \leq K, \quad \|A_p u_j\|_p \leq K .$$

Then we have, by estimate (9.1),

$$\|u_j\|_{2,p} \leq C\|A_p u_j\|_p \leq CK . \tag{14.20}$$

Therefore, by Rellich's theorem we may assume that the sequence $\{u_j\}$ itself is a Cauchy sequence in the space $H^{1,p}(D)$. Then, applying estimate (14.17) to the sequence $\{u_j - u_k\}$ and using estimate (14.20) we obtain that

$$\begin{aligned} \|Su_j - Su_k\|_p &\leq \eta |u_j - u_k|_{2,p} + C_\eta \left(\|u_j - u_k\|_p + |u_j - u_k|_{1,p} \right) \\ &\leq 2\eta CK + C_\eta \|u_j - u_k\|_{1,p} . \end{aligned}$$

Hence we have

$$\limsup_{j,k \to \infty} \|Su_j - Su_k\|_p \leq 2\eta CK .$$

This proves that the sequence $\{Su_j\}$ is a Cauchy sequence in the space $L^p(D)$, since η is arbitrary. □

In view of Lemma 14.8, assertion (14.19) follows from an application of [GK, Theorem 2.6]. Indeed, we have, by Theorem 1.3,

$$\operatorname{ind}(W_p - \lambda I) = \operatorname{ind}(A_p - \lambda I + S) = \operatorname{ind}(A_p - \lambda I) = 0 .$$

(3) Summing up, we have proved that the operator $W_p - \lambda I$ is bijective for all $\lambda \in \Sigma_p(\varepsilon)$ and its inverse $(W_p - \lambda I)^{-1}$ satisfies estimate (1.9).

The proof of Theorem 14.6 is now complete. □

Step 2: Finally, the proof of part (ii) of Theorem 1.8 needs an (easy and by now standard) additional argument as in Sect. 13.4. □

14.5 Proof of Theorem 1.9

This final section is devoted to the proof of Theorem 1.9. Namely, we prove that Theorem 1.8 remains valid with $L^p(D)$ and W_p replaced by $C_0(\overline{D} \setminus M)$ and $\mathfrak{W}$, respectively. Recall that $\mathfrak{W}$ is a linear operator from $C_0(\overline{D} \setminus M)$ into itself defined as follows:

(a) The domain of definition $\mathcal{D}(\mathfrak{W})$ is the set

$$\begin{aligned} \mathcal{D}(\mathfrak{W}) = \{ u \in C_0(\overline{D} \setminus M) \cap H^{2,p}(D) : Wu \in C_0(\overline{D} \setminus M), \\ L_0 u = 0 \text{ on } \partial D \} , \qquad N < p < \infty . \end{aligned}$$

(b) $\mathfrak{W}u = Wu$, $u \in \mathcal{D}(\mathfrak{W})$.

Theorem 1.9 follows from Theorem 1.8 by using Sobolev's imbedding theorems and a λ-dependent localization argument (Lemma 14.10). The proof is carried out in a chain of auxiliary lemmas.

Step 1: We begin with a version of estimate (14.15):

Lemma 14.9. *Let $N < p < \infty$. Assume that condition (H) is satisfied:*
(H) $\mu(x') + |\gamma(x')| > 0$ on ∂D.
Then, for every $\varepsilon > 0$, there exists a constant $r_p(\varepsilon) > 0$ such that if $\lambda = r^2 e^{i\vartheta}$ with $r \geq r_p(\varepsilon)$ and $-\pi + \varepsilon \leq \vartheta \leq \pi - \varepsilon$, we have, for all $u \in \mathcal{D}(W_p)$,

$$|\lambda|^{1/2}\|u\|_{C^1(\overline{D})} + |\lambda|\,\|u\|_{C(\overline{D})} \leq C_p(\varepsilon)|\lambda|^{N/2p}\|(W - \lambda I)u\|_p\,, \tag{14.21}$$

with a constant $C_p(\varepsilon) > 0$.

Proof. First, it follows from an application of the Gagliardo-Nirenberg inequality (12.3) with $p = r > N$, $q := \infty$ and $\theta := N/p$ that

$$\|u\|_{C(\overline{D})} \leq C|u|_{1,p}^{N/p}\|u\|_p^{1-N/p}, \quad u \in H^{1,p}(D)\,. \tag{14.22}$$

Here and in the following the letter C denotes a generic positive constant depending on p and ε, but independent of u and λ.

Combining inequality (14.22) with inequality (14.15), we obtain that

$$\begin{aligned}\|u\|_{C(\overline{D})} &\leq C\left(|\lambda|^{-1/2}\|(W-\lambda)u\|_p\right)^{N/p}\left(|\lambda|^{-1}\|(W-\lambda)u\|_p\right)^{1-N/p}\\ &= C|\lambda|^{-1+N/2p}\|(W-\lambda)u\|_p\,,\end{aligned}$$

so that

$$|\lambda|\,\|u\|_{C(\overline{D})} \leq C|\lambda|^{N/2p}\|(W-\lambda)u\|_p, \quad u \in \mathcal{D}(W_p)\,. \tag{14.23}$$

Similarly, applying inequality (14.22) to the functions $D_i u \in H^{1,p}(D)$, $1 \leq i \leq n$, we obtain that

$$\begin{aligned}\|\nabla u\|_{C(\overline{D})} &\leq C|\nabla u|_{1,p}^{N/p}\|\nabla u\|_p^{1-N/p}\\ &\leq C|u|_{2,p}^{N/p}|u|_{1,p}^{1-N/p}\\ &\leq C\left(\|(W-\lambda)u\|_p\right)^{N/p}\left(|\lambda|^{-1/2}\|(W-\lambda)u\|_p\right)^{1-N/p}\\ &= C|\lambda|^{-1/2+N/2p}\|(W-\lambda)u\|_p\,.\end{aligned}$$

This proves that

$$|\lambda|^{1/2}\|u\|_{C^1(\overline{D})} \leq C|\lambda|^{N/2p}\|(W-\lambda)u\|_p, \quad u \in \mathcal{D}(W_p)\,. \tag{14.24}$$

Therefore, the desired inequality (14.21) follows by combining inequalities (14.23) and (14.24). □

Step 2: The next lemma proves the resolvent estimate

$$\|(\mathfrak{W} - \lambda I)^{-1}\| \leq \frac{c(\varepsilon)}{|\lambda|}, \quad \lambda \in \Sigma(\varepsilon) . \tag{1.10}$$

Lemma 14.10. *Let $N < p < \infty$. If condition (H) is satisfied, then, for every $\varepsilon > 0$, there exists a constant $r(\varepsilon) > 0$ such that if $\lambda = r^2 e^{i\vartheta}$ with $r \geq r(\varepsilon)$ and $-\pi + \varepsilon \leq \vartheta \leq \pi - \varepsilon$, we have, for all $u \in \mathcal{D}(\mathfrak{W})$,*

$$|\lambda|^{1/2} \|u\|_{C^1(\overline{D})} + |\lambda| \, \|u\|_{C(\overline{D})} \leq c(\varepsilon) \|(\mathfrak{W} - \lambda I)u\|_{C(\overline{D})} , \tag{14.25}$$

with a constant $c(\varepsilon) > 0$.

Proof. The proof is divided into four steps.

(1) First, we show that the domain

$$\mathcal{D}(\mathfrak{W}) = \left\{u \in C_0(\overline{D} \setminus M) \cap H^{2,p}(D) : Wu \in C_0(\overline{D} \setminus M), \ L_0 u = 0 \text{ on } \partial D\right\}$$

is *independent* of $N < p < \infty$.

We let

$$\mathcal{D}_p = \left\{u \in H^{2,p}(D) \cap C_0(\overline{D} \setminus M) : Wu \in C_0(\overline{D} \setminus M), \ L_0 u = 0 \text{ on } \partial D\right\} .$$

Since we have $L^{p_1}(D) \subset L^{p_2}(D)$ for $p_1 > p_2$, it follows that

$$\mathcal{D}_{p_1} \subset \mathcal{D}_{p_2} \quad \text{if } p_1 > p_2 .$$

Conversely, let v be an arbitrary element of $\mathcal{D}_{p_2}$:

$$v \in H^{2,p_2}(D) \cap C_0(\overline{D} \setminus M), \quad Wv \in C_0(\overline{D} \setminus M), \quad L_0 v = 0 .$$

Then, since we have v, $Wv \in C_0(\overline{D} \setminus M) \subset L^{p_1}(D)$, it follows from an application of Theorem 14.6 with $p := p_1$ that there exists a unique function $u \in H^{2,p_1}(D)$ such that

$$\begin{cases} (W - \lambda)u = (W - \lambda)v & \text{in } D , \\ L_0 u = 0 & \text{on } \partial D , \end{cases}$$

if we choose λ sufficiently large. Hence we have $u - v \in H^{2,p_2}(D)$ and

$$\begin{cases} (W - \lambda)(u - v) = 0 & \text{in } D , \\ L_0(u - v) = 0 & \text{on } \partial D . \end{cases}$$

Therefore, by applying again Theorem 14.6 with $p := p_2$ we obtain that $u - v = 0$, so that $v = u \in H^{2,p_1}(D)$. This proves that $v \in \mathcal{D}_{p_1}$.

(2) We shall make use of a λ-*dependent localization* argument in order to adjust the term $\|(W - \lambda)u\|_p$ in inequality (14.21) to obtain inequality (14.25), just as in Sect. 12.2.

(2–a) If x'_0 is a point of ∂D, then we introduce a local coordinate system (x', x_N) near x'_0 (see Fig. 12.2) such that $x' = (x_1, x_2, \ldots, x_{N-1})$ give local coordinates for the boundary ∂D and that

$$\begin{aligned} D &= \{(x', x_N) : x_N > 0\} , \\ \partial D &= \{(x', x_N) : x_N = 0\} , \\ x'_0 &= (0, \ldots, 0, 0) , \\ \mathbf{n} &= (0, \ldots, 0, 1) , \end{aligned}$$

and let

$$\begin{aligned} G_0 &= B(x'_0, \eta_0) \cap D , \\ G' &= B(x'_0, \eta) \cap D, \ 0 < \eta < \eta_0 , \\ G'' &= B(x'_0, \eta/2) \cap D, \ 0 < \eta < \eta_0 . \end{aligned}$$

Similarly, if x_0 is a point of D, then we let (see Fig. 12.3)

$$\begin{aligned} G_0 &= B(x_0, \eta_0) \subset D , \\ G' &= B(x_0, \eta), \ 0 < \eta < \eta_0 , \\ G'' &= B(x_0, \eta/2), \ 0 < \eta < \eta_0 . \end{aligned}$$

(2–b) We take a function $\Theta(t) \in C_0^\infty(\mathbf{R})$ such that $\Theta(t)$ equals one near the origin, and define a localizing function

$$\varphi_0(x) := \Theta\left(\frac{|x' - x'_0|^2}{\eta^2}\right) \Theta\left(\frac{x_N - t_0}{\eta}\right), \quad x_0 = (x'_0, t_0) . \tag{14.26}$$

Then we have the following claim, analogous to Claim 12.5:

Claim 14.11. *If $u \in \mathcal{D}(\mathfrak{W})$, then it follows that $\varphi_0 u \in \mathcal{D}(W_p)$.*

Proof. First, we recall that

$$u \in H^{2,p}(D) \quad \text{for all } 1 < p < \infty .$$

Hence we have

$$\varphi_0 u \in H^{2,p}(D) .$$

Furthermore, it follows from the proof of Claim 12.5 that the function $\varphi_0 u$ satisfies the boundary condition

$$L_0(\varphi_0 u) = 0 \quad \text{on } \partial D .$$

Summing up, we have proved that

$$\varphi_0 u \in \mathcal{D}(W_p) \quad \text{for all } 1 < p < \infty . \quad \square$$

(3) Now let u be an arbitrary element of $\mathcal{D}(\mathfrak{W})$. Then, by Claim 14.11 we can apply inequality (14.21) to the function $\varphi_0 u$ to obtain that

$$\begin{aligned} &|\lambda|^{1/2}\|u\|_{C^1(\overline{G''})} + |\lambda|\,\|u\|_{C(\overline{G''})} \\ &\le |\lambda|^{1/2}\,\|\varphi_0 u\|_{C^1(\overline{G'})} + |\lambda|\,\|\varphi_0 u\|_{C(\overline{G'})} \\ &= |\lambda|^{1/2}\,\|\varphi_0 u\|_{C^1(\overline{D})} + |\lambda|\,\|\varphi_0 u\|_{C(\overline{D})} \\ &\le C|\lambda|^{N/2p}\,\|(W-\lambda)(\varphi_0 u)\|_{L^p(D)} \ . \end{aligned} \tag{14.27}$$

(3–a) We estimate the last term $\|(W-\lambda)(\varphi_0 u)\|_{L^p(D)}$ in terms of the supremum norm of $C(\overline{D})$.

First, we write the term $(W-\lambda)(\varphi_0 u)$ in the form

$$(W-\lambda)(\varphi_0 u) = \varphi_0\left((W-\lambda)u\right) + [A,\varphi_0]u + [S,\varphi_0]u \ ,$$

where $[A,\varphi_0]$ and $[S,\varphi_0]$ are the commutators of A and φ_0 and of S and φ_0, respectively:

$$\begin{aligned} [A,\varphi_0]u &= A(\varphi_0 u) - \varphi_0 Au \ , \\ [S,\varphi_0]u &= S(\varphi_0 u) - \varphi_0 Su \ . \end{aligned}$$

Since we have, for some constant $c > 0$,

$$|G'| \le |B(x_0,\eta)| \le c\eta^N \ ,$$

it follows from an application of Claim 12.6 that

$$\begin{aligned} \|\varphi_0(W-\lambda)u\|_{L^p(D)} &= \|\varphi_0(W-\lambda)u\|_{L^p(G')} \\ &\le c^{1/p}\eta^{N/p}\,\|(W-\lambda)u\|_{C(\overline{G'})} \\ &\le c^{1/p}\eta^{N/p}\,\|(W-\lambda)u\|_{C(\overline{D})} \ . \end{aligned} \tag{14.28}$$

On the other hand, we can estimate the commutators $[A,\varphi_0]u$ and $[S,\varphi_0]u$ as follows:

Lemma 14.12. *There exists a constant $C > 0$, independent of η, such that we have, as $\eta \downarrow 0$,*

$$\|[A,\varphi_0]u\|_{L^p(D)} \le C\left(\eta^{-1+N/p}\|u\|_{C^1(\overline{D})} + \eta^{-2+N/p}\|u\|_{C(\overline{D})}\right) , \tag{14.29a}$$

$$\|[S,\varphi_0]u\|_{L^p(D)} \le C\left(\eta^{-1+N/p}\|u\|_{C^1(\overline{D})} + \eta^{-2+N/p}\|u\|_{C(\overline{D})}\right) . \tag{14.29b}$$

Proof. Estimate (14.29a) is already proved in Lemma 12.7 (see estimate (12.14))

To prove estimate (14.29b), it should be noticed that

$$
\begin{aligned}
&S(\varphi_0 u)(x) \\
&= \int_{\mathbf{R}^N \setminus \{0\}} (\varphi_0(x+z)u(x+z) - \varphi_0(x)u(x) - z \cdot \nabla(\varphi_0 u)(x))\, s(x,z)\, m(dz) \\
&= \varphi_0(x) \int_{\mathbf{R}^N \setminus \{0\}} (u(x+z) - u(x) - z \cdot \nabla u(x))\, s(x,z)\, m(dz) \\
&\quad + \left(\int_{\mathbf{R}^N \setminus \{0\}} (u(x+z) - u(x)) z\, s(x,z)\, m(dz) \right) \cdot \nabla \varphi_0(x) \\
&\quad + \int_{\mathbf{R}^N \setminus \{0\}} (\varphi_0(x+z) - \varphi_0(x) - z \cdot \nabla \varphi_0(x))\, u(x+z)\, s(x,z)\, m(dz) \\
&= \varphi_0(x) S u(x) + \left(\int_{\mathbf{R}^N \setminus \{0\}} (u(x+z) - u(x)) z\, s(x,z)\, m(dz) \right) \cdot \nabla \varphi_0(x) \\
&\quad + \int_{\mathbf{R}^N \setminus \{0\}} (\varphi_0(x+z) - \varphi_0(x) - z \cdot \nabla \varphi_0(x))\, u(x+z)\, s(x,z)\, m(dz) .
\end{aligned}
$$

Hence we can write the commutator $[S, \varphi_0]u$ in the form

$$
\begin{aligned}
&[S, \varphi_0] u(x) \\
&= \left(\int_{\mathbf{R}^N \setminus \{0\}} (u(x+z) - u(x)) z\, s(x,z)\, m(dz) \right) \cdot \nabla \varphi_0(x) \\
&\quad + \int_{\mathbf{R}^N \setminus \{0\}} (\varphi_0(x+z) - \varphi_0(x) - z \cdot \nabla \varphi_0(x))\, u(x+z)\, s(x,z)\, m(dz) \\
&:= S_0^{(1)} u(x) + S_0^{(2)} u(x) .
\end{aligned}
$$

First, just as in Lemma 14.2 we can estimate the term $S_0^{(1)} u$ as follows:

$$
\begin{aligned}
\|S_0^{(1)} u\|_{L^p(D)} &= \|S_0^{(1)} u\|_{L^p(G')} \\
&\leq 2 \left(\sigma(\eta) \|u\|_{C^1(\overline{D})} + \delta(\eta) \|u\|_{C(\overline{D})} \right) \|\nabla \varphi_0\|_{L^p(G')} \\
&\leq 2 \left(\sigma(\eta) \|u\|_{C^1(\overline{D})} + \left(\frac{C_1}{\eta} + C_2 \right) \|u\|_{C(\overline{D})} \right) \|\nabla \varphi_0\|_{L^p(G')} .
\end{aligned}
$$

However, it follows from an application of Claim 12.6 that

$$
\begin{aligned}
\|\nabla \varphi_0\|_{L^p(G')} &\leq C \eta^{N/p} \|\nabla \varphi_0\|_{C(\overline{G'})} \leq C' \eta^{-1+N/p} , \\
\|\nabla^2 \varphi_0\|_{L^p(G')} &\leq C \eta^{N/p} \|\nabla^2 \varphi_0\|_{C(\overline{G'})} \leq C' \eta^{-2+N/p} ,
\end{aligned}
$$

since we have, by formula (14.26),

$$
|\nabla \varphi_0(x)| = O(\eta^{-1}), \quad |\nabla^2 \varphi_0(x)| = O(\eta^{-2})
$$

as $\eta \downarrow 0$. Hence we obtain that

$$\|S_0^{(1)}u\|_{L^p(D)} \leq C\left(\eta^{-1+N/p}\|u\|_{C^1(\overline{D})} + \eta^{-2+N/p}\|u\|_{C(\overline{D})}\right) . \tag{14.30}$$

Similarly, arguing as in the proof of Lemma 14.7 we can estimate the term $S_0^{(2)}u$ as follows:

$$\begin{aligned} \|S_0^{(2)}u\|_{L^p(D)} &\leq C\|u\|_{C(\overline{D})}\|\nabla^2\varphi_0\|_{L^p(G')} \\ &\leq C\|u\|_{C(\overline{D})}\, \eta^{N/p}\|\nabla^2\varphi_0\|_{C(\overline{G'})} \\ &\leq C\eta^{-2+N/p}\|u\|_{C(\overline{D})} . \end{aligned} \tag{14.31}$$

Thus the desired estimate (14.29b) follows by combining estimates (14.30) and (14.31), since $[S, \varphi_0]u = S_0^{(1)}u + S_0^{(2)}u$. □

Therefore, combining estimates (14.27), (14.28), (14.29a) and (14.29b) we obtain that

$$\begin{aligned} &|\lambda|^{1/2}\|u\|_{C^1(\overline{G''})} + |\lambda|\,\|u\|_{C(\overline{G''})} \\ &\leq C|\lambda|^{N/2p}\,\|(W-\lambda)(\varphi_0 u)\|_{L^p(D)} \\ &= C|\lambda|^{N/2p}\,\|\varphi_0\left((W-\lambda)u\right) + [A,\varphi_0]u + [S,\varphi_0]u\|_{L^p(D)} \\ &\leq C|\lambda|^{N/2p}\Big(\eta^{N/p}\,\|(W-\lambda)u\|_{C(\overline{G'})} + \eta^{-1+N/p}\|u\|_{C^1(\overline{G'})} \\ &\quad + \eta^{-2+N/p}\|u\|_{C(\overline{G'})}\Big) \\ &\leq C|\lambda|^{N/2p}\Big(\eta^{N/p}\,\|(W-\lambda)u\|_{C(\overline{D})} + \eta^{-1+N/p}\|u\|_{C^1(\overline{D})} \\ &\quad + \eta^{-2+N/p}\|u\|_{C(\overline{D})}\Big) . \end{aligned} \tag{14.32}$$

(3–b) We recall (see Fig. 12.3) that the closure $\overline{D} = D \cup \partial D$ can be covered by a finite number of sets of the forms

$$\begin{cases} B(x_0, \eta/2), & x_0 \in D \ , \\ B(x_0', \eta/2) \cap \overline{D}, & x_0' \in \partial D \ . \end{cases}$$

Therefore, taking the supremum of inequality (14.32) over $x \in \overline{D}$ we find that

$$\begin{aligned} &|\lambda|^{1/2}\|u\|_{C^1(\overline{D})} + |\lambda|\,\|u\|_{C(\overline{D})} \\ &\leq C|\lambda|^{N/2p}\eta^{N/p}\left(\|(W-\lambda)u\|_{C(\overline{D})} + \eta^{-1}\|u\|_{C^1(\overline{D})} + \eta^{-2}\|u\|_{C(\overline{D})}\right) . \end{aligned} \tag{14.33}$$

(4) We now choose the localization parameter η. We let

$$\eta = \frac{\eta_0}{|\lambda|^{1/2}}K ,$$

where K is a positive constant (to be chosen later on) satisfying the condition

$$0 < \eta = \frac{\eta_0}{|\lambda|^{1/2}} K < \eta_0 \,,$$

that is,

$$0 < K < |\lambda|^{1/2} \,.$$

Then we obtain from inequality (14.33) that

$$\begin{aligned} &|\lambda|^{1/2} \|u\|_{C^1(\overline{D})} + |\lambda| \, \|u\|_{C(\overline{D})} \\ &\leq C \eta_0^{N/p} K^{N/p} \|(W - \lambda)u\|_{C(\overline{D})} + \left(C \eta_0^{N/p-1} K^{-1+N/p} \right) |\lambda|^{1/2} \|u\|_{C^1(\overline{D})} \\ &\quad + \left(C \eta_0^{N/p-2} K^{-2+N/p} \right) |\lambda| \, \|u\|_{C(\overline{D})} \,. \end{aligned} \tag{14.34}$$

However, since the exponents $-1 + N/p$ and $-2 + N/p$ are negative, we can choose the constant K so large that

$$C \eta_0^{N/p-1} K^{-1+N/p} < 1 \,,$$

and

$$C \eta_0^{N/p-2} K^{-2+N/p} < 1 \,.$$

Then the desired inequality (14.25) follows from inequality (14.34). The proof of Lemma 14.10 is complete. □

Step 3: The next lemma, together with Lemma 14.10, proves part (i) of Theorem 1.9. Namely, the resolvent set of $\mathfrak{W}$ contains the set

$$\Sigma(\varepsilon) = \left\{ \lambda = r^2 e^{i\vartheta} : r \geq r(\varepsilon), \; -\pi + \varepsilon \leq \vartheta \leq \pi - \varepsilon \right\} \,.$$

Lemma 14.13. *If $\lambda \in \Sigma(\varepsilon)$, then, for any $f \in C_0(\overline{D} \setminus M)$ there exists a unique function $u \in \mathcal{D}(\mathfrak{W})$ such that $(\mathfrak{W} - \lambda I)u = f$.*

Proof. Since we have, for all $1 < p < \infty$,

$$f \in C_0(\overline{D} \setminus M) \subset L^p(D) \,,$$

it follows from an application of Theorem 1.8 that if $\lambda \in \Sigma_p(\varepsilon)$, there exists a unique function $u \in H^{2,p}(D)$ such that

$$(W - \lambda)u = f \quad \text{in } D \,, \tag{14.35}$$

and

$$L_0 u = \mu(x') \frac{\partial u}{\partial \mathbf{n}} + \gamma(x') u = 0 \quad \text{on } \partial D \,. \tag{14.36}$$

However, by Sobolev's imbedding theorem it follows that

$$u \in H^{2,p}(D) \subset C^{2-N/p}(\overline{D}) \subset C^1(\overline{D}) \quad \text{if} \quad N < p < \infty \,.$$

Hence we have, by formula (14.36) and condition (H),

$$u = 0 \quad \text{on} \quad M = \{x' \in \partial D : \mu(x') = 0\} ,$$

so that

$$u \in C_0(\overline{D} \setminus M) .$$

Furthermore, in view of equation (14.35) we find that

$$Wu = f + \lambda u \in C_0(\overline{D} \setminus M) .$$

Summing up, we have proved that

$$\begin{cases} u \in \mathcal{D}(\mathfrak{W}) , \\ (\mathfrak{W} - \lambda I)u = f . \end{cases}$$

This proves Lemma 14.13. □

Step 4: Finally, part (ii) of Theorem 1.9 is proved just as in Sect. 13.4. □

14.6 Concluding Remarks

By combining Theorems 1.7 and 1.9, we can prove that the operator $\mathfrak{W}$ coincides with the minimal closed extension $\overline{\mathcal{W}}$:

$$\mathfrak{W} = \overline{\mathcal{W}} . \tag{14.37}$$

Indeed, we recall that

$$\mathcal{D}(\mathcal{W}) = \{u \in C^2(\overline{D}) \cap C_0(\overline{D} \setminus M) : Wu \in C_0(\overline{D} \setminus M),\ L_0 u = 0 \text{ on } \partial D\} ,$$
$$\mathcal{D}(\mathfrak{W}) = \{u \in H^{2,p}(D) \cap C_0(\overline{D} \setminus M) : Wu \in C_0(\overline{D} \setminus M),\ L_0 u = 0 \text{ on } \partial D\} .$$

Then we have, by Theorems 1.7 and 1.9,

$$\overline{\mathcal{W}} \subset \mathfrak{W} .$$

However, since $\overline{\mathcal{W}}$ and $\mathfrak{W}$ generate semigroups of class (C_0) on $C_0(\overline{D} \setminus M)$, arguing as in Step (5) of the proof of Theorem 2.9 we obtain that $\mathcal{D}(\overline{\mathcal{W}}) = \mathcal{D}(\mathfrak{W})$. This proves formula (14.37).

Summing up, we have proved the following result:

Theorem 14.14. *If condition (H) is satisfied, then the operator $\mathfrak{W}$ is the infinitesimal generator of some Feller semigroup $\{e^{t\mathfrak{W}}\}_{t \geq 0}$ on $\overline{D} \setminus M$. Moreover, it generates a semigroup $e^{z\mathfrak{W}}$ on $C_0(\overline{D} \setminus M)$ which is* analytic *in the sector $\Delta_\varepsilon = \{z = t + is : z \neq 0, |\arg z| < \pi/2 - \varepsilon\}$ for any $0 < \varepsilon < \pi/2$ (see Fig. 1.6).*

A

Boundedness of Pseudo-Differential Operators

In this appendix we prove an L^p boundedness theorem for pseudo-differential operators — a global version of Theorem 5.14 — which plays a fundamental role throughout the book. Bourdaud [Bo] proved the L^p boundedness by using the multiplier theorem of Marcinkiewicz, just as in Coifman-Meyer [CM]: The method of proof consists of the following two steps:

(1) A characterization of L^p functions in terms of the Littlewood-Paley series.
(2) A reduction of the problem to elementary symbols of the form

$$\sigma(x,\xi) = \sum_{j=0}^{\infty} M_j(2^{j\delta}x)\,\psi(2^{-j}\xi)\,,$$

where $\psi(\xi)$ is a smooth function with compact support which does not contain the origin and $\{M_j\}$ is a bounded sequence in an appropriate Hölder space.

For a general study of L^p theory of pseudo-differential operators, the reader might refer to Coifman-Meyer [CM], Kumano-go [Ku] and Taylor [Ty].

A.1 The Littlewood-Paley Series

In this subsection we characterize L^p functions by their Littlewood-Paley series. We begin with the following elementary lemma (cf. [BL, Lemma 6.1.7]):

Lemma A.1. *For a given constant $a > 1$, there exists a function $\varphi(\xi) \in C_0^\infty(\mathbf{R}^n)$ such that*

$$\operatorname{supp}\varphi = \left\{\xi \in \mathbf{R}^n : \frac{1}{a} \leq |\xi| \leq a\right\}\,, \tag{A.1a}$$

$$\sum_{j\in\mathbf{Z}} \varphi(a^{-j}\,\xi) = 1 \quad \textit{for all } \xi \neq 0\,. \tag{A.1b}$$

Proof. If $h(\xi)$ is a non-negative function in $C_0^\infty(\mathbf{R}^n)$ such that

$$\operatorname{supp} h = \left\{\xi \in \mathbf{R}^n : \frac{1}{a} \le |\xi| \le a\right\},$$

then we have, for all $k \in \mathbf{Z}$,

$$\operatorname{supp} h(a^{-k}\,\cdot) = \{\xi \in \mathbf{R}^n : a^{k-1} \le |\xi| \le a^{k+1}\}.$$

Thus we can define a function $H(\xi) \in C_0^\infty(\mathbf{R}^n)$ by the formula

$$H(\xi) = \sum_{k\in\mathbf{Z}} h(a^{-k}\,\xi), \quad \xi \in \mathbf{R}^n.$$

It should be noticed that the sum is locally finite, and that

$$\begin{cases} H(0) = 0, \\ H(\xi) > 0 \quad \text{for all } \xi \neq 0. \end{cases}$$

Therefore, it is easy to verify that the function

$$\varphi(\xi) = \frac{h(\xi)}{H(\xi)}, \quad \xi \in \mathbf{R}^n,$$

enjoys the desired properties (A.1a) and (A.1b). □

Furthermore, if we define a function $\psi_0(\xi)$ by the formula

$$\psi_0(\xi) = \sum_{j=-\infty}^{0} \varphi(a^{-j}\,\xi), \quad \xi \in \mathbf{R}^n,$$

then it follows from properties (A.1a) and (A.1b) that

$$\psi_0 \in C_0^\infty(\mathbf{R}^n), \tag{A.2a}$$
$$\operatorname{supp} \psi_0 = \{\xi \in \mathbf{R}^n : |\xi| \le a\}, \tag{A.2b}$$
$$\psi_0(\xi) = 1 - \sum_{j=1}^{\infty} \varphi(a^{-j}\,\xi) \quad \text{for all } \xi \neq 0. \tag{A.2c}$$

Now, using the Fourier transform we can introduce a family of linear operators

$$\Delta_0 : \mathcal{S}'(\mathbf{R}^n) \longrightarrow \mathcal{S}'(\mathbf{R}^n),$$
$$\Delta_j : \mathcal{S}'(\mathbf{R}^n) \longrightarrow \mathcal{S}'(\mathbf{R}^n), \quad j = 1, 2, \dots,$$

by the formulas

$$\widehat{\Delta_0 f}(\xi) = \psi_0(\xi)\widehat{f}(\xi)\,,$$
$$\widehat{\Delta_j f}(\xi) = \varphi(a^{-j}\xi)\widehat{f}(\xi). \quad j = 1,\ 2, \dots\,.$$

Then, by properties (A.1) and (A.2) it is easy to see that

$$f = \sum_{j=0}^{\infty} \Delta_j f, \quad f \in \mathcal{S}'(\mathbf{R}^n)\,. \tag{A.3}$$

Indeed, it suffices to note that, for all $\xi \neq 0$,

$$\begin{aligned}\sum_{j=0}^{\infty} \widehat{\Delta_j f}(\xi) &= \widehat{\Delta_0 f}(\xi) + \sum_{j=1}^{\infty} \widehat{\Delta_j f}(\xi) \\ &= \left(\psi_0(\xi) + \sum_{j=1}^{\infty} \varphi(a^{-j}\xi)\right)\widehat{f}(\xi) \\ &= \left(\sum_{j\in\mathbf{Z}} \varphi(a^{-j}\xi)\right)\widehat{f}(\xi) \\ &= \widehat{f}(\xi)\,.\end{aligned}$$

The series (A.3) is called the *Littlewood-Paley series* of f.

A.2 Definition of Sobolev and Besov Spaces

In this subsection, following Bergh-Löfström [BL] we give Peetre's equivalent definition of Besov spaces and generalized Sobolev spaces.

To do this, we choose $a = 2$ in Lemma A.1. Namely, $\varphi(\xi)$ is a function in $C_0^\infty(\mathbf{R}^n)$ which satisfies the conditions

$$\operatorname{supp}\varphi = \left\{\xi \in \mathbf{R}^n : \frac{1}{2} \leq |\xi| \leq 2\right\}\,, \tag{A.4a}$$

$$\sum_{k\in\mathbf{Z}} \varphi(2^{-k}\xi) = 1 \quad \text{for all } \xi \neq 0\,. \tag{A.4b}$$

Then we can define functions $\varphi_k(x)$, $\psi(x) \in \mathcal{S}(\mathbf{R}^n)$ by the formulas

$$\widehat{\varphi_k}(\xi) = \varphi(2^{-k}\xi), \quad k \in \mathbf{Z}\,,$$
$$\widehat{\psi}(\xi) = \psi_0(\xi) = 1 - \sum_{j=1}^{\infty} \varphi(2^{-j}\xi)\,.$$

It should be noticed that

$$\operatorname{supp}\widehat{\varphi_k} = \{\xi \in \mathbf{R}^n : 2^{k-1} \leq |\xi| \leq 2^{k+1}\}, \quad k \in \mathbf{Z}\,,$$
$$\operatorname{supp}\widehat{\psi} = \{\xi \in \mathbf{R}^n : |\xi| \leq 2\}\,.$$

Furthermore, if $s \in \mathbf{R}$ we define the Bessel potential J^s by the formula

$$J^s = (I - \Delta)^{s/2} \colon \mathcal{S}'(\mathbf{R}^n) \longrightarrow \mathcal{S}'(\mathbf{R}^n) .$$

More precisely, we let

$$J^s f = \mathcal{F}^*((1 + |\xi|^2)^{s/2} \widehat{f}(\xi)), \quad f \in \mathcal{S}'(\mathbf{R}^n) .$$

Then we have the following basic properties of the objects just defined (see [BL, Lemma 6.2.1]):

Lemma A.2. *(i) Assume that a distribution $f \in \mathcal{S}'(\mathbf{R}^n)$ satisfies the condition*

$$\varphi_k * f \in L^p(\mathbf{R}^n), \quad k \in \mathbf{Z} ,$$

with $1 \leq p \leq \infty$. Then we have, for all $s \in \mathbf{R}$,

$$\|J^s(\varphi_k * f)\|_{L^p} \leq C\, 2^{sk} \, \|\varphi_k * f\|_{L^p}, \quad k = 1,\ 2, \ldots ,$$

with a constant $C > 0$ independent of p and k.
(ii) If a distribution $f \in \mathcal{S}'(\mathbf{R}^n)$ satisfies the condition

$$\psi * f \in L^p(\mathbf{R}^n) ,$$

with $1 \leq p \leq \infty$, then we have, for all $s \in \mathbf{R}$,

$$\|J^s(\psi * f)\|_{L^p} \leq C' \, \|\psi * f\|_{L^p} ,$$

with a constant $C' > 0$ independent of p and k.

Now, by virtue of Lemma A.2 we can make the following definition of the Besov and generalized Sobolev spaces (see [BL, Definition 6.2.2]):

Definition A.1. Let $s \in \mathbf{R}$, and $1 \leq p,\ q \leq \infty$. If $f \in \mathcal{S}'(\mathbf{R}^n)$, we let

$$\|f\|_{B^s_{p,q}} = \begin{cases} \|\psi * f\|_{L^p} + \left(\displaystyle\sum_{k=1}^{\infty} \left(2^{sk} \, \|\varphi_k * f\|_{L^p}\right)^q \right)^{1/q} & \text{if } 1 \leq q < \infty , \\ \|\psi * f\|_{L^p} + \sup_{k \geq 1} \left(2^{sk} \, \|\varphi_k * f\|_{L^p}\right) & \text{if } q = \infty , \end{cases}$$

$$\|f\|_{H^s_p} = \|J^s f\|_{L^p} .$$

Then the *Besov space* $B^s_{p,q}(\mathbf{R}^n)$ and the generalized *Sobolev space* $H^s_p(\mathbf{R}^n)$ are defined respectively as follows:

$$B^s_{p,q}(\mathbf{R}^n) = \{f \in \mathcal{S}'(\mathbf{R}^n) : \|f\|_{B^s_{p,q}} < \infty\} ,$$
$$H^s_p(\mathbf{R}^n) = \{f \in \mathcal{S}'(\mathbf{R}^n) : \|f\|_{H^s_p} < \infty\} .$$

Remark A.1. It is known (see Bergh-Löfström [BL], Triebel [Tr]) that the Sobolev space $H^{s,p}(\mathbf{R}^n)$ and the Besov space $B^{s,p}(\mathbf{R}^{n-1})$ introduced in Sect. 5.2 are respectively equivalent to the following:

$$H^{s,p}(\mathbf{R}^n) = H_p^s(\mathbf{R}^n), \quad s \in \mathbf{R},\, 1 < p < \infty\,,$$
$$B^{s,p}(\mathbf{R}^{n-1}) = B_{p,p}^s(\mathbf{R}^{n-1}), \quad s \in \mathbf{R},\, 1 \le p \le \infty\,.$$

It should be noticed that

$$H_p^0(\mathbf{R}^n) = L^p(\mathbf{R}^n), \quad 1 \le p \le \infty\,.$$

Furthermore, it is easy to verify the following two assertions:

(I) The spaces $B_{p,q}^s(\mathbf{R}^n)$ and $H_p^s(\mathbf{R}^n)$ are Banach spaces with norms $\|\cdot\|_{B_{p,q}^s}$ and $\|\cdot\|_{H_p^s}$, respectively.
(II) The Bessel potential J^σ is a topological isomorphism of $B_{p,q}^s(\mathbf{R}^n)$ onto $B_{p,q}^{s-\sigma}(\mathbf{R}^n)$ for each $\sigma \in \mathbf{R}$, and is also a topological isomorphism of $H_p^s(\mathbf{R}^n)$ onto $H_p^{s-\sigma}(\mathbf{R}^n)$ for each $\sigma \in \mathbf{R}$, respectively.

The next theorem characterizes the spaces $B_{p,q}^s(\mathbf{R}^n)$ and $H_p^s(\mathbf{R}^n)$ in terms of the Littlewood-Paley series (see [BL, Theorems 6.4.3]):

Theorem A.3. *(i) Let $s \in \mathbf{R}$ and $1 < p < \infty$. Then we have, for any $f \in \mathcal{S}'(\mathbf{R}^n)$,*

$$f \in H_p^s(\mathbf{R}^n) \Longleftrightarrow \left(\sum_{j=0}^\infty 2^{2sj}\, |\Delta_j f|^2\right)^{1/2} \in L^p(\mathbf{R}^n)\,.$$

(ii) Let $s \in \mathbf{R}$ and $1 \le p,\, q \le \infty$. Then we have, for any $f \in \mathcal{S}'(\mathbf{R}^n)$,

$$f \in B_{p,q}^s(\mathbf{R}^n) \Longleftrightarrow \sum_{j=0}^\infty 2^{sqj}\, \|\Delta_j f\|_{B_{p,q}^s} < \infty\,.$$

Remark A.2. Theorem A.3 remains valid with the constant 2 replaced by a general constant $a > 1$.

A.3 Non-Regular Symbols

In this subsection we introduce a class of non-regular symbols $\sigma(x, \xi)$ which are Hölder continuous with respect to the variable x and belong to the Hörmander class $S_{1,\delta}^0$ with respect to the variable ξ.

Let $0 \le \delta \le 1$, $N \in \mathbf{N}$ and r a *non-integral* positive number. A function $\sigma(x, \xi)$ defined on $\mathbf{R}^n \times \mathbf{R}^n$ belongs to the class $S_{1,\delta}^0(N, r)$ if it satisfies the following two conditions:

(a) For each $|\alpha| \le N$ and $|\beta| < r$, there exists a constant $C_{\alpha,\beta} > 0$ such that

$$\left|\partial_\xi^\alpha \partial_x^\beta \sigma(x,\xi)\right| \le C_{\alpha,\beta}(1+|\xi|)^{-|\alpha|+\delta|\beta|}, \quad x,\ \xi \in \mathbf{R}^n .$$

(b) For each $|\alpha| \le N$ and $|\beta| = [r]$, there exists a constant $C'_{\alpha,\beta} > 0$ such that

$$\left|\partial_\xi^\alpha \partial_x^\beta \sigma(x+h,\xi) - \partial_\xi^\alpha \partial_x^\beta \sigma(x,\xi)\right| \le C'_{\alpha,\beta}|h|^{r-[r]}(1+|\xi|)^{-|\alpha|+r\delta}, \quad x,\ \xi,\ h \in \mathbf{R}^n . \tag{A.5}$$

Here $[r]$ is the integral part of r.

Remark A.3. If r is integral, then condition (A.5) should be replaced by the *Zygmund condition* for $|\beta| = r-1$

$$\begin{aligned} &\left|\partial_\xi^\alpha \partial_x^\beta \sigma(x+h,\xi) + \partial_\xi^\alpha \partial_x^\beta \sigma(x-h,\xi) - 2\partial_\xi^\alpha \partial_x^\beta \sigma(x,\xi)\right| \\ &\le C'_{\alpha,\beta}|h|(1+|\xi|)^{-|\alpha|+r\delta}, \quad x,\ \xi,\ h \in \mathbf{R}^n . \end{aligned}$$

It is easy to verify that the class $S^0_{1,\delta}(N,r)$ is a Banach space with respect to the norm

$$\begin{aligned} \|\sigma\| = &\sum_{\substack{|\alpha|\le N \\ |\beta|<r}} \sup_{x,\xi} \left\{ \frac{|\partial_\xi^\alpha \partial_x^\beta \sigma(x,\xi)|}{(1+|\xi|)^{-|\alpha|+\delta|\beta|}} \right\} \\ &+ \sum_{\substack{|\alpha|\le N \\ |\beta|=[r]}} \sup_{x,\xi,h} \left\{ \frac{|\partial_\xi^\alpha \partial_x^\beta \sigma(x+h,\xi) - \sigma(x,\xi)|}{|h|^{r-[r]}(1+|\xi|)^{-|\alpha|+r\delta}} \right\} . \end{aligned}$$

The next lemma asserts that every symbol $\sigma(x,\xi)$ in $S^0_{1,\delta}(N,r)$ can be decomposed into elementary symbols (cf. [CM, Chap. II, Proposition 5]):

Lemma A.4. *Let $0 \le \delta \le 1$, $r > 0$ and let N be an* even *integer greater than n. Then every symbol $\sigma(x,\xi) \in S^0_{1,\delta}(N,r)$ can be decomposed into the form*

$$\sigma(x,\xi) = \widetilde{\sigma}(x,\xi) + \sum_{k\in\mathbf{Z}^n} (1+|k|^2)^{-(n+1)/2}\sigma_k(x,\xi) . \tag{A.6}$$

Here:

(i) The symbol $\widetilde{\sigma}(x,\xi) \in S^0_{1,\delta}(N,r)$ satisfies the condition

$$\widetilde{\sigma}(x,\xi) = 0 \quad \textit{for all } |\xi| \ge 2 , \tag{A.7}$$

and there exists a constant $C > 0$, depending on n and N, such that

$$\|\partial_\xi^\alpha \widetilde{\sigma}(\cdot,\xi)\|_{\Lambda_r} \le C\|\sigma\|, \quad |\alpha| \le N .$$

(ii) Every elementary symbol $\sigma_k(x,\xi)$ is written in the form

$$\sigma_k(x,\xi)=\sum_{j=0}^{\infty} M_{k,j}(2^{j\delta}x)\,\psi_k(2^{-j}\xi)\,, \tag{A.8}$$

where

$$\psi_k(\xi)=(1+|k|^2)^{(n+1-N)/2}e^{ik\xi}\,\theta(\xi)$$
$$\theta(\xi)\in C_0^\infty(\mathbf{R}^n),\quad \operatorname{supp}\theta=\left\{\xi\in\mathbf{R}^n:\frac{1}{3}\le|\xi|\le 3\right\}\,,$$
$$M_{k,j}(x)=(1+|k|^2)^{N/2}C_{k,j}(x)\in\Lambda_r(\mathbf{R}^n)\,.$$

Moreover, we have, with some constant $C'>0$ independent of k and j,

$$\|\psi_k\|_{L^\infty_{N-n-1}}\le C'\,, \tag{A.9a}$$
$$\|M_{k,j}\|_{\Lambda_r}\le C'\|\sigma\|\,. \tag{A.9b}$$

Here $\Lambda_r(\mathbf{R}^n)=B^r_{\infty,\infty}(\mathbf{R}^n)$ is the classical Hölder space and $L^\infty_m(\mathbf{R}^n)$ is the space of functions on $\mathbf{R}^n$ whose distribution derivatives of order $\le m$ belong to $L^\infty(\mathbf{R}^n)$, respectively.

Proof. The proof is divided into three steps.

Step 1: First, we take a non-negative function $\mu(\xi)\in C_0^\infty(\mathbf{R}^n)$ which satisfies the condition

$$\mu(\xi)=\begin{cases}1 & \text{if } |\xi|\le 1\,,\\ 0 & \text{if } |\xi|\ge 2\,,\end{cases} \tag{A.10}$$

and let

$$\begin{aligned}\sigma(x,\xi)&=\mu(\xi)\sigma(x,\xi)+(1-\mu(\xi))\sigma(x,\xi)\\ &:=\widetilde{\sigma}(x,\xi)+\tau(x,\xi)\,.\end{aligned} \tag{A.11}$$

By condition (A.10) and the Leibniz rule, it is easy to verify that

$$\widetilde{\sigma}(x,\xi)=\mu(\xi)\sigma(x,\xi)=0 \quad \text{for all } |\xi|\ge 2\,,$$

and that we have, for all $|\alpha|\le N$,

$$\begin{aligned}\|\partial_\xi^\alpha\widetilde{\sigma}(\cdot,\xi)\|_{\Lambda_r}&\le C_1\|\sigma\|(1+|\xi|)^{-|\alpha|+r\delta}\\ &\le C_1\|\sigma\|(1+|\xi|)^{r\delta}\\ &\le 3^{r\delta}\,C_1\|\sigma\|\,,\end{aligned}$$

with a constant $C_1>0$.

Similarly we have,

$$\|\tau\|\le C_2\|\sigma\|\,,$$

with a constant $C_2 > 0$.

Step 2: Secondly, we take a non-negative function $\lambda(\xi) \in C_0^\infty(\mathbf{R}^n)$ which satisfies the conditions

$$\operatorname{supp} \lambda = \left\{ \xi \in \mathbf{R}^n : \frac{1}{2} \leq |\xi| \leq 2 \right\} ,$$

$$\sum_{j=0}^{\infty} \lambda(2^{-j}\,\xi) = 1 \quad \text{for all } |\xi| \geq 1 ,$$

and let

$$\begin{aligned}\tau_j(x,\xi) &= \lambda(\xi)\,\tau(2^{-j\delta}\,x, 2^j\,\xi) \\ &= \lambda(\xi)\,(1-\mu(2^j\,\xi))\sigma(2^{-j\delta}\,x, 2^j\,\xi), \quad j = 0,\ 1,\ 2, \ldots .\end{aligned}$$

Then the symbols $\tau_j(x,\xi)$ are estimated as follows:

Claim A.5. *For each $|\alpha| \leq n$, there exists a constant $C > 0$, independent of j, such that*

$$\|D_\xi^\alpha \tau_j(\cdot,\xi)\|_{\Lambda_r} \leq C\,\|\sigma\|, \quad j = 0,\ 1,\ 2, \ldots . \tag{A.12}$$

Proof. For example, we consider the case where $\alpha = 0$: It should be noticed that we have, for $|\beta| = [r]$,

$$\begin{aligned}&\left(\frac{1}{2^{j\delta}}\right)^{|\beta|} \left| \partial_x^\beta \sigma\left(\frac{x+h}{2^{j\delta}}, 2^j\,\xi\right) - \partial_x^\beta \sigma\left(\frac{x}{2^{j\delta}}, 2^j\,\xi\right)\right| \\ &\leq \|\sigma\| \left(\frac{1}{2^{j\delta}}\right)^{|\beta|} \left(\frac{|h|}{2^{j\delta}}\right)^{r-[r]} (1+|2^j\xi|)^{r\delta} \\ &= \|\sigma\| |h|^{r-[r]} \left(\frac{1+|2^j\xi|}{2^j}\right)^{r\delta} .\end{aligned}$$

However, it follows that

$$1 + 2^{j-1} \leq 1 + |2^j\xi| \leq 1 + 2^{j+1}, \quad \frac{1}{2} \leq |\xi| \leq 2 ,$$

so that

$$\frac{1}{2} \leq \frac{1}{2^j} + \frac{1}{2} \leq \frac{1+|2^j\xi|}{2^j} \leq \frac{1}{2^j} + 2 \leq 3, \quad \xi \in \operatorname{supp} \lambda .$$

Hence we have estimate (A.12) for $\alpha = 0$. □

Step 3: Thirdly, we take a non-negative function $\theta(\xi) \in C_0^\infty(\mathbf{R}^n)$ which satisfies the conditions

$$\operatorname{supp} \theta = \left\{ \xi \in \mathbf{R}^n : \frac{1}{3} \leq |\xi| \leq 3 \right\} ,$$

$$\theta(\xi) = 1 \quad \text{on } \operatorname{supp} \lambda .$$

Then we can expand the symbol $\tau_j(x,\xi) = \tau_j(x,\xi)\,\theta(\xi)$ in the Fourier series form

$$\tau_j(x,\xi) = \left(\sum_{k\in\mathbf{Z}^n} C_{k,j}(x)\, e^{ik\cdot\xi}\right)\theta(\xi)\,,$$

where

$$C_{k,j}(x) = \frac{1}{(2\pi)^n}\int_{\mathbf{R}^n} \tau_j(x,\eta)\, e^{-ik\cdot\eta}\, d\eta\,.$$

Moreover, we can rewrite the symbol $\tau_j(x,\xi)$ in the form (see decomposition (A.8)):

$$\begin{aligned}\tau_j(x,\xi) &= \sum_{k\in\mathbf{Z}^n} (1+|k|^2)^{-(n+1)/2}\left\{(1+|k|^2)^{N/2}C_{k,j}(x)\right\}\\ &\quad\times\left\{(1+|k|^2)^{(n+1-N)/2}e^{ik\xi}\,\theta(\xi)\right\}\\ &:= \sum_{k\in\mathbf{Z}^n} u_k\, M_{k,j}(x)\,\psi_k(\xi)\,. \end{aligned}\tag{A.13}$$

Here it should be noticed that

$$N \text{ is an } \textit{even} \text{ integer greater than } n\,,$$
$$\sum_{k\in\mathbf{Z}^n} u_k = \sum_{k\in\mathbf{Z}^n}(1+|k|^2)^{-(n+1)/2} < \infty\,.$$

Hence we have, by decomposition (A.13),

$$\begin{aligned}\tau(x,\xi) &= \left(\sum_{j=0}^{\infty}\lambda(2^{-j}\,\xi)\right)\tau(x,\xi)\\ &= \sum_{j=0}^{\infty}\tau_j(2^{j\delta}\,x, 2^{-j}\,\xi)\\ &= \sum_{j=0}^{\infty}\sum_{k\in\mathbf{Z}^n} u_k\, M_{k,j}(2^{j\delta}\,x)\,\psi_k(2^{-j}\,\xi)\\ &= \sum_{k\in\mathbf{Z}^n} u_k\left\{\sum_{j=0}^{\infty} M_{k,j}(2^{j\delta}\,x)\,\psi_k(2^{-j}\,\xi)\right\}\\ &:= \sum_{k\in\mathbf{Z}^n}(1+|k|^2)^{-(n+1)/2}\,\sigma_k(x,\xi)\,. \end{aligned}\tag{A.14}$$

The desired decomposition (A.6) follows by combining decompositions (A.11) and (A.14).

It remains to prove estimates (A.9a) and (A.9b) for the functions $\psi_k(\xi)$ and $M_{k,j}(x)$.

(1) The estimate (A.9a) of $\psi_k(\xi)$: Since we have, for $|\alpha| \leq N-n-1$,

$$\left|\partial_\xi^\alpha \psi_k(\xi)\right| \leq C_3 \|\theta\|_{L^\infty_{N-n-1}} ,$$

it follows that

$$\|\psi_k\|_{L^\infty_{N-n-1}} \leq C_4 .$$

(2) The estimate (A.9b) of $M_{k,j}(x)$: By integration by parts, it follows that

$$\begin{aligned} M_{k,j}(x) &= \frac{(1+|k|^2)^{N/2}}{(2\pi)^n} \int_{\mathbf{R}^n} e^{-ik\xi}\, \tau_j(x,\xi)\, d\xi \\ &= \frac{1}{(2\pi)^n} \int_{\mathbf{R}^n} \tau_j(x,\xi) \cdot (I - \Delta_\xi)^{N/2} e^{-ik\xi}\, d\xi \\ &= \frac{1}{(2\pi)^n} \int_{\mathbf{R}^n} e^{-ik\xi} (I - \Delta_\xi)^{N/2} \tau_j(x,\xi)\, d\xi . \end{aligned}$$

Hence, by Claim A.5 it is easy to see that

$$\|M_{k,j}\|_{\Lambda_r} \leq C_5 \|\sigma\| .$$

Now the proof of Lemma A.4 is complete. □

A.4 The L^p Boundedness Theorem

In this subsection we formulate an L^p boundedness theorem for pseudo-differential operators with non-regular symbols due to Bourdaud [Bo, Theorem 1]:

Theorem A.6. *Let $0 \leq \delta \leq 1$, $r > 0$ and let N be an* even *integer greater than $(3n/2)+1$. If $\sigma(x,\xi)$ is a symbol in the class $S^0_{1,\delta}(N,r)$, then the pseudo-differential operator $\sigma(x,D)$, defined by the formula*

$$\sigma(x,D)f(x) = \frac{1}{(2\pi)^n} \int_{\mathbf{R}^n} e^{ix\cdot\xi} \sigma(x,\xi) \widehat{f}(\xi)\, d\xi ,$$

is bounded on the Sobolev space $H^s_p(\mathbf{R}^n)$ for all $(\delta-1)r < s < r$ and $1 < p < \infty$, and is bounded on the Besov space $B^s_{p,q}(\mathbf{R}^n)$ for all $(\delta-1)r < s < r$ and $1 \leq p$, $q \leq \infty$, respectively.

By virtue of Lemma A.4, the proof of Theorem A.6 is reduced to the proof of the following two Propositions A.7 and A.8:

Proposition A.7. *Let $0 \leq \delta \leq 1$, $r > 0$ and let N be an* even *integer greater than n. Assume that a symbol $\sigma(x,\xi) \in S^0_{1,\delta}(N,r)$ satisfies the condition*

$$\sigma(x,\xi) = 0 \quad \text{for all } |\xi| \geq 2 . \tag{A.15}$$

Then the pseudo-differential operator

$$\sigma(x,D)f(x) = \frac{1}{(2\pi)^n} \int_{\mathbf{R}^n} e^{ix\cdot\xi} \sigma(x,\xi) \widehat{f}(\xi)\, d\xi$$

is bounded on the Sobolev space $H_p^s(\mathbf{R}^n)$ *for all* $(\delta - 1)r < s < r$ *and* $1 < p < \infty$, *and is bounded on the Besov space* $B_{p,q}^s(\mathbf{R}^n)$ *for all* $(\delta - 1)r < s < r$ *and* $1 \leq p,\ q \leq \infty$. *More precisely, we have the estimate for the operator norm*

$$\|\sigma(x,D)\| \leq C\|\sigma\| ,$$

with a constant $C > 0$ *depending on* n, p, q, r, s *and* δ.

Proposition A.8. *Let* $0 \leq \delta \leq 1$, $r > 0$ *and let* N *be an* even *integer greater than* n. *Assume that a symbol* $\sigma(x,\xi) \in S_{1,\delta}^0(N,r)$ *is an elementary symbol of the form*

$$\sigma(x,\xi) = \sum_{j=0}^{\infty} M_j(2^{j\delta}x)\, \psi(2^{-j}\xi) , \tag{A.16}$$

where

$$\psi(\xi) \in C_0^\infty(\mathbf{R}^n), \quad \operatorname{supp}\psi = \left\{\xi \in \mathbf{R}^n : \frac{1}{3} \leq |\xi| \leq 3\right\} ,$$

The sequence $\{M_j\}$ *is bounded in the Hölder space* $\Lambda_r(\mathbf{R}^n)$.

Then the pseudo-differential operator

$$\sigma(x,D)f(x) = \frac{1}{(2\pi)^n} \int_{\mathbf{R}^n} e^{ix\cdot\xi} \sigma(x,\xi) \widehat{f}(\xi)\, d\xi$$

is bounded on the Sobolev space $H_p^s(\mathbf{R}^n)$ *for all* $(\delta - 1)r < s < r$ *and* $1 < p < \infty$, *and is bounded on the Besov space* $B_{p,q}^s(\mathbf{R}^n)$ *for all* $(\delta - 1)r < s < r$ *and* $1 \leq p,\ q \leq \infty$. *More precisely, we have the estimate for the operator norm*

$$\|\sigma(x,D)\| \leq C \left(\sup_{j\geq 0} \|M_j\|_{\Lambda_r}\right) \|\psi\|_{L^\infty_{N-n-1}} ,$$

with a constant $C > 0$ *depending on* n, p, q, r, s *and* δ.

In this appendix we shall only prove the Sobolev space case. The proof of the Besov space case is left to the reader.

A.5 Proof of Proposition A.7

First, it should be noticed that

$$\begin{aligned} \|\sigma(x,D)u\|_{H_p^s} &= \|J^s(\sigma(x,D)u)\|_{L^p} \\ &= \|\sigma(x,D)J^{-s}v\|_{L^p} , \\ u = J^{-s}v &\in H_p^s(\mathbf{R}^n),\ v \in L^p(\mathbf{R}^n) . \end{aligned}$$

In other words, we have the following diagram:

$$\begin{array}{ccc} H_p^s & \xrightarrow{\sigma(x,D)} & H_p^s \\ {\scriptstyle J^{-s}}\uparrow & & \downarrow{\scriptstyle J^s} \\ L^p & \xrightarrow[J^s\sigma(x,D)J^{-s}]{} & L^p \end{array}$$

Furthermore, we find that the symbol $\tau(x,\xi)$ of $J^s\sigma(x,D)J^{-s}$ is written in the form

$$\tau(x,\xi) = \sigma(x,\xi) + r(x,\xi) ,$$

where $r(x,\xi)$ belongs to the Hörmander class $S^{-1}_{1,\delta}$ with respect to the variable ξ.

Therefore, the proof of Proposition A.7 is reduced to the proof of the L^p boundedness of the two operators $\sigma(x,D)$ and $r(x,D)$ which will be proved in Lemma A.9 and Lemma A.13, respectively.

Lemma A.9. *Let $0 \leq \delta \leq 1$, $r > 0$ and let N be an* even *integer greater than n. Assume that a symbol $\sigma(x,\xi) \in S^0_{1,\delta}(N,r)$ satisfies the condition*

$$\sigma(x,\xi) = 0 \quad \textit{for all } |\xi| \geq 2 . \tag{A.17}$$

Then there exists a constant $C > 0$ such that

$$\|\sigma(x,D)f\|_{L^p} \leq C\|f\|_{L^p}, \quad f \in L^p(\mathbf{R}^n) . \tag{A.18}$$

Proof. First, we have, by condition (A.17),

$$\begin{aligned} \sigma(x,D)f(x) &= \frac{1}{(2\pi)^n}\int_{\mathbf{R}^n} e^{ix\cdot\xi}\sigma(x,\xi)\widehat{f}(\xi)\,d\xi \\ &= \frac{1}{(2\pi)^n}\int_{\mathbf{R}^n}\left(\int_{\mathbf{R}^n} e^{i(x-y)\cdot\xi}\sigma(x,\xi)\,d\xi\right) f(y)\,dy \\ &= \int_{\mathbf{R}^n} K(x-y,y)f(y)\,dy , \end{aligned}$$

where

$$K(x,z) = \frac{1}{(2\pi)^n}\int_{\mathbf{R}^n} e^{iz\cdot\xi}\sigma(x,\xi)\,d\xi .$$

However, by integration by parts it follows that, for any integer $\ell \in \mathbf{N}$,

$$\begin{aligned} (1+|z|^2)^\ell\, K(x,z) &= \frac{1}{(2\pi)^n}\int_{\mathbf{R}^n} e^{iz\cdot\xi}(1+|z|^2)^\ell\,\sigma(x,\xi)\,d\xi \\ &= \frac{1}{(2\pi)^n}\int_{\mathbf{R}^n} (I-\Delta_\xi)^\ell e^{iz\cdot\xi}\cdot\sigma(x,\xi)\,d\xi \\ &= \frac{1}{(2\pi)^n}\int_{\mathbf{R}^n} e^{iz\cdot\xi}\,(I-\Delta_\xi)^\ell\sigma(x,\xi)\,d\xi . \end{aligned}$$

Hence we can find a constant $C_1 > 0$ such that

$$(1 + |z|^2)^\ell \, |K(x,z)| \le C_1 \|\sigma\| \, .$$

Now, if we take the integer ℓ satisfying the condition

$$n < 2\ell \le N \, ,$$

then we have

$$|K(x,z)| \le \frac{C_1 \|\sigma\|}{(1 + |z|^2)^\ell} \, .$$

Thus it follows that

$$\begin{aligned} |\sigma(x,D)f(x)| &\le \int_{\mathbf{R}^n} |K(x-y,y)| \, |f(y)| \, dy \\ &\le C_1 \|\sigma\| \int_{\mathbf{R}^n} \frac{|f(y)|}{(1 + |x-y|^2)^\ell} \, dy \\ &:= K_0 * |f|(x) \, , \end{aligned}$$

where

$$K_0(x) = \frac{C_1 \|\sigma\|}{(1 + |x|^2)^\ell} \in L^1(\mathbf{R}^n) \, .$$

Therefore, applying Young's inequality (Corollary 4.3) we obtain that

$$\|\sigma(x,D)f\|_{L^p} \le \|K_0 * |f|\|_{L^p} \le \|K_0\|_{L^1} \, \|f\|_{L^p} \, .$$

This proves inequality (A.14) with $C = \|K_0\|_{L^1}$. □

The next lemma, due to Nagase [Na, Theorem 2], plays an essential role in the proof of an L^p boundedness theorem for pseudo-differential operators $r(x,D)$ of *negative order*:

Lemma A.10. *Assume that a symbol $r(x,\xi) \in S^0_{1,\delta}(N,r)$ satisfies the condition:*

There exists a constant $\rho > 0$ such that, for each $|\alpha| \le n+1$ we have, with some constant $C > 0$,

$$|\partial_\xi^\alpha r(x,\xi)| \le C(1 + |\xi|)^{-|\alpha|-\rho} \, . \tag{A.19}$$

Then there exists a function $K(x,z) \in C(\mathbf{R}^n \times (\mathbf{R}^n \setminus \{0\}))$ such that

$$r(x,D)f = \int_{\mathbf{R}^n} K(x, x-y) f(y) \, dy, \quad f \in L^p(\mathbf{R}^n) \, . \tag{A.20}$$

Moreover, for each $0 < \rho' < \min(1,\rho)$, there exists a constant $C' > 0$ such that

$$|K(x,z)| \le C' \frac{1}{1 + |z|} \frac{1}{|z|^{n-\rho'}} \, . \tag{A.21}$$

Proof. The proof is divided into five steps.

Step 1: First, we take a function $\chi(\xi) \in C_0^\infty(\mathbf{R}^n)$ such that

$$\chi(\xi) = \begin{cases} 1 & \text{if } |\xi| \leq \dfrac{1}{2} , \\ 0 & \text{if } |\xi| \geq 1 , \end{cases}$$

and let

$$r_\varepsilon(x,\xi) = \chi(\varepsilon\xi) r(x,\xi), \quad 0 < \varepsilon < 1 .$$

Then it is clear that the support of $r_\varepsilon(x,\xi)$ is compact with respect to the variable ξ.

Step 2: If we let

$$K_\varepsilon(x,z) = \frac{1}{(2\pi)^n} \int_{\mathbf{R}^n} e^{iz\cdot\xi} r_\varepsilon(x,\xi)\, d\xi, \quad 0 < \varepsilon < 1 ,$$

then it follows from an application of the Lebesgue convergence theorem that

$$\begin{aligned} r(x,D)f(x) &= \frac{1}{(2\pi)^n} \int_{\mathbf{R}^n} e^{ix\cdot\xi} r(x,\xi) \widehat{f}(\xi)\, d\xi \\ &= \frac{1}{(2\pi)^n} \lim_{\varepsilon\downarrow 0} \int_{\mathbf{R}^n} e^{ix\cdot\xi} \chi(\varepsilon\xi) r(x,\xi) \widehat{f}(\xi)\, d\xi \\ &= \frac{1}{(2\pi)^n} \lim_{\varepsilon\downarrow 0} \iint_{\mathbf{R}^n\times\mathbf{R}^n} e^{i(x-y)\cdot\xi} r_\varepsilon(x,\xi) f(y)\, dy d\xi \\ &= \lim_{\varepsilon\downarrow 0} \left(\frac{1}{(2\pi)^n} \int_{\mathbf{R}^n} e^{i(x-y)\cdot\xi} r_\varepsilon(x,\xi)\, d\xi \right) f(y)\, dy , \end{aligned}$$

that is,

$$r(x,D)f(x) = \lim_{\varepsilon\downarrow 0} \int_{\mathbf{R}^n} K_\varepsilon(x, x-y) f(y)\, dy, \quad f \in L^p(\mathbf{R}^n) . \tag{A.22}$$

However, by integration by parts it follows that

$$\begin{aligned} z^\alpha K_\varepsilon(x,z) &= \frac{1}{(2\pi)^n} \int_{\mathbf{R}^n} z^\alpha e^{iz\cdot\xi} r_\varepsilon(x,\xi)\, d\xi \\ &= \frac{1}{(2\pi)^n} \int_{\mathbf{R}^n} D_\xi^\alpha \left(e^{iz\cdot\xi}\right) \cdot r_\varepsilon(x,\xi)\, d\xi \\ &= \frac{(-1)^{|\alpha|}}{(2\pi)^n} \int_{\mathbf{R}^n} e^{iz\cdot\xi} D_\xi^\alpha r_\varepsilon(x,\xi)\, d\xi \\ &= \frac{(-1)^{|\alpha|}}{(2\pi)^n} \int_{\mathbf{R}^n} (e^{iz\cdot\xi} - 1) D_\xi^\alpha r_\varepsilon(x,\xi)\, d\xi , \end{aligned} \tag{A.23}$$

since we have

$$\int_{\mathbf{R}^n} D_\xi^\alpha r_\varepsilon(x,\xi)\, d\xi = \int_{\mathbf{R}^n} D_\xi^\alpha \left(\chi(\varepsilon\xi)\, r(x,\xi)\right) d\xi = 0 .$$

Here we remark the following elementary claim:

Claim A.11. *(a)* $D_\xi^\alpha r_\varepsilon(x,\xi) \to D_\xi^\alpha r(x,\xi)$ *as* $\varepsilon \downarrow 0$.
(b) For each $|\alpha| = n$, *we have*

$$|D_\xi^\alpha r_\varepsilon(x,\xi)| \leq \|r\|(1+|\xi|)^{-n-\rho}, \tag{A.24}$$

where

$$\|r\| = \sum_{|\alpha|\leq n} \sup_{x,\xi} \left\{ \frac{|\partial_\xi^\alpha r(x,\xi)|}{(1+|\xi|)^{-|\alpha|-\rho}} \right\}.$$

Therefore, by letting $\varepsilon \downarrow 0$ in formula (A.23) it follows from an application of the Lebesgue convergence theorem that

$$\lim_{\varepsilon\downarrow 0} z^\alpha K_\varepsilon(x,z) = \frac{(-1)^{|\alpha|}}{(2\pi)^n} \int_{\mathbf{R}^n} (e^{iz\cdot\xi}-1)\, D_\xi^\alpha r(x,\xi)\, d\xi. \tag{A.25}$$

If we let

$$K(x,z) = \frac{(-1)^{|\alpha|}}{(2\pi)^n z^\alpha} \int_{\mathbf{R}^n} (e^{iz\cdot\xi}-1)\, D_\xi^\alpha r(x,\xi)\, d\xi, \quad z \neq 0, \tag{A.26}$$

we obtain from formula (A.25) that

$$\lim_{\varepsilon\downarrow 0} z^\alpha K_\varepsilon(x,z) = z^\alpha K(x,z).$$

Step 3: To prove inequality (A.21), we estimate the function $z^\alpha K(x,z)$ for $|\alpha| = n$ and $|\alpha| = n+1$. To do this, we need the following elementary inequality:

Claim A.12. *We have, for* $0 < \rho' < 1$,

$$|e^{iz\cdot\xi} - 1| \leq 2|z|^{\rho'}|\xi|^{\rho'} \tag{A.27}$$

Proof. By the mean value theorem, it follows that

$$|e^{iz\cdot\xi} - 1| \leq \min\{2, |z|\,|\xi|\} \leq 2\min\{1, |z|\,|\xi|\}. \tag{A.28}$$

But we have

$$\begin{cases} |z|\,|\xi| \leq 1 \Longrightarrow |z|\,|\xi| \leq (|z|\,|\xi|)^{\rho'}, \\ |z|\,|\xi| \geq 1 \Longrightarrow 1 \leq (|z|\,|\xi|)^{\rho'}, \end{cases} \tag{A.29}$$

since $0 < \rho' < 1$.

The desired inequality (A.27) follows by combining inequalities (A.28) and (A.29). □

Case (a): $|\alpha| = n$. By using formula (A.25) and inequalities (A.24) and (A.27), we obtain that

$$
\begin{aligned}
|z^\alpha K_\varepsilon(x,z)| &= \frac{1}{(2\pi)^n}\int_{\mathbf{R}^n} \left|e^{iz\cdot\xi}-1\right| |D_\xi^\alpha r_\varepsilon(x,\xi)|\,d\xi \\
&\leq \frac{2\|r\||z|^{\rho'}}{(2\pi)^n}\int_{\mathbf{R}^n} |\xi|^{\rho'}(1+|\xi|)^{-n-\rho}\,d\xi \\
&\leq \frac{2\|r\||z|^{\rho'}}{(2\pi)^n}\int_{\mathbf{R}^n} \frac{1}{(1+|\xi|)^{n+(\rho-\rho')}}\,d\xi \\
&\leq C_1\|r\||z|^{\rho'}, \qquad \text{(A.30)}
\end{aligned}
$$

with some constant $C_1 > 0$.

Case (b): $|\alpha| = n+1$. Just as in the case (a), we have, with some constant $C_2 > 0$,

$$
\begin{aligned}
|z^\alpha K_\varepsilon(x,z)| &\leq \frac{2\|r\||z|^{\rho'}}{(2\pi)^n}\int_{\mathbf{R}^n} \frac{1}{(1+|\xi|)^{n+\rho+(1-\rho')}}\,d\xi \\
&\leq C_2\|r\||z|^{\rho'}. \qquad \text{(A.31)}
\end{aligned}
$$

Therefore, by letting $\varepsilon \downarrow 0$ in inequalities (A.30) and (A.31) it follows that, for $|\alpha| = n$ and $|\alpha| = n+1$, there exists a constant $C_3 > 0$ such that

$$|z^\alpha K(x,z)| \leq C_3|z|^{\rho'}.$$

This proves inequality (A.21).

Step 4: On the other hand, by combining Claims A.11 and A.12 we obtain that, for $|\alpha| = n$ and $|\alpha| = n+1$,

$$\left|\left(e^{ix\cdot\xi}-1\right) D_\xi^\alpha r(x,\xi)\right| \leq \frac{C_3|z|^{\rho'}}{(1+|\xi|)^{n+(\rho-\rho')}}.$$

Thus, by formula (A.26) it follows from an application of the Lebesgue convergence theorem that the function

$$z^\alpha K(x,z) = \frac{(-1)^{|\alpha|}}{(2\pi)^n}\int_{\mathbf{R}^n} \left(e^{iz\cdot\xi}-1\right) D_\xi^\alpha r(x,\xi)\,d\xi$$

is continuous on $\mathbf{R}^n \times \mathbf{R}^n$. In particular we have

$$K(x,z) \in C(\mathbf{R}^n \times (\mathbf{R}^n \setminus \{0\})).$$

Step 5: Finally, by inequality (A.21) we can let $\varepsilon \downarrow 0$ in formula (A.22) to obtain the integral formula (A.20). □

As a corollary of Lemma A.10, we can prove an L^p boundedness theorem for pseudo-differential operators $r(x,D)$ of negative order (see Nagase [Na, Theorem 3]):

Lemma A.13. *If a bounded continuous symbol $r(x,\xi)$ satisfies condition (A.19), then there exists a constant $C>0$ such that*

$$\|r(x,D)f\|_{L^p} \leq C\|f\|_{L^p}, \quad f \in L^p(\mathbf{R}^n). \tag{A.32}$$

Proof. By inequality (A.21), it follows that

$$\begin{aligned}
\int_{\mathbf{R}^n} |K(x,x-y)|\,dx &\leq C' \int_{\mathbf{R}^n} \frac{1}{1+|x|}\frac{1}{|x|^{n-\rho'}}\,dx \\
&\leq C'\left(\int_{|x|\leq 1} \frac{1}{|x|^{n-\rho'}}\,dx + \int_{|x|>1} \frac{1}{|x|^{n+(1-\rho')}}\,dx\right) \\
&< \infty .
\end{aligned}$$

Similarly we have the estimate

$$\begin{aligned}
\int_{\mathbf{R}^n} |K(x,x-y)|\,dy &\leq C'\left(\int_{|y|\leq 1} \frac{1}{|y|^{n-\rho'}}\,dy + \int_{|y|>1} \frac{1}{|y|^{n+(1-\rho')}}\,dy\right) \\
&< \infty .
\end{aligned}$$

Therefore, the desired inequality (A.32) follows by applying generalized Young's inequality (Theorem 4.2). □

A.6 Proof of Proposition A.8

The next lemma plays an essential role in the proof of Proposition A.8:

Lemma A.14. *Let $1<p<\infty$, $s\in\mathbf{R}$, a, $\gamma>1$ and $m>n/2$.*

(i) If $\psi(\xi)$ is a function in $C_0^\infty(\mathbf{R}^n)$ which satisfies the condition

$$\operatorname{supp}\psi \subset \left\{\xi \in \mathbf{R}^n : \frac{1}{\gamma} \leq |\xi| \leq \gamma\right\},$$

we let

$$\begin{aligned}
f_j(x) &= \psi(a^{-j}D)(x) \\
&:= \frac{1}{(2\pi)^n}\int_{\mathbf{R}^n} e^{ix\cdot\xi}\psi\left(a^{-j}\xi\right)\widehat{f}(\xi)\,d\xi, \quad f \in H_p^s(\mathbf{R}^n).
\end{aligned}$$

Then there exists a constant $C>0$, independent of f, such that

$$\left\|\left(\sum_{j=0}^{\infty} a^{2sj}|f_j|^2\right)^{1/2}\right\|_{L^p} \leq C\|\psi\|_{L^\infty_{N-n-1}}\|f\|_{H_p^s}. \tag{A.33}$$

(ii) If $\{f_j\}$ is a sequence in $\mathcal{S}'(\mathbf{R}^n)$ which satisfies the condition

$$\operatorname{supp}\widehat{f_j} \subset \left\{\xi \in \mathbf{R}^n : \frac{a^j}{\gamma} \le |\xi| \le \gamma a^j\right\} ,$$

then there exists a constant $C > 0$, independent of $\{f_j\}$, such that

$$\left\|\sum_{j=0}^{\infty} f_j\right\|_{H_p^s} \le C \left\|\left(\sum_{j=0}^{\infty} a^{2sj}|f_j|^2\right)^{1/2}\right\|_{L^p} . \tag{A.34}$$

(iii) If $s > 0$, then inequality (A.34) remains valid for every sequence $\{f_j\}$ in $S'(\mathbf{R}^n)$ which satisfies the condition

$$\operatorname{supp}\widehat{f_j} \subset \left\{\xi \in \mathbf{R}^n : |\xi| \le \gamma a^j\right\} .$$

Parts (i) and (ii) are essentially proved in Bergh-Löfström [BL, Theorem 6.4.3], while part (iii) is proved in Meyer [Me, Lemme 5].

A.6.1 Proof of the Case $\delta = 1$

First, we prove the case where $\delta = 1$ and $0 < s < r$. The proof is divided into five steps.

Step 1: Let $\{M_j\}$ be a bounded sequence in the Hölder space $\Lambda_r(\mathbf{R}^n) = B^r_{\infty,\infty}(\mathbf{R}^n)$ (see [Tr, Theorem 2.5.7]). We take $a = 2$ and let

$$M_j(x) = \sum_{\ell=0}^{\infty} M_{j,\ell}(x) \tag{A.35}$$

be the Littlewood-Paley series of the function $M_j(x)$, that is,

$$\widehat{M_{j,0}}(\xi) = \psi_0(\xi)\widehat{M_j}(\xi) ,$$
$$\widehat{M_{j,k}}(\xi) = \varphi(2^{-k}\xi)\widehat{M_j}(\xi), \quad k = 1,\ 2, \ldots ,$$

where

$$\operatorname{supp}\varphi = \left\{\xi \in \mathbf{R}^n : \frac{1}{2} \le |\xi| \le 2\right\} , \tag{A.36a}$$

$$\sum_{j\in\mathbf{Z}} \varphi(2^{-j}\xi) = 1 \quad \text{for all } \xi \ne 0 \tag{A.36b}$$

and

$$\psi_0 \in C_0^\infty(\mathbf{R}^n) , \tag{A.37a}$$

$$\operatorname{supp}\psi_0 = \{\xi \in \mathbf{R}^n : |\xi| \le 2\} , \tag{A.37b}$$

$$\psi_0(\xi) = 1 - \sum_{j=1}^{\infty} \varphi(2^{-j}\xi) \quad \text{for all } \xi \ne 0 . \tag{A.37c}$$

If we introduce functions $\varphi_\ell(x) \in \mathcal{S}(\mathbf{R}^n)$ by the formulas

$$\widehat{\varphi_\ell}(\xi) = \begin{cases} \psi_0(\xi) & \text{if } \ell = 0 \ , \\ \varphi(2^{-k}\xi), & \text{if } \ell = 1,\ 2, \ldots \ , \end{cases}$$

then the functions $M_{j,\ell}(x)$ can be expressed in the convolution form

$$M_{j,\ell}(x) = \varphi_\ell * M_j(x), \quad \ell = 0,\ 1,\ 2, \ldots \ .$$

Moreover, the functions $M_{j,\ell}(x)$ are estimated as follows:

Claim A.15. *Let $r' > 0$ be an arbitrary number such that*

$$[r] < r' < r, \quad [r'] = [r] \ . \tag{A.38}$$

If we let

$$A = \sup_{j \geq 0} \|M_j\|_{\Lambda^r} \ ,$$

then we have, for all $\ell = 0,\ 1,\ 2,\ \ldots,$

$$\|M_{j,\ell}\|_{L^\infty} \leq C\, A\, 2^{-r'\ell} \ , \tag{A.39}$$

with some constant $C > 0$ independent of ℓ.

Proof. By Part (i) of Lemma A.2, it follows that

$$\begin{aligned} \|M_{j,\ell}\|_{L^\infty} &= \|\varphi_\ell * M_j\|_{L^\infty} \\ &= \|J^{-r'} \cdot J^{r'}(\varphi_\ell * M_j)\|_{L^\infty} \\ &\leq C_1\, 2^{-r'\ell} \|\varphi_\ell * (J^{r'} M_j)\|_{L^\infty} \ . \end{aligned} \tag{A.40}$$

However, we have, by Young's inequality (Corollary 4.3),

$$\begin{aligned} \|\varphi_\ell * (J^{r'} M_j)\|_{L^\infty} &\leq \|\varphi_\ell\|_{L^1}\, \|J^{r'} M_j\|_{L^\infty} \\ &= \|\widehat{\varphi}\|_{L^1}\, \|M_j\|_{H^{r'}_\infty} \ . \end{aligned} \tag{A.41}$$

Since $[r'] = [r] < r' < r$, we have the inclusions

$$\Lambda_r = B^r_{\infty,\infty} \subset B^{r'}_{\infty,1} \subset H^{r'}_\infty \ ,$$

and hence, with some constants C_2, $C_3 > 0$,

$$\|M_j\|_{H^{r'}_\infty} \leq C_2\, \|M_j\|_{B^{r'}_{\infty,1}} \leq C_3\, \|M_j\|_{\Lambda_r} \leq C_3\, A \ . \tag{A.42}$$

Therefore, combining inequalities (A.40), (A.41) and (A.42) we obtain that

$$\begin{aligned} \|M_{j,\ell}\|_{L^\infty} &\leq C_1\, 2^{-r'\ell} \|J^{r'} M_j\|_{L^\infty} \\ &\leq (C_1\, C_3\, \|\widehat{\varphi}\|_{L^1})\, 2^{-r'\ell} \|M_j\|_{\Lambda_r} \\ &\leq (C_1\, C_3\, \|\widehat{\varphi}\|_{L^1})\, A\, 2^{-r'\ell} \ . \end{aligned}$$

This proves inequality (A.39). □

Step 2: If we let

$$N_{j,h}(x) = M_{j,h-j}(2^j x), \quad h \geq j\,, \tag{A.43}$$

then we have the following two assertions:

Claim A.16. *The functions $N_{j,h}(x)$ are estimated as follows:*

$$\|N_{j,h}\|_{L^\infty} \leq C\,A\,2^{-(h-j)r'}, \quad h \geq j\,.$$

Proof. Indeed, it follows from an application of Claim A.15 that

$$\begin{aligned}\|N_{j,h}\|_{L^\infty} &= \|M_{j,h-j}\|_{L^\infty} \\ &\leq C\,A\,2^{-(h-j)r'}\,. \quad \square\end{aligned}$$

Claim A.17. *The spectra of the functions $N_{j,h}(x)$ are estimated as follows:*

$$\operatorname{supp} \widehat{N_{j,h}} \subset \{\xi \in \mathbf{R}^n : 2^{h-1} \leq |\xi| \leq 2^{h+1}\}, \quad h \geq j\,. \tag{A.44}$$

Proof. By formula (A.43), we can rewrite the function $\widehat{N_{j,h}}(\xi)$ as follows:

$$\begin{aligned}\widehat{N_{j,h}}(\xi) &= \int_{\mathbf{R}^n} e^{-ix\cdot\xi} M_{j,h-j}(2^j x)\,dx \\ &= \int_{\mathbf{R}^n} e^{-iy\cdot(\xi/2j)} M_{j,h-j}(y)\,\frac{dy}{2^{jn}} \\ &= \frac{1}{2^{jn}} \widehat{M_{j,h-j}}(2^{-j}\xi) \\ &= \frac{1}{2^{jn}} \varphi(2^{-h}\xi)\,\widehat{M_j}(2^{-j}\xi)\,.\end{aligned}$$

However, since we have

$$\operatorname{supp} \varphi = \left\{\xi \in \mathbf{R}^n : \frac{1}{2} \leq |\xi| \leq 2\right\},$$

it follows that

$$\operatorname{supp} \widehat{N_{j,h}} \subset \{\xi \in \mathbf{R}^n : 2^{h-1} \leq |\xi| \leq 2^{h+1}\}\,. \quad \square$$

Step 3: Now, using decomposition (A.8) with $\delta = 1$ we express the function $\sigma(x,D)f$ in terms of the functions $N_{j,h}$ and f_j.

By formulas (A.35) and (A.43), it follows that

$$\begin{aligned}\sigma(x,D)f(x) &= \frac{1}{(2\pi)^n} \int_{\mathbf{R}^n} e^{ix\cdot\xi}\,\sigma(x,\xi)\widehat{f}(\xi)\,d\xi \\ &= \sum_{j=0}^{\infty} \frac{1}{(2\pi)^n} \int_{\mathbf{R}^n} e^{ix\cdot\xi}\,M_j(2^j x)\,\psi(2^{-j}\xi)\,\widehat{f}(\xi)\,d\xi\end{aligned}$$

$$
\begin{aligned}
&= \sum_{j=0}^{\infty} \frac{M_j(2^j x)}{(2\pi)^n} \int_{\mathbf{R}^n} e^{ix\cdot\xi}\, \psi(2^{-j}\xi)\, \widehat{f}(\xi)\, d\xi \\
&= \sum_{j=0}^{\infty} M_j(2^j x)\, \psi(2^{-j} D) f(x) \\
&= \sum_{j=0}^{\infty} M_j(2^j x)\, f_j(x) \\
&= \sum_{j=0}^{\infty} \left(\sum_{\ell=0}^{\infty} M_{j,\ell}(2^j x) \right) f_j(x) \\
&= \sum_{j=0}^{\infty} \left(\sum_{h=j}^{\infty} M_{j,h-j}(2^j x) \right) f_j(x) \\
&= \sum_{j=0}^{\infty} \left(\sum_{h=j}^{\infty} N_{j,h}(x) \right) f_j(x) \,, \qquad \text{(A.45)}
\end{aligned}
$$

where

$$
f_j(x) = \psi(2^{-j} D)(x) = \frac{1}{(2\pi)^n} \int_{\mathbf{R}^n} e^{ix\cdot\xi} \psi\left(2^{-j}\,\xi\right) \widehat{f}(\xi)\, d\xi \,.
$$

However, the spectra of the functions $N_{j,h}(x) f_j(x)$ are estimated as follows:

Claim A.18. *We have, for all $h \geq j$,*

$$
\operatorname{supp} \widehat{N_{j,h} f_j} \subset \begin{cases} \left\{ \xi \in \mathbf{R}^n : \dfrac{1}{3} \cdot 2^h \leq |\xi| \leq 3 \cdot 2^h \right\} & \text{if } h \geq j+5 \ , \\ \{\xi \in \mathbf{R}^n : |\xi| \leq 35 \cdot 2^j\} \,. & \text{if } j \leq h \leq j+4 \ . \end{cases}
$$

Proof. First, it should be noticed that

$$
\widehat{N_{j,h} f_j}(\xi) = \frac{1}{(2\pi)^n} \int_{\mathbf{R}^n} \widehat{N_{j,h}}(\xi - \eta)\, \psi(2^{-j}\eta)\, \widehat{f}(\eta)\, d\eta \,,
$$

and that

$$
\begin{aligned}
&\operatorname{supp} \widehat{N_{j,h}}(\cdot - \eta) \subset \{\xi \in \mathbf{R}^n : 2^{h-1} \leq |\xi - \eta| \leq 2^{h+1}\} \,, \\
&\operatorname{supp} \psi(2^{-j} \cdot) = \left\{ \eta \in \mathbf{R}^n : \frac{1}{3} \cdot 2^j \leq |\eta| \leq 3 \cdot 2^j \right\} \,.
\end{aligned}
$$

Therefore, we can estimate the support of $\widehat{N_{j,h} f_j}$ as follows.

Case (A): $h \geq j + 5$. In this case, we have the estimate

$$
|\xi| \leq |\xi - \eta| + |\eta|
$$

$$
\begin{aligned}
&\leq 2^{h+1} + 3 \cdot 2^j \\
&= 2^h (2 + 3 \cdot 2^{j-h}) \\
&\leq 2^h \left(2 + 3 \cdot 2^{-5}\right) \\
&\leq 3 \cdot 2^h .
\end{aligned}
$$

Similarly we have the estimate

$$
\begin{aligned}
|\xi| &\geq |\xi - \eta| - |\eta| \\
&\geq 2^{h-1} - 3 \cdot 2^j \\
&= 2^h \left(\frac{1}{2} - 3 \cdot 2^{j-h}\right) \\
&\geq 2^h \left(\frac{1}{2} - 3 \cdot \frac{1}{32}\right) \\
&\geq \frac{1}{3} \cdot 2^h .
\end{aligned}
$$

This proves that

$$
\operatorname{supp} \widehat{N_{j,h} f_j} \subset \left\{ \xi \in \mathbf{R}^n : \frac{1}{3} \cdot 2^h \leq |\xi| \leq 3 \cdot 2^h \right\}
$$

if $h \geq j + 5$.

Case (B): $j \leq h \leq j + 4$. In this case, we have the estimates

$$
\begin{aligned}
&|\xi - \eta| \leq 2^{h+1} \leq 2^{j+5}, \\
&|\eta| \leq 3 \cdot 2^j ,
\end{aligned}
$$

and hence

$$
|\xi| \leq |\xi - \eta| + |\eta| \leq 2^{j+5} + 3 \cdot 2^j = 35 \cdot 2^j .
$$

This proves that

$$
\operatorname{supp} \widehat{N_{j,h} f_j} \subset \left\{ \xi \in \mathbf{R}^n : |\xi| \leq 35 \cdot 2^j \right\}
$$

if $j \leq h \leq j + 4$.

The proof of Claim A.18 is complete. □

Step 4: By virtue of Claim A.18, we can rewrite formula (A.45) in the form

$$
\begin{aligned}
\sigma(x, D) f &= \sum_{j=0}^{\infty} \left(\sum_{h=j}^{\infty} N_{j,h} \right) f_j \\
&= \sum_{\nu=0}^{4} \left(\sum_{j=0}^{\infty} N_{j,j+\nu} f_j \right) + \sum_{h=5}^{\infty} \left(\sum_{j=0}^{h-5} N_{j,h} f_j \right) \\
&:= g_1 + g_2 .
\end{aligned}
$$

(i) The *estimate* of $g_1(x)$: Since $0 < s < r$, applying part (iii) of Lemma A.14 with $\gamma := 35$ and $a := 2$ and then Claim A.16 we obtain that

$$\begin{aligned}\|g_1\|_{H_p^s} &\leq \sum_{\nu=0}^{4}\left\|\sum_{j=0}^{\infty} N_{j,j+\nu} f_j\right\|_{H_p^s} \\ &\leq \sum_{\nu=0}^{4}\left\|\left(\sum_{j=0}^{\infty} 4^{sj}|N_{j,j+\nu} f_j|^2\right)^{1/2}\right\|_{L^p} \\ &\leq C_1\, A\left\|\left(\sum_{j=0}^{\infty} 4^{sj}|f_j|^2\right)^{1/2}\right\|_{L^p} . \qquad \text{(A.46)}\end{aligned}$$

However, we have, by inequality (A.33) with $\gamma := 3$, $a := 2$ and $m := N - n - 1$,

$$\left\|\left(\sum_{j=0}^{\infty} 4^{sj}|f_j|^2\right)^{1/2}\right\|_{L^p} \leq C_2\, \|\psi\|_{L^\infty_{N-n-1}} \|f\|_{H_p^s} .$$

Therefore, combining this inequality with inequality (A.46) we obtain that

$$\|g_1\|_{H_p^s} \leq C_1\, C_2\, A \|\psi\|_{L^\infty_{N-n-1}} \|f\|_{H_p^s} . \qquad \text{(A.47)}$$

(ii) The *estimate* of $g_2(x)$: By applying inequality (A.34) with $\gamma := 3$ and $a := 2$ and then Claim A.16, we obtain that

$$\begin{aligned}\|g_2\|_{H_p^s} &\leq C_3\left\|\left(\sum_{h=5}^{\infty} 4^{hs}\left|\sum_{j=0}^{h-5} N_{j,h} f_j\right|^2\right)^{1/2}\right\|_{L^p} \\ &\leq C_4\, A\left\|\left(\sum_{h=5}^{\infty} 4^{hs}\left(\sum_{j=0}^{h-5} 2^{(j-h)r'}|f_j|\right)^2\right)^{1/2}\right\|_{L^p} \\ &= C_4\, A\left\|\left(\sum_{h=5}^{\infty} 4^{h(s-r')}\left(\sum_{j=0}^{h-5} 2^{j(r'-s)}\left(2^{sj}|f_j|\right)\right)^2\right)^{1/2}\right\|_{L^p} \qquad \text{(A.48)}\end{aligned}$$

To estimate the function $g_2(x)$, we need the following elementary result:

Lemma A.19. *If $b > 1$ and $\xi = (\xi_j)$ is an element of the Hilbert space ℓ^2, we let*

$$\eta = (\eta_k) ,$$
$$\eta_k = \frac{\sum_{j=0}^{k} b^j \xi_j}{b^k}, \quad k = 0,\ 1,\ 2, \ldots .$$

Then it follows that $\eta \in \ell^2$ and there exists a constant $C > 1$, independent of ξ, such that

$$\|\eta\|_{\ell^2} \leq C\|\xi\|_{\ell^2} . \tag{A.49}$$

For example, we may take

$$C = \frac{\sqrt{2}\,\sqrt{b^2+1}}{b^2-1}\, b .$$

Proof. (1) First, we have, for any given $\varepsilon > 0$, the elementary inequality

$$\begin{aligned}(a_1 + a_2 + \cdots + a_n)^2 \leq (1+\varepsilon^2)^{n-1} {a_1}^2 &+ (1+\varepsilon^2)^{n-2}\left(1+\frac{1}{\varepsilon^2}\right) {a_2}^2 \\ &+ \cdots + (1+\varepsilon^2)\left(1+\frac{1}{\varepsilon^2}\right) {a_{n-1}}^2 \\ &+ \left(1+\frac{1}{\varepsilon^2}\right) {a_n}^2, \quad a_1,\, a_2, \ldots, a_n \in \mathbf{R} . \end{aligned} \tag{A.50}$$

(2) Now, applying inequality (A.50) we obtain that

$$\begin{aligned}\eta_k^2 &= \left(\sum_{j=0}^{k} b^{j-k} \xi_j\right)^2 \\ &\leq (1+\varepsilon^2)^k \left(\frac{\xi_0}{b^k}\right)^2 + (1+\varepsilon^2)^k \left(\frac{1}{\varepsilon^2}\right)\left(\frac{\xi_1}{b^{k-1}}\right)^2 \\ &\quad + \cdots + (1+\varepsilon^2)^3 \left(\frac{1}{\varepsilon^2}\right)\left(\frac{\xi_{k-2}}{b^2}\right)^2 + (1+\varepsilon^2)^2 \left(\frac{1}{\varepsilon^2}\right)\left(\frac{\xi_{k-1}}{b}\right)^2 \\ &\quad + \left(\frac{1+\varepsilon^2}{\varepsilon^2}\right) {\xi_k}^2 \\ &= \left(\frac{1+\varepsilon^2}{b^2}\right)^k {\xi_0}^2 + \left(\frac{1+\varepsilon^2}{\varepsilon^2}\right)\left\{\left(\frac{1+\varepsilon^2}{b^2}\right)^{k-1} {\xi_1}^2 \right. \\ &\quad \left. + \cdots + \left(\frac{1+\varepsilon^2}{b^2}\right)^2 {\xi_{k-2}}^2 + \left(\frac{1+\varepsilon^2}{b^2}\right)^2 {\xi_{k-1}}^2 + {\xi_k}^2 \right\} \\ &= \left(\frac{1+\varepsilon^2}{b^2}\right)^k {\xi_0}^2 + \left(\frac{1+\varepsilon^2}{\varepsilon^2}\right) \sum_{j=1}^{k} \left(\frac{1+\varepsilon^2}{b^2}\right)^{k-j} {\xi_j}^2 . \end{aligned}$$

Hence it follows that

$$\begin{aligned}&\sum_{k=0}^{\infty} \eta_k^2 \\ &\leq \sum_{k=0}^{\infty} \left(\frac{1+\varepsilon^2}{b^2}\right)^k {\xi_0}^2 + \left(\frac{1+\varepsilon^2}{\varepsilon^2}\right) \sum_{k=1}^{\infty}\sum_{j=1}^{k} \left(\frac{1+\varepsilon^2}{b^2}\right)^{k-j} {\xi_j}^2 \\ &= \sum_{k=0}^{\infty} \left(\frac{1+\varepsilon^2}{b^2}\right)^k {\xi_0}^2 + \left(\frac{1+\varepsilon^2}{\varepsilon^2}\right) \sum_{j=1}^{\infty} \left(\sum_{k=j}^{\infty} \left(\frac{1+\varepsilon^2}{b^2}\right)^{k-j}\right) {\xi_j}^2 . \end{aligned} \tag{A.51}$$

If we take

$$\varepsilon = \left(\frac{b^2-1}{2}\right)^{1/2},$$

then we have, by inequality (A.51),

$$\begin{aligned}
\sum_{k=0}^{\infty} \eta_k^2 &\leq \left(\frac{1}{1-((1+\varepsilon^2)/b^2)}\right) {\xi_0}^2 + \left(\frac{1+\varepsilon^2}{\varepsilon^2}\right)\left(\frac{1}{1-((1+\varepsilon^2)/b^2)}\right) \sum_{j=1}^{\infty} {\xi_j}^2 \\
&= \left(\frac{2b^2}{b^2-1}\right) {\xi_0}^2 + \left(\frac{b^2+1}{b^2-1}\right)\left(\frac{2b^2}{b^2-1}\right) \sum_{j=1}^{\infty} {\xi_j}^2 \\
&\leq \left(\frac{b^2+1}{b^2-1}\right)\left(\frac{2b^2}{b^2-1}\right) \sum_{j=0}^{\infty} {\xi_j}^2 .
\end{aligned}$$

This proves the desired inequality (A.49). □

Now, applying Lemma A.19 with

$$\begin{aligned}
&b := 2^{r'-s} > 1, \quad 0 < s < r', \\
&\xi_j := 2^{sj}\,|f_j|,
\end{aligned}$$

we obtain that

$$\begin{aligned}
&\sum_{h=5}^{\infty} 4^{h(s-r')} \left(\sum_{j=0}^{h-5} 2^{j(r'-s)} \left(2^{sj}|f_j|\right)\right)^2 \\
&= \sum_{h=5}^{\infty} \left(b^{-(h-5)} \left(\sum_{j=0}^{h-5} b^j \xi_j\right)\right)^2 \times b^{-10} \\
&= \sum_{\ell=0}^{\infty} \left(b^{-\ell} \left(\sum_{j=0}^{\ell} b^j \xi_j\right)\right)^{1/2} \times b^{-10} \\
&\leq {C_5}^2 \sum_{j=0}^{\infty} |\xi_j|^2 \\
&= {C_5}^2 \sum_{j=0}^{\infty} 4^{sj}\,|f_j|^2 .
\end{aligned}$$

Thus we have, by inequality (A.48),

$$\begin{aligned}
\|g_2\|_{H_p^s} &\leq C_4\, C_5\, A \left\| \left(\sum_{j=0}^{\infty} 4^{sj} |f_j|^2\right)^{1/2} \right\|_{L^p} \\
&\leq C_6\, A \|\psi\|_{L^\infty_{N-n-1}} \|f\|_{H_p^s} . \qquad \text{(A.52)}
\end{aligned}$$

Step 5: By combining inequalities (A.47) and (A.52), we have proved that, for $0 < s < r'$,

$$\begin{aligned}\|\sigma(x,D)f\|_{H_p^s} &= \left\|\sum_{j=0}^{\infty}\sum_{h=j}^{\infty} N_{j,h}\, f_j\right\|_{H_p^s} \\ &= \|g_1 + g_2\|_{H_p^s} \\ &\leq C_7\, A\|\psi\|_{L^\infty_{N-n-1}}\|f\|_{H_p^s} \, .\end{aligned}$$

Now the proof of Proposition A.8 for $\delta = 1$ and $0 < s < r$ is complete, since $r' > 0$ may be chosen arbitrarily close to r. □

A.6.2 Proof of the Case $0 \leq \delta < 1$

Secondly, we prove the case where $0 \leq \delta < 1$ and $(\delta - 1)r < s < r$: To do this, it should be noticed that

$$S^0_{1,\delta}(N,r) \subset S^0_{1,1}(N,r) \quad \text{for } 0 \leq \delta < 1 \, .$$

Hence it follows from the proof of the case $\delta = 1$ that if a symbol $\sigma(x,\xi)$ is in the class $S^0_{1,\delta}(N,r)$, then the operator $\sigma(x,D)$ is bounded on $H_p^s(\mathbf{R}^n)$ for all $0 < s < r$ and $1 < p < \infty$.

Therefore, we have only to consider the case where $(\delta - 1)r < s < 0$, since Proposition A.8 for $s = 0$ follows from an interpolation argument. Indeed, it suffices to note that the space $H^0_p(\mathbf{R}^n) = L^p(\mathbf{R}^n)$ is a complex interpolation space between the spaces $H^\sigma_p(\mathbf{R}^n)$ and $H^{-\sigma}_p(\mathbf{R}^n)$ with $0 < \sigma < (1-\delta)r$ (see [BL, Theorem 6.4.5], [Tr, Theorem 2.4.7]). The proof is divided into four steps.

Step 1: Now we assume that $\sigma(x,\xi) \in S^0_{1,\delta}(N,r)$ is an elementary symbol of the form

$$\sigma(x,\xi) = \sum_{j=0}^{\infty} M_j(2^{j\delta}x)\,\psi(2^{-j}\xi) \, , \tag{A.53}$$

where

$$\psi(\xi) \in C_0^\infty(\mathbf{R}^n), \quad \operatorname{supp}\psi = \left\{\xi \in \mathbf{R}^n : \frac{1}{3} \leq |\xi| \leq 3\right\} ,$$

The sequence $\{M_j\}$ is bounded in the Hölder space $\Lambda_r(\mathbf{R}^n)$.

We take

$$a = 2^{1-\delta} > 1, \quad 0 \leq \delta < 1 \, ,$$

and let

$$M_j(x) = \sum_{\ell=0}^{\infty} M_{j,\ell}(x) \tag{A.54}$$

be the Littlewood-Paley series of $M_j(x)$, that is,

$$\widehat{M_{j,0}}(\xi) = \psi_0(\xi)\widehat{M_j}(\xi) \,,$$
$$\widehat{M_{j,k}}(\xi) = \varphi(a^{-k}\,\xi)\widehat{M_j}(\xi), \quad k = 1,\ 2, \dots \,,$$

where

$$\operatorname{supp}\varphi = \left\{\xi \in \mathbf{R}^n : \frac{1}{a} \leq |\xi| \leq a\right\} , \tag{A.55a}$$
$$\sum_{j\in\mathbf{Z}} \varphi(a^{-j}\,\xi) = 1 \quad \text{for all } \xi \neq 0 \tag{A.55b}$$

and

$$\psi_0 \in C_0^\infty(\mathbf{R}^n) \,, \tag{A.56a}$$
$$\operatorname{supp}\psi_0 = \{\xi \in \mathbf{R}^n : |\xi| \leq a\} \,, \tag{A.56b}$$
$$\psi_0(\xi) = 1 - \sum_{j=1}^{\infty} \varphi(2^{-j}\,\xi) \quad \text{for all } \xi \neq 0 \,. \tag{A.56c}$$

If we introduce functions $\varphi_\ell(x) \in \mathcal{S}(\mathbf{R}^n)$ by the formulas

$$\widehat{\varphi_\ell}(\xi) = \begin{cases} \psi_0(\xi) & \text{if } \ell = 0 \,, \\ \varphi(a^{-\ell}\,\xi), & \text{if } \ell = 1,\ 2, \dots \,, \end{cases}$$

then the functions $M_{j,\ell}$ can be expressed in the convolution form

$$M_{j,\ell}(x) = \varphi_\ell * M_j(x), \quad \ell = 0,\ 1,\ 2, \dots \,.$$

Moreover, the functions $M_{j,\ell}$ are estimated as follows:

Claim A.20. *If we let*

$$A = \sup_{j\geq 0} \|M_j\|_{\Lambda^r} \,,$$

then we have, for all $\ell = 0,\ 1,\ 2,\ \dots$,

$$\|M_{j,\ell}\|_{L^\infty} \leq C\,A\,a^{-r\ell} \,, \tag{A.57}$$

with some constant $C > 0$ *independent of* ℓ.

Proof. By Part (i) of Lemma A.2, it follows that

$$\begin{aligned} \|M_{j,\ell}\|_{L^\infty} &= \|\varphi_\ell * M_j\|_{L^\infty} \\ &= \left\| J^{(\delta-1)r} \cdot J^{(1-\delta)r}(\varphi_\ell * M_j) \right\|_{L^\infty} \\ &\leq C_1\,a^{-r\ell} \|\varphi_\ell * (J^{(1-\delta)r} M_j)\|_{L^\infty} \,. \end{aligned} \tag{A.58}$$

However, we have, by Young's inequality (Corollary 4.3),

$$\begin{aligned}\|\varphi_\ell * (J^{(1-\delta)r} M_j)\|_{L^\infty} &\leq \|\varphi_\ell\|_{L^1} \, \|J^{(1-\delta)r} M_j\|_{L^\infty} \\ &= \|\widehat{\varphi}\|_{L^1} \, \|M_j\|_{H_\infty^{(1-\delta)r}} \,. \end{aligned} \tag{A.59}$$

Since $0 \leq \delta < 1$, we have the inclusions

$$\Lambda_r = B^r_{\infty,\infty} \subset B^{(1-\delta)r}_{\infty,1} \subset H^{(1-\delta)r}_\infty \,,$$

and hence, with some constants C_2, $C_3 > 0$,

$$\|M_j\|_{H_\infty^{(1-\delta)r}} \leq C_2 \, \|M_j\|_{B^{(1-\delta)r}_{\infty,1}} \leq C_3 \, \|M_j\|_{\Lambda_r} \leq C_3 \, A \,. \tag{A.60}$$

Therefore, combining inequalities (A.58), (A.59) and (A.60) we obtain that

$$\begin{aligned}\|M_{j,\ell}\|_{L^\infty} &\leq C_1 \, a^{-r\ell} \|J^{(1-\delta)r} M_j\|_{L^\infty} \\ &\leq (C_1 \, C_3 \, \|\widehat{\varphi}\|_{L^1}) \, a^{-r\ell} \|M_j\|_{\Lambda_r} \\ &\leq (C_1 \, C_3 \, \|\widehat{\varphi}\|_{L^1}) \, A \, a^{-r\ell} \,.\end{aligned}$$

This proves inequality (A.57). □

Step 2: By formulas (A.8) and (A.54), it follows that

$$\begin{aligned}\sigma(x, D) f &= \sum_{j=0}^\infty M_j(2^{j\delta} x) \, \psi(2^{-j} D) f \\ &= \sum_{j=0}^\infty \left(\sum_{\ell=0}^\infty M_{j,\ell}(2^{j\delta} x) \right) f_j \,. \end{aligned} \tag{A.61}$$

However, since $0 \leq \delta < 1$, we can find a number $\nu > 2$ such that

$$2^{(1-\nu)(1-\delta)} \leq \frac{1}{12} \,.$$

Then the spectra of the functions $M_{j,\ell}(2^{j\delta} x) f_j(x)$ are estimated as follows:

Claim A.21. *We have, for all $\ell \geq 0$,*

$$\operatorname{supp} \widehat{M_{j,\ell} f_j} \subset \begin{cases} \left\{ \xi \in \mathbf{R}^n : \dfrac{1}{4} \cdot 2^j \leq |\xi| \leq 4 \cdot 2^j \right\} & \text{if } j \geq \ell + \nu \,, \\ \{ \xi \in \mathbf{R}^n : |\xi| \leq \gamma \cdot 2^\ell \} . & \text{if } 0 \leq j \leq \ell + \nu - 1 \,, \end{cases}$$

where

$$\gamma = 3 \cdot 2^{(\nu-1)} + 2^{(\nu-2)\delta+1} \,.$$

Proof. First, it should be noticed that

$$\begin{aligned}\widehat{M_{j,\ell}f_j}(\xi) &= \int_{\mathbf{R}^n} e^{-ix\cdot\xi} M_{j,\ell}(2^{j\delta}\,x)\, f_j(x)\, dx \\ &= \widehat{M_{j,\ell}f_j} * \widehat{f}_j(\xi) \\ &= \frac{1}{(2^{j\delta})^n}\int_{\mathbf{R}^n} \widehat{M_{j,\ell}}(2^{-j\delta}(\xi-\eta))\, \psi(2^{-j}\eta)\, \widehat{f}(\eta)\, d\eta \,,\end{aligned}$$

and that

$$\begin{aligned}&2^{(1-\delta)(\ell-1)} \leq \frac{|\xi-\eta|}{2^{j\delta}} \leq 2^{(1-\delta)(\ell+1)} \,, \\ &\frac{1}{3} \leq \frac{|\eta|}{2^j} \leq 3 \,.\end{aligned}$$

Therefore, we can estimate the support of $\widehat{M_{j,h}f_j}$ as follows.

Case (A): $j \geq \ell + \nu$. In this case, we have the estimate

$$\begin{aligned}|\xi| &\leq |\xi-\eta| + |\eta| \\ &\leq 2^{(1-\delta)(\ell+1)+j\delta} + 3\cdot 2^j \\ &= 2^j\left(3 + 2^{(1-\delta)(\ell+1)+j(\delta-1)}\right) \\ &\leq 2^j\left(3 + 2^{(1-\delta)(1-\nu)}\right) \\ &\leq 2^j\left(3+\frac{1}{12}\right) \\ &\leq 4\cdot 2^j \,.\end{aligned}$$

Similarly we have the estimate

$$\begin{aligned}|\xi| &\geq |\xi-\eta| - |\eta| \\ &\geq \frac{1}{3}\cdot 2^j - 2^{(1-\delta)(\ell+1)+j\delta} \\ &= 2^j\left(\frac{1}{3} - 2^{(1-\delta)(\ell-j)+1-\delta}\right) \\ &\geq 2^j\left(\frac{1}{3} - 2^{(1-\delta)(1-\nu)}\right) \\ &\geq 2^j\left(\frac{1}{3}-\frac{1}{12}\right) \\ &= \frac{1}{4}\cdot 2^j \,.\end{aligned}$$

This proves that

$$\operatorname{supp}\widehat{M_{j,\ell}f_j} \subset \left\{\xi\in\mathbf{R}^n : \frac{1}{4}\cdot 2^j \leq |\xi| \leq 4\cdot 2^j\right\} \quad \text{if } j \geq \ell+\nu \,,$$

with $\gamma = 3\cdot 2^{(\nu-1)} + 2^{(\nu-2)\delta+1}$.

Case (B): $0 \le j \le \ell + \nu - 1$. In this case, we have the estimate

$$\begin{aligned}|\xi| &\le |\xi - \eta| + |\eta| \\ &\le 2^{(1-\delta)(\ell+1)+j\delta} + 3 \cdot 2^j \\ &\le 2^\ell \left(3 \cdot 2^{(\nu-1)} + 2^{(\nu-2)\delta+1}\right) .\end{aligned}$$

This proves that

$$\operatorname{supp} \widehat{M_{j,\ell} f_j} \subset \left\{\xi \in \mathbf{R}^n : |\xi| \le \gamma \cdot 2^\ell\right\} \quad \text{if } 0 \le j \le \ell + \nu - 1 .$$

The proof of Claim A.21 is complete. □

Step 3: By virtue of Claim A.21, we can rewrite formula (A.61) in the form

$$\begin{aligned}\sigma(x, D)f &= \sum_{j=\nu}^{\infty} \left(\sum_{\ell=0}^{j-\nu} M_{j,\ell}(2^{j\delta} x)\right) f_j \\ &\quad + \sum_{\ell=0}^{\infty} \left(\sum_{j=0}^{\ell+\nu-1} M_{j,\ell}(2^{j\delta} x) f_j\right) \\ &:= g_1 + g_2 .\end{aligned}$$

(i) The *estimate* of $g_1(x)$: By applying inequality (A.34) with $\gamma := 4$ and $a := 2$, we obtain that

$$\|g_1\|_{H_p^s} \le C_1 \left\| \left(\sum_{j=\nu}^{\infty} 4^{sj} \left| \sum_{\ell=0}^{j-\nu} M_{j,\ell}(2^{j\delta} x) \right|^2 |f_j|^2 \right)^{1/2} \right\|_{L^p} .$$

However, by Claim A.20 it follows that

$$\begin{aligned}\left| \sum_{\ell=0}^{j-\nu} M_{j,\ell}(2^{j\delta} x) \right| &\le \sum_{\ell=0}^{j-\nu} \|M_{j,\ell}\|_{L^\infty} \\ &\le C_2 A \sum_{\ell=0}^{j-\nu} \left(\frac{1}{2^{(1-\delta)r}} \right)^\ell \\ &\le C_3 A .\end{aligned}$$

Hence we have, by inequality (A.33) with $\gamma := 3$, $a := 2$ and $m := N - n - 1$,

$$\begin{aligned}\|g_1\|_{H_p^s} &\le C_1 C_3 A \left\| \left(\sum_{j=\nu}^{\infty} 4^{sj} |f_j|^2 \right)^{1/2} \right\|_{L^p} \\ &\le C_4 A \|\psi\|_{L^\infty_{N-n-1}} \|f\|_{H_p^s} . \end{aligned} \tag{A.62}$$

(ii) The *estimate* of $g_2(x)$: First, it follows that

$$\begin{aligned}
&\sum_{\ell=0}^{\infty} 4^{(s+(1-\delta)r)\ell} \left| \sum_{j=0}^{\ell+\nu-1} M_{j,\ell}(2^{j\delta}\, x) f_j(x) \right|^2 \\
&\le \sum_{\ell=0}^{\infty} 4^{(s+(1-\delta)r)\ell} \left(\sum_{j=0}^{\ell+\nu-1} |M_{j,\ell}(2^{j\delta}\, x)|\, |f_j(x)|^2 \right)^2 .
\end{aligned}$$

However, we have, by Claim A.20,

$$\begin{aligned}
\sum_{j=0}^{\ell+\nu-1} |M_{j,\ell}(2^{j\delta}\, x)|\, |f_j(x)|^2 &\le \sum_{j=0}^{\ell+\nu-1} \|M_{j,\ell}\|_{L^\infty}\, |f_j(x)| \\
&\le C_5\, A\, 2^{(\delta-1)r\ell} \sum_{j=0}^{\ell+\nu-1} |f_j(x)| .
\end{aligned}$$

Hence we obtain that

$$\begin{aligned}
&\sum_{\ell=0}^{\infty} 4^{(s+(1-\delta)r)\ell} \left| \sum_{j=0}^{\ell+\nu-1} M_{j,\ell}(2^{j\delta}\, x) f_j(x) \right|^2 \\
&\le {C_5}^2\, A^2 \sum_{\ell=0}^{\infty} 4^{(s+(1-\delta)r)\ell}\, 4^{(\delta-1)r\ell} \left(\sum_{j=0}^{\ell+\nu-1} |f_j(x)| \right)^2 \\
&= {C_5}^2\, A^2 \sum_{\ell=0}^{\infty} 4^{s\ell} \left(\sum_{j=0}^{\ell+\nu-1} |f_j(x)| \right)^2 . \qquad \text{(A.63)}
\end{aligned}$$

If we let

$$\begin{aligned}
&b := 2^{-s} > 1, \quad (\delta-1)r < s < 0 , \\
&\xi_j := 2^{sj}\, |f_j| ,
\end{aligned}$$

then we can write the last term in inequality (A.63) in the form

$$\begin{aligned}
&\sum_{\ell=0}^{\infty} 4^{s\ell} \left(\sum_{j=0}^{\ell+\nu-1} |f_j(x)| \right)^2 \\
&= \sum_{\ell=0}^{\infty} b^{-2(\ell+\nu-1)} \left(\sum_{j=0}^{\ell+\nu-1} b^j \xi_j| \right)^2 \times 4^{s(1-\nu)} .
\end{aligned}$$

Therefore, by applying Lemma A.19 we obtain from inequality (A.63) that

$$\sum_{\ell=0}^{\infty} 4^{(s+(1-\delta)r)\ell} \left| \sum_{j=0}^{\ell+\nu-1} M_{j,\ell}(2^{j\delta}\, x) f_j(x) \right|^2 \le {C_5}^2\, A^2 \sum_{j=0}^{\infty} 4^{sj}\, |f_j|^2 .$$

By raising this inequality to the power $p/2$ and then integrating, it follows that

$$\left\| \left(\sum_{\ell=0}^{\infty} 4^{(s+(1-\delta)r)\ell} \left| \sum_{j=0}^{\ell+\nu-1} M_{j,\ell}(2^{j\delta} x) f_j(x) \right|^2 \right)^{1/2} \right\|_{L^p}^p$$

$$\leq C_5{}^p A^p \left\| \sum_{j=0}^{\infty} 4^{sj} |f_j|^2 \right\|_{L^p}^p . \tag{A.64}$$

On the other hand, since we have, by Claim A.21,

$$g_2(x) = \sum_{\ell=0}^{\infty} \left(\sum_{j=0}^{\ell+\nu-1} M_{j,\ell}(2^{j\delta} x) f_j(x) \right),$$

$$\operatorname{supp} \widehat{M_{j,\ell} f_j} \subset \left\{ \xi \in \mathbf{R}^n : |\xi| \leq \gamma \cdot 2^{\ell} \right\} \quad \text{if } 0 \leq j \leq \ell + \nu - 1,$$

it follows from an application of part (iii) of Lemma A.14 with $s := s + (1-\delta)r > 0$ that

$$\|g_2\|_{H_p^{s+(1-\delta)r}} \leq C_6 \left\| \left(\sum_{\ell=0}^{\infty} 4^{(s+(1-\delta)r)\ell} \left| \sum_{j=0}^{\ell+\nu-1} M_{j,\ell}(2^{j\delta} x) f_j \right|^2 \right)^{1/2} \right\|_{L^p} . \tag{A.65}$$

By combining inequalities (A.65) and (A.64) and using inequality (A.33) with $a := 2$ and $\gamma := 3$, we obtain that

$$\begin{aligned} \|g_2\|_{H_p^{s+(1-\delta)r}} &\leq C_5\, C_6\, A \left\| \left(\sum_{j=0}^{\infty} 4^{sj} |f_j|^2 \right)^{1/2} \right\|_{L^p} \\ &\leq C_7\, A \|\psi\|_{L^\infty_{N-n-1}} \|f\|_{H_p^s} . \end{aligned}$$

This implies the estimate

$$\begin{aligned} \|g_2\|_{H_p^s} &\leq C_8\, \|g_2\|_{H_p^{s+(1-\delta)r}} \\ &\leq C_7\, C_8\, A \|\psi\|_{L^\infty_{N-n-1}} \|f\|_{H_p^s} , \end{aligned} \tag{A.66}$$

since for $H_p^{s+(1-\delta)r} \subset H_p^s$ for $0 \leq \delta < 1$ and $r > 0$.

Step 4: Summing up, we obtain from inequalities (A.62) and (A.66) that

$$\begin{aligned} \|\sigma(x, D) f\|_{H_p^s} &= \left\| \sum_{j=0}^{\infty} \sum_{\ell=0}^{\infty} M_{j,\ell}(2^{j\delta})\, f_j \right\|_{H_p^s} \\ &= \|g_1 + g_2\|_{H_p^s} \\ &\leq C_9\, A \|\psi\|_{L^\infty_{N-n-1}} \|f\|_{H_p^s} . \end{aligned}$$

The proof of Proposition A.8 for $0 \leq \delta < 1$ and $(\delta - 1)r < s < 0$ is now complete. □

B

Unique Solvability of Pseudo-Differential Operators

In this appendix, we give a sketch of the proof of a unique solvability theorem for pseudo-differential operators (Theorem 7.16) which plays an essential role in the construction of Feller semigroups in Chap. 7:

Theorem B.1. *Let T be a second-order, classical pseudo-differential operator on an n-dimensional, compact smooth manifold M without boundary such that*

$$T = P + S\,,$$

where:

(a) The operator P is a second-order, degenerate elliptic differential operator *on M with non-positive principal symbol, and $P1 \leq 0$ on M.*
(b) The operator S is a classical pseudo-differential operator of order $2-\kappa$, $\kappa > 0$, on M and its distribution kernel $s(x,y)$ is non-negative *off the diagonal $\Delta_M = \{(x,x) : x \in M\}$ in $M \times M$.*
(c) $T1 = P1 + S1 \leq 0$ on M.

Then, for each integer $k \geq 1$ there exists a constant $\lambda = \lambda(k) > 0$ such that, for any $f \in C^{k+\theta}(M)$, we can find a function $\varphi \in C^{k+\theta}(M)$ satisfying the equation

$$(T - \lambda I)\varphi = f \quad \text{on } M\,,$$

and the estimate

$$\|\varphi\|_{C^{k+\theta}(M)} \leq C(\lambda)\|f\|_{C^{k+\theta}(M)}\,.$$

Here $C(\lambda) > 0$ is a constant independent of f.

Proof. We prove Theorem B.1 by using a method of *elliptic regularizations* essentially due to Oleĭnik-Radkevič [OR, Chap. I]), just as in the proof of Cancelier [Cn, Théorème 4.5]. The proof is divided into four steps.

Step 1: We begin with the following results:

Theorem B.2. *Let $T = P + S$ be a second-order, classical pseudo-differential operator on M as in Theorem B.1. Assume that*

$$T1 = P1 + S1 < 0 \quad on\ M\ .$$

Then we have, for all $\varphi \in C^2(M)$,

$$\|\varphi\|_{C(M)} \leq \left(\frac{1}{\min_M(-T1)}\right) \|T\varphi\|_{C(M)}\ .$$

Theorem B.2 is a compact manifold version of Theorem C.2 which will be proved in Appendix C.

Theorem B.3. *Let $T = P + S$ be a second-order, classical pseudo-differential operator on M as in Theorem B.1. Assume that the operator T is* elliptic *on M and satisfies the condition*

$$T1 = P1 + S1 < 0 \quad on\ M\ .$$

Then, for each integer $k \geq 0$, the operator

$$T: C^{k+2+\theta}(M) \longrightarrow C^{k+\theta}(M)$$

is bijective.

Proof. Since T is elliptic and its principal symbol is *real*, it follows from an application of [Ta2, Corollary 6.7.12] that

$$\operatorname{ind} T = \dim N(T) - \operatorname{codim} R(T) = 0\ .$$

However, Theorem B.2 tells us that T is injective, that is, $\dim N(T) = 0$. Hence we obtain that $\operatorname{codim} R(T) = 0$, which proves that T is surjective. □

Step 2: First, we prove Theorem B.1 for the Sobolev space $W^{1,\infty}(M)$:

Claim B.4. *There exists a constant $\lambda = \lambda(1) > 0$ such that, for any $f \in W^{1,\infty}(M)$, we can find a function $\varphi \in W^{1,\infty}(M)$ satisfying the equation*

$$(T - \lambda I)\varphi = f \quad on\ M\ ,$$

and the estimate

$$\|\varphi\|_{1,\infty} \leq C_1(\lambda)\|f\|_{1,\infty}\ .$$

Here $C_1(\lambda) > 0$ is a constant independent of f.

Proof. The proof is divided into three steps.

(1) Let $\{(U_\alpha, \chi_\alpha)\}_{\alpha=1}^N$ be a finite open covering of M by local charts, and let $\{\sigma_\alpha\}_{\alpha=1}^N$ be a family of non-negative functions in $C^\infty(M \times M)$ such that

$$\operatorname{supp} \sigma_\alpha \subset U_\alpha \times U_\alpha\ ,$$

and

$$\sum_{\alpha=1}^{N} \sigma_\alpha(x,y) = 1 \text{ in a neighborhood of the diagonal } \Delta_M = \{(x,x) : x \in M\}\,.$$

Then it is easy to see that the operator $T = P + S$ can be written in local coordinates $(x_1, x_2, \ldots, x_n)$ in the form (see Sect. 3.4)

$$\begin{aligned} T\varphi(x) &= \sum_{i,j=1}^{n} \alpha^{ij}(x) \frac{\partial^2 \varphi}{\partial x_i \partial x_j}(x) + \sum_{i=1}^{n} \beta^i(x) \frac{\partial \varphi}{\partial x_i}(x) + \gamma(x)\varphi(x) \\ &\quad + \int_M s(x,y) \left[\varphi(y) - \sigma(x,y) \left(\varphi(x) + \sum_{i=1}^{n} (y_i - x_i) \frac{\partial \varphi}{\partial x_i}(x) \right) \right] dy. \end{aligned}$$

Here:

(a) The operator $\sum_{i,j=1}^{n} \alpha^{ij}(x) \partial^2 / \partial x_i \partial x_j$ is the principal part of P; more precisely, the $\alpha^{ij}(x)$ are the components of a smooth symmetric contravariant tensor of type $\binom{2}{0}$ on M satisfying the condition

$$\sum_{i,j=1}^{n} \alpha^{ij}(x) \xi_i \xi_j \geq 0, \quad x \in M,\ \xi = \sum_{j=1}^{n} \xi_j dx_j \in T_x^*(M)\,,$$

where $T_x^*(M)$ is the cotangent space of M at x.

(b) The function $\sigma(x,y) = \sum_{\alpha=1}^{N} \sigma_\alpha(x,y)$ is a local unity function on M.

(c) The density dy is a strictly positive density on M.

(d) $P1(x) = \gamma(x) \leq 0$ on M and $T1(x) = \gamma(x) + \int_M s(x,y)[1 - \sigma(x,y)]\, dy \leq 0$ on M.

Furthermore, it should be emphasized that there exists a constant $C > 0$ such that the distribution kernel $s(x,y)$ of $S \in L_{cl}^{2-\kappa}(M)$, $\kappa > 0$, satisfies the estimate

$$|s(x,y)| \leq \frac{C}{|x-y|^{n+2-\kappa}}, \quad (x,y) \in (M \times M) \setminus \Delta_M\,,$$

where $|x - y|$ is the geodesic distance between x and y with respect to the Riemannian metric of M (cf. Coifman-Meyer [CM, Chap. IV, Proposition 1]). Hence we find that the integral

$$\int_M s(x,y) \left[\varphi(y) - \sigma(x,y) \left(\varphi(x) + \sum_{i=1}^{n} (y_i - x_i) \frac{\partial \varphi}{\partial x_i}(x) \right) \right] dy$$

is absolutely convergent, since $\kappa > 0$ and $\sigma(x,y) = 1$ in a neighborhood of the diagonal Δ_M.

Now, if $\varphi \in C^1(M)$, we define a continuous function $B_T(\varphi, \varphi)$ on M by the formula

$$B_T(\varphi,\varphi)(x) = 2\sum_{i,j=1}^{n} \alpha^{ij}(x)\frac{\partial\varphi}{\partial x_i}(x)\frac{\partial\varphi}{\partial x_j}(x)$$
$$+ \int_M s(x,y)\,(\varphi(y)-\varphi(x))^2\,dy - T1(x)\cdot\varphi(x)^2, \quad x \in M.$$

It should be noticed that the function $B_T(\varphi,\varphi)$ is *non-negative* on M for all $\varphi \in C^1(M)$.

The next result may be proved just as in the proof of Cancelier [Cn, Théorème 4.1].

Lemma B.5. *Let $\{X_j\}_{j=1}^r$ be a family of real smooth vector fields on M such that the X_j span the tangent space $T_x(M)$ at each point x of M. If $\varphi \in C^\infty(M)$, we let*

$$p_1(x) = \sum_{j=1}^{r} |X_j\varphi(x)|^2, \quad x \in M,$$

and

$$R_1(x) = Tp_1(x) - \sum_{j=1}^{r} B_T(X_j\varphi, X_j\varphi)(x), \quad x \in M.$$

Then, for each $\eta > 0$, there exist constants $\beta_0 > 0$ and $\beta_1 > 0$ such that we have, for all $\varphi \in C^\infty(M)$,

$$|R_1(x)| \le \eta\sum_{j=1}^{r} B_T(X_j\varphi, X_j\varphi)(x) + \beta_0\|\varphi\|^2_{C(M)}$$
$$+ \beta_1\|\varphi\|^2_{C^1(M)} + \frac{1}{2}\|T\varphi\|^2_{C^1(M)}, \quad x \in M. \tag{B.1}$$

Remark B.1. The constants β_0 and β_1 are *uniform* for the operators $T+\varepsilon\Lambda-\lambda I$, $0 \le \varepsilon \le 1$, $\lambda \ge 0$, where Λ is a second-order, *elliptic* differential operator on M defined by the formula

$$\Lambda = -\sum_{j=1}^{r} X_j^* X_j = \sum_{j=1}^{r} X_j^2 + \sum_{j=1}^{r} \operatorname{div} X_j \cdot X_j .$$

(2) First, let f be an arbitrary element of $C^\infty(M)$. Since the operator $T+\varepsilon\Lambda-\lambda I$ is elliptic for all $\varepsilon > 0$ and $(T+\varepsilon\Lambda-\lambda I)1 = T1-\lambda \le -\lambda < 0$ on M for $\lambda > 0$, it follows from an application of Theorem B.3 that we can find a unique function $\varphi_\varepsilon \in C^\infty(M)$ such that

$$(T+\varepsilon\Lambda-\lambda I)\varphi_\varepsilon = f \quad \text{on } M.$$

Furthermore, applying Theorem B.2 to the operator $T+\varepsilon\Lambda-\lambda I$ we obtain that

$$\|\varphi_\varepsilon\|_{C(M)} \le \frac{1}{\lambda}\|f\|_{C(M)}, \tag{B.2}$$

since $\min_M(-(T+\varepsilon\Lambda-\lambda I)1) \geq \lambda$.

Let x_0 be a point of M at which the function

$$p_1^\varepsilon(x) = \sum_{j=1}^{r} |X_j \varphi_\varepsilon(x)|^2$$

attains its positive maximum. Then we have

$$\Lambda p_1^\varepsilon(x_0) = \left(\sum_{j=1}^{r} X_j^2\right) p_1^\varepsilon(x_0) \leq 0\,,$$

and also

$$\begin{aligned}
Tp_1^\varepsilon(x_0) &= \sum_{i,j=1}^{n} \alpha^{ij}(x_0) \frac{\partial^2 p_1^\varepsilon}{\partial x_i \partial x_j}(x_0) + \gamma(x_0) p_1^\varepsilon(x_0) \\
&\quad + \int_M s(x_0,y)[p_1^\varepsilon(y) - \sigma(x_0,y) p_1^\varepsilon(x_0)]\, dy \\
&\leq \left(\gamma(x_0) + \int_M s(x_0,y)[1-\sigma(x_0,y)]\, dy\right) p_1^\varepsilon(x_0) \\
&\quad + \int_M s(x_0,y)[p_1^\varepsilon(y) - p_1^\varepsilon(x_0)]\, dy \\
&\leq T1(x_0) \cdot p_1^\varepsilon(x_0).
\end{aligned}$$

Hence, using inequality (B.1) with $\eta = 1/2$ and inequality (B.2) we obtain that

$$\begin{aligned}
\lambda p_1^\varepsilon(x_0) &\leq (\lambda - T1(x_0)) p_1^\varepsilon(x_0) - \varepsilon \Lambda p_1^\varepsilon(x_0) \\
&\leq (\lambda - T - \varepsilon\Lambda) p_1^\varepsilon(x_0) \\
&= -\left((T + \varepsilon\Lambda - \lambda) p_1^\varepsilon(x_0) - \sum_{j=1}^{r} B_{T+\varepsilon\Lambda-\lambda I}(X_j\varphi_\varepsilon, X_j\varphi_\varepsilon)(x_0)\right) \\
&\quad - \sum_{j=1}^{r} B_{T+\varepsilon\Lambda-\lambda I}(X_j\varphi_\varepsilon, X_j\varphi_\varepsilon)(x_0) \\
&\leq -\frac{1}{2} \sum_{j=1}^{r} B_{T+\varepsilon\Lambda-\lambda I}(X_j\varphi_\varepsilon, X_j\varphi_\varepsilon)(x_0) + \beta_0 \|\varphi_\varepsilon\|_{C(M)}^2 \\
&\quad + \beta_1 \|\varphi_\varepsilon\|_{C^1(M)}^2 + \frac{1}{2} \|f\|_{C^1(M)}^2 \\
&\leq \frac{\beta_0}{\lambda^2} \|f\|_{C(M)}^2 + \beta_1 \|\varphi_\varepsilon\|_{C^1(M)}^2 + \frac{1}{2} \|f\|_{C^1(M)}^2.
\end{aligned}$$

This proves that

$$\begin{aligned}
(\lambda - \beta_1)\|\varphi_\varepsilon\|_{C^1(M)}^2 &\leq \lambda \left(\|\varphi_\varepsilon\|_{C(M)}^2 + p_1^\varepsilon(x_0)\right) - \beta_1 \|\varphi_\varepsilon\|_{C^1(M)}^2 \\
&\leq \frac{1}{\lambda} \|f\|_{C(M)}^2 + \frac{\beta_0}{\lambda^2} \|f\|_{C(M)}^2 + \frac{1}{2} \|f\|_{C^1(M)}^2 . \qquad \text{(B.3)}
\end{aligned}$$

Here it should be emphasized (see Remark B.1) that the constants β_0 and β_1 are independent of $\varepsilon > 0$ and $\lambda > 0$. Therefore, if $\lambda > 0$ is so large that

$$\lambda > \beta_1 ,$$

then it follows from inequality (B.3) that

$$\|\varphi_\varepsilon\|^2_{C^1(M)} \leq C(\lambda)\|f\|^2_{C^1(M)} , \tag{B.4}$$

where $C(\lambda) > 0$ is a constant *independent* of $\varepsilon > 0$.

(3) Now let f be an arbitrary element of $W^{1,\infty}(M)$. Then we can find a sequence $\{f_p\}_{p=1}^\infty$ in $C^\infty(M)$ such that

$$\begin{cases} f_p \longrightarrow f & \text{in } C(M) , \\ \|f_p\|_{C^1(M)} \leq \|f\|_{1,\infty} . \end{cases}$$

If $\varphi_{\varepsilon p}$ is a unique solution in $C^\infty(M)$ of the equation

$$(T + \varepsilon\Lambda - \lambda I)\varphi_{\varepsilon p} = f_p \quad \text{on M} , \tag{B.5}$$

it follows from an application of inequality (B.4) that

$$\|\varphi_{\varepsilon p}\|^2_{C^1(M)} \leq C(\lambda)\|f_p\|^2_{C^1(M)} \leq C(\lambda)\|f\|^2_{1,\infty} .$$

This proves that the sequence $\{\varphi_{\varepsilon p}\}$ is uniformly bounded and equicontinuous on M. Hence, by virtue of the Ascoli-Arzelà theorem we can choose a subsequence $\{\varphi_{\varepsilon' p'}\}$ which converges uniformly to a function $\varphi \in C(M)$, as $\varepsilon' \downarrow 0$ and $p' \to \infty$. Furthermore, since the unit ball in the Hilbert space $L^2(M)$ is *sequentially weakly compact* (see Yosida [Yo, Chap. V, Sect. 2, Theorem 1]), we may assume that the sequence $\{\partial_j \varphi_{\varepsilon' p'}\}$ converges weakly to a function ψ_j in $L^2(M)$, for each $1 \leq j \leq n$. Then it follows that

$$\partial_j \varphi = \psi_j \in L^2(M), \quad 1 \leq j \leq n .$$

On the other hand, it is easy to verify that the set

$$K = \{g \in L^2(M) : \|g\|_\infty \leq \sqrt{C(\lambda)}\, \|f\|_{1,\infty}\}$$

is convex and strongly closed in $L^2(M)$. Moreover, we obtain that the set K is *weakly closed* in $L^2(M)$, if we apply the following Mazur's theorem (see [Yo, Chap. IV, Sect. 6, Theorem 3]) with

$$X := L^2(M), \quad M := K .$$

Theorem B.6 (Mazur). *Let X be a real or complex, normed linear space and M a closed convex, balanced subset of X. Then, for any $x_0 \notin M$, there exists a continuous linear functional f_0 on X such that $f_0(x_0) > 1$ and $|f_0(x)| \leq 1$ on M.*

However, we have

$$\begin{cases} \partial_j \varphi_{\varepsilon' p'} \in K , \\ \partial_j \varphi_{\varepsilon' p'} \longrightarrow \psi_j \quad \text{weakly in } L^2(M) \text{ for each } 1 \leq j \leq n . \end{cases}$$

Therefore, since the set K is weakly closed in $L^2(M)$, it follows that

$$\partial_j \varphi = \psi_j \in K, \quad 1 \leq j \leq n ,$$

that is,

$$\|\partial_j \varphi\|_\infty \leq \sqrt{C(\lambda)}\, \|f\|_{1,\infty}, \quad 1 \leq j \leq n .$$

Summing up, we have proved that

$$\begin{cases} \varphi \in W^{1,\infty}(M) , \\ \|\varphi\|_{1,\infty} \leq C_1(\lambda) \|f\|_{1,\infty} , \end{cases}$$

where $C_1(\lambda) > 0$ is a constant *independent* of f.

Finally, by letting $\varepsilon' \downarrow 0$ and $p' \to \infty$ in the equation

$$(T + \varepsilon' \Lambda - \lambda I)\varphi_{\varepsilon' p'} = f_{p'} \quad \text{on M} , \tag{B.5$'$}$$

we obtain that

$$(T - \lambda I)\varphi = f \quad \text{on } M .$$

The proof of Claim B.4 is complete. □

Step 3: Similarly we can prove Theorem B.1 for the Sobolev spaces $W^{m,\infty}(M)$ where $m \geq 2$:

Claim B.7. *For each integer $m \geq 2$, there exists a constant $\lambda = \lambda(m) > 0$ such that, for any $f \in W^{m,\infty}(M)$, we can find a function $\varphi \in W^{m,\infty}(M)$ satisfying the equation*

$$(T - \lambda I)\varphi = f \quad \text{on } M ,$$

and the estimate

$$\|\varphi\|_{m,\infty} \leq C_m(\lambda) \|f\|_{m,\infty} .$$

Here $C_m(\lambda) > 0$ is a constant independent of f.

Step 4: Theorem B.1 follows from Claims B.4 and B.7 by a well-known interpolation argument. Indeed, it suffices to note that the Hölder space $C^{k+\theta}(M)$ is a real interpolation space between the Sobolev spaces $W^{k,\infty}(M)$ and $W^{k+1,\infty}(M)$ (see Bergh-Löfström [BL, Theorem 6.4.5], Triebel [Tr, Theorem 2.4.2]):

$$C^{k+\theta}(M) = B^{k+\theta}_{\infty,\infty}(M) = (W^{k,\infty}(M), W^{k+1,\infty}(M))_{\theta,\infty}$$

The proof of Theorem B.1 is now complete. □

C

The Maximum Principle

In this appendix, we prove various maximum principles for second-order elliptic Waldenfels operators such as the weak and strong maximum principles and the boundary point lemma which play an important role in Chaps. 13 and 14. For a general study of maximum principles, the reader might refer to Protter-Weinberger [PW], Bony-Courrège-Priouret [BCP] and also Taira [Ta2].

Let W be a second-order, elliptic *Waldenfels* operator with real coefficients such that

$$\begin{aligned} Wu(x) &= Au(x) + Su(x) \\ &:= \sum_{i,j=1}^{N} a^{ij}(x) \frac{\partial^2 u}{\partial x_i \partial x_j}(x) + \sum_{i=1}^{N} b^i(x) \frac{\partial u}{\partial x_i}(x) + c(x)u(x) \\ &\quad + \int_{\mathbf{R}^N \setminus \{0\}} \left(u(x+z) - u(x) - \sum_{j=1}^{N} z_j \frac{\partial u}{\partial x_j}(x) \right) s(x,z)\, m(dz) \,. \end{aligned}$$

Here:

(1) $a^{ij}(x) \in C^\infty(\overline{D})$, $a^{ij}(x) = a^{ji}(x)$ and there exists a constant $a_0 > 0$ such that

$$\sum_{i,j=1}^{N} a^{ij}(x)\xi_i\xi_j \geq a_0|\xi|^2, \quad x \in \overline{D},\ \xi \in \mathbf{R}^N \,.$$

(2) $b^i(x) \in C^\infty(\overline{D})$.

(3) $c(x) \in C^\infty(\overline{D})$, and $c(x) \leq 0$ in D, but $c(x) \not\equiv 0$ in D.

(4) $s(x,z) \in C(\overline{D} \times \mathbf{R}^N)$ and $0 \leq s(x,z) \leq 1$ on $\overline{D} \times \mathbf{R}^N$, and there exist constants $C_0 > 0$ and $0 < \theta < 1$ such that

$$|s(x,z) - s(y,z)| \leq C_0|x-y|^\theta, \quad x, y \in \overline{D},\ z \in \mathbf{R}^N \,,$$

and

$$s(x,z) = 0 \quad \text{if } x + z \notin \overline{D} \,.$$

(5) The measure $m(dz)$ is a Radon measure on $\mathbf{R}^N \setminus \{0\}$ which satisfies the moment condition

$$\int_{\{|z| \leq 1\}} |z|^2 \, m(dz) + \int_{\{|z|>1\}} |z| \, m(dz) < \infty .$$

C.1 The Weak Maximum Principle

First, we prove the weak maximum principle:

Theorem C.1. *Let W be a second-order elliptic Waldenfels operator. Then we have the following two assertions:*

(i) If a function $u(x) \in C(\overline{D}) \cap C^2(D)$ satisfies the condition: $Wu(x) \geq 0$ in D and if $W1(x) < 0$ in D, then the function $u(x)$ may take its positive *maximum only on the boundary ∂D.*

(ii) If a function $u(x) \in C(\overline{D}) \cap C^2(D)$ satisfies the condition: $Wu(x) > 0$ in D and if $W1(x) \leq 0$ in D, then the function $u(x)$ may take its non-negative *maximum only on the boundary ∂D.*

Proof. Assume, to the contrary, that there exists a point x_0 of D such that

$$u(x_0) = \max_{x \in \overline{D}} u(x) .$$

Then we have

$$\frac{\partial u}{\partial x_i}(x_0) = 0, \ 1 \leq i \leq N ;$$
$$\sum_{i,j=1}^{N} a^{ij}(x_0) \frac{\partial^2 u}{\partial x_i \partial x_j}(x_0) \leq 0 ,$$

and hence

$$Au(x_0) = \sum_{i,j=1}^{N} a^{ij}(x_0) \frac{\partial^2 u}{\partial x_i \partial x_j}(x_0) + c(x_0)u(x_0) \leq c(x_0)u(x_0) , \quad \text{(C.1)}$$

$$Su(x_0) = \int_{\mathbf{R}^N \setminus \{0\}} (u(x_0 + z) - u(x_0)) \, s(x_0, z) \, m(dz) \leq 0 . \quad \text{(C.2)}$$

Assertion (i): If $Wu(x) \geq 0$, $c(x) = W1(x) < 0$ in D and if $u(x_0) = \max_{\overline{D}} u > 0$, then it follows from inequalities (C.1) and (C.2) that

$$0 \leq Wu(x_0) = Au(x_0) + Su(x_0) \leq \max_{x \in \overline{D}} u(x) \cdot W1(x_0) < 0 .$$

This is a contradiction.

Assertion (ii): Similarly, if $Wu(x) > 0$, $c(x) = W1(x) \leq 0$ in D and if $u(x_0) = \max_{\overline{D}} u \geq 0$, then it follows from inequalities (C.1) and (C.2) that

$$0 < Wu(x_0) = Au(x_0) + Su(x_0) \leq \max_{x \in \overline{D}} u(x) \cdot W1(x_0) \leq 0 .$$

This is also a contradiction.

The proof of Theorem C.1 is complete. □

As an application of the weak maximum principle, we can obtain a pointwise estimate for solutions of the non-homogeneous equation $Wu = f$:

Theorem C.2. *Let W be a second-order elliptic Waldenfels operator. Assume that*

$$W1(x) < 0 \quad on\ \overline{D} = D \cup \partial D .$$

Then we have, for all $u \in C(\overline{D}) \cap C^2(D)$,

$$\max_{\overline{D}} |u| \leq \max \left\{ \left(\frac{1}{\min_{\overline{D}}(-W1)} \right) \sup_{D} |Wu|, \ \max_{\partial D} |u| \right\} . \tag{C.3}$$

Proof. We let

$$M = \max \left\{ \left(\frac{1}{\min_{\overline{D}}(-W1)} \right) \sup_{D} |Wu|, \ \max_{\partial D} |u| \right\} ,$$

and consider two functions

$$v_{\pm}(x) = M \pm u(x) .$$

Then it follows that

$$Wv_{\pm}(x) = MW1(x) \pm Wu(x) \leq 0 \quad \text{in } D .$$

Hence, applying part (i) of Theorem C.1 to the functions $-v_{\pm}(x)$ we obtain that the functions $v_{\pm}(x)$ may take their negative minimums only on the boundary ∂D. However, we have

$$v_{\pm}(x) = M \pm u(x) \geq 0 \quad \text{on } \partial D ,$$

and hence

$$v_{\pm}(x) \geq 0 \quad \text{on } \overline{D} .$$

This proves estimate (C.3). □

C.2 The Strong Maximum Principle

The next theorem is a generalization of the strong maximum principle for the Laplacian to the integro-differential operator case:

Theorem C.3. *Let W be a second-order elliptic Waldenfels operator. Assume that a function $u(x) \in C^2(\overline{D})$ satisfies the conditions: $Wu(x) \geq 0$ in D and $\max_{\overline{D}} u \geq 0$. If the function $u(x)$ takes its maximum at a point of D, then it is a constant.*

Proof. The proof is divided into three steps.

Step 1: We let

$$M = \max_{x \in \overline{D}} u(x) ,$$
$$F = \{x \in D : u(x) = M\} ,$$

and assume, to the contrary, that

$$F \subsetneqq D .$$

Since F is closed in D, we can find a point x_0 of F and an open ball V contained in the set $D \setminus F$, centered at x_1, such that (see Fig. C.1)

(a) $V \subset D \setminus F$;
(b) x_0 is on the boundary ∂V of V.

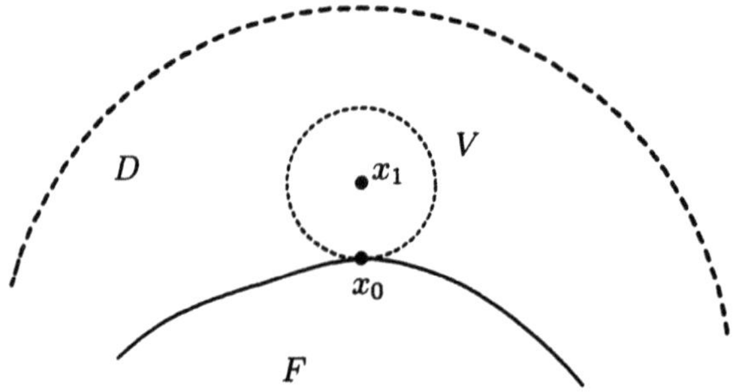

Fig. C.1.

Step 2: The next claim on the existence of "barriers" is an essential step in the proof of Theorem C.3:

Claim C.4. *There exists a function $v(x) \in C^\infty(\overline{D})$ which satisfies the following four properties:*

(i) $v(x) > 0$ in V;

(ii) $v(x) < 0$ *on* $\overline{D} \setminus \overline{V}$;
(iii) $v(x) = 0$ *on* ∂V;
(iv) $Wv(x_0) > 0$.

Proof. We define a function $v(x)$ by the formula

$$v(x) = \exp\left[-q|x - x_1|^2\right] - \exp\left[-q\rho^2\right], \quad \rho = |x_0 - x_1|,$$

where q is a positive constant to be chosen later on. Then it is easy to see that the function $v(x)$ satisfies conditions (i) through (iii). Hence it suffices to show that $v(x)$ satisfies condition (iv) for q sufficiently large.

First, we estimate the function $Av(x_0)$ from below. To do this, it should be noticed that

$$v(x_0) = 0, \tag{C.4a}$$
$$\nabla v(x_0) = 2q(x_1 - x_0) \exp\left[-q\rho^2\right] \neq 0. \tag{C.4b}$$

Hence we have

$$\begin{aligned} Av(x_0) = \Bigg\{ & 4q^2 \sum_{i,j=1}^{N} a^{ij}(x_0)(x_1^i - x_0^i)(x_1^j - x_0^j) \\ & - 2q \sum_{i=1}^{N} \left(a^{ii}(x_0) + b^i(x_0)(x_0^i - x_1^i)\right) \Bigg\} \exp\left[-q\rho^2\right]. \end{aligned}$$

Since the matrix (a^{ij}) is positive definite, we can estimate the function $Av(x_0)$ from below as follows:

$$Av(x_0) \geq \left(4a_0\rho^2 q^2 - Cq\right) \exp\left[-q\rho^2\right], \tag{C.5}$$

where $C > 0$ is a constant independent of q.

Secondly, to estimate the function $Sv(x_0)$ we study the kernel $s(x_0, z)$: Recall that

$$u(x_0) = M = \max_{x \in \overline{D}} u(x).$$

Hence we have

$$Au(x_0) = \sum_{i,j=1}^{N} a^{ij}(x_0) \frac{\partial^2 u}{\partial x_i \partial x_j}(x_0) + c(x_0)u(x_0) \leq c(x_0)u(x_0) \leq 0,$$

and also

$$Su(x_0) = \int_{\mathbf{R}^N \setminus \{0\}} s(x_0, z)\left(u(x_0 + z) - u(x_0)\right) m(dz) \leq 0.$$

This implies that

$$Au(x_0) = 0 ,$$
$$Su(x_0) = 0 .$$

Indeed, it suffices to note that

$$0 \le Wu(x_0) = Au(x_0) + Su(x_0) \le 0 .$$

Thus we obtain that

$$\begin{aligned} 0 &= Su(x_0) \\ &= \int_{\mathbf{R}^N \setminus \{0\}} s(x_0, z)\,(u(x_0 + z) - u(x_0))\, m(dz) \\ &= \int_{x_0 + z \notin F} s(x_0, z)\,(u(x_0 + z) - u(x_0))\, m(dz) , \end{aligned}$$

so that

$$s(x_0, z) = 0 \quad \text{if } x_0 + z \notin F .$$

Therefore, we can write the function $Sv(x_0)$ in the form (see Fig. C.2)

$$\begin{aligned} &Sv(x_0) \\ &= \int_{\substack{x_0 + z \in F \\ z \neq 0}} s(x_0, z) \left(v(x_0 + z) - v(x_0) - \sum_{j=1}^{N} z_j \frac{\partial v}{\partial x_j}(x_0) \right) m(dz) . \end{aligned}$$

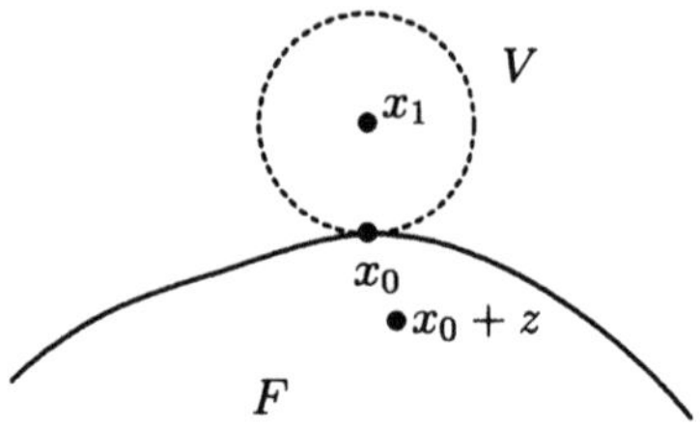

Fig. C.2.

For each $\varepsilon > 0$, we decompose the function $Sv(x_0)$ into the two terms

$$\begin{aligned} &Sv(x_0) \\ &= \int_{\substack{x_0 + z \in F \\ 0 < |z| \le \varepsilon}} s(x_0, z) \left(v(x_0 + z) - v(x_0) - \sum_{j=1}^{N} z_j \frac{\partial v}{\partial x_j}(x_0) \right) m(dz) \\ &\quad + \int_{\substack{x_0 + z \in F \\ |z| > \varepsilon}} s(x_0, z) \left(v(x_0 + z) - v(x_0) - \sum_{j=1}^{N} z_j \frac{\partial v}{\partial x_j}(x_0) \right) m(dz) \\ &:= S_1^{(\varepsilon)} v(x_0) + S_2^{(\varepsilon)} v(x_0) . \end{aligned}$$

Then, by using formulas (C.4) and Claim 14.1 we can estimate the second term $S_2^{(\varepsilon)}v(x_0)$ as follows:

$$\begin{aligned}
&S_2^{(\varepsilon)}v(x_0)\\
&= \int_{\substack{x_0+z\in F\\ |z|>\varepsilon}} s(x_0,z)\left(v(x_0+z)-v(x_0)-\sum_{j=1}^{N} z_j\frac{\partial v}{\partial x_j}(x_0)\right) m(dz)\\
&\leq \left\{\int_{|z|>\varepsilon} m(dz) + 2q\rho\int_{|z|>\varepsilon} |z| m(dz)\right\}\exp\left[-q\rho^2\right]\\
&= \{\tau(\varepsilon)+2q\rho\delta(\varepsilon)\}\exp\left[-q\rho^2\right]\\
&\leq \left\{\left(\frac{C_1}{\varepsilon^2}+C_2\right)+2q\rho\left(\frac{C_1}{\varepsilon}+C_2\right)\right\}\exp\left[-q\rho^2\right] ,
\end{aligned} \tag{C.6}$$

where

$$C_1 = \int_{\{|z|\leq 1\}} |z|^2\, m(dz), \quad C_2 = \int_{\{|z|>1\}} |z|\, m(dz) .$$

Similarly, by using the mean value theorem we can estimate the first term $S_1^{(\varepsilon)}v(x_0)$ as follows:

$$\begin{aligned}
&S_1^{(\varepsilon)}v(x_0)\\
&= \int_{\substack{x_0+z\in F\\ 0<|z|\leq\varepsilon}} s(x_0,z)\left(v(x_0+z)-v(x_0)-\sum_{j=1}^{N} z_j\frac{\partial v}{\partial x_j}(x_0)\right) m(dz)\\
&\leq \left\{\int_{\substack{x_0+z\in F\\ 0<|z|\leq\varepsilon}} m(dz)\left(C_3q^2+C_4q\right)|z|^2 m(dz)\right\}\exp\left[-q\rho^2\right]\\
&\leq \sigma(\varepsilon)\left(C_3q^2+C_4q\right)\exp\left[-q\rho^2\right] ,
\end{aligned} \tag{C.7}$$

where C_3 and C_4 are positive constants independent of q.

Hence, by estimates (C.6) and (C.7) and Claim 14.1 it follows that, for every $\eta > 0$ there exists a constant $C_\eta > 0$ such that

$$|Sv(x_0)| \leq \left(\eta q^2 + C_\eta q\right)\exp\left[-q\rho^2\right] . \tag{C.8}$$

Therefore, if we let

$$\eta = 2a_0\rho^2 ,$$
$$q > \frac{C+C_\eta}{2a_0\rho^2} ,$$

then we obtain from estimates (C.5) and (C.8) that

$$\begin{aligned}
Wv(x_0) &= Av(x_0) + Sv(x_0)\\
&\geq Av(x_0) - |Sv(x_0)|
\end{aligned}$$

$$
\begin{aligned}
&\geq \left(4a_0\rho^2 q^2 - Cq - \eta q^2 - C_\eta q\right) \exp\left[-q\rho^2\right] \\
&= \left(2a_0\rho^2 q - (C + C_\eta)\right) q \exp\left[-q\rho^2\right] \\
&> 0.
\end{aligned}
$$

The proof of Claim C.4 is complete. □

Step 3: Now we introduce a function

$$u_\lambda(x) = u(x) + \lambda v(x) ,$$

where λ is a positive constant to be chosen later on. Then, by Claim C.4 we have the following four assertions (see Fig. C.3):

(a) There exists a neighborhood V' of x_0 such that

$$Wv > 0 \quad \text{in } V' ,$$

since $Wv(x)$ is a continuous function.

(b) $Wu_\lambda = Wu + \lambda Wv \geq \lambda Wv > 0$ in V'.

(c) $u_\lambda = u + \lambda v \leq u \leq M$ on $\overline{D} \setminus V$, since $v \leq 0$ on $\overline{D} \setminus V$.

(d) $u_\lambda = u + \lambda v \leq M$ on $\overline{V} \setminus V'$ for λ sufficiently small, since $u < M$ on $\overline{V} \setminus V'$.

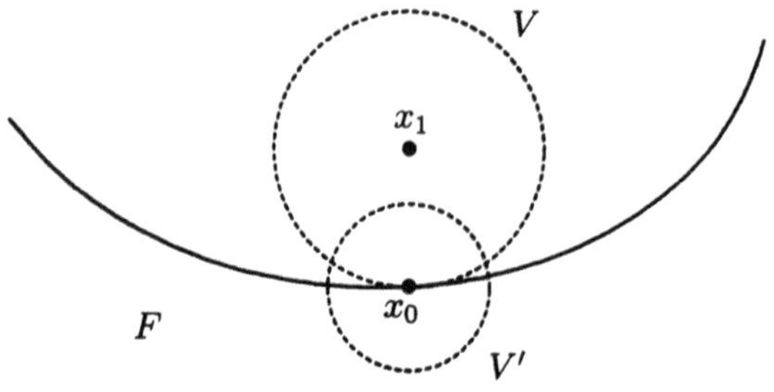

Fig. C.3.

Therefore, we obtain that (see Fig. C.4 below)

$$
\begin{aligned}
&Wu_\lambda > 0 \quad \text{in } V' , \\
&u_\lambda \leq M \quad \text{on } \partial V' , \\
&u_\lambda(x_0) = M .
\end{aligned}
$$

However, this contradicts part (ii) of Theorem C.1 (the weak maximum principle).

Summing up, we have proved that

$$F = D ,$$

that is,

$$u(x) \equiv M \quad \text{in } D .$$

Now the proof of Theorem C.3 is complete. □

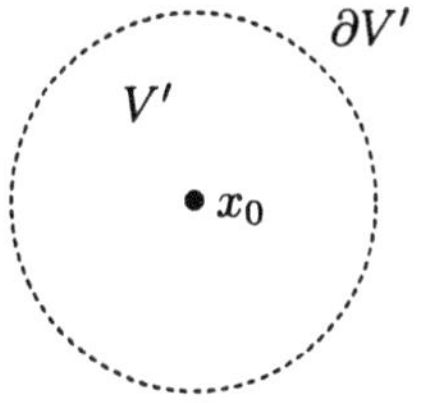

Fig. C.4.

C.3 The Boundary Point Lemma

Finally, we consider the interior normal derivative $(\partial u)/(\partial \mathbf{n})$ at a boundary point where the function $u(x)$ takes its non-negative maximum.

The boundary point lemma reads as follows:

Lemma C.5. *Let W be a second-order elliptic Waldenfels operator. Assume that a function $u(x) \in C^2(\overline{D})$ satisfies the conditions: $Wu(x) \geq 0$ in D, $\max_{\overline{D}} u \geq 0$ and further that there exists a point $x_0' \in \partial D$ such that $u(x_0') = \max_{\overline{D}} u$. Then the interior normal derivative $(\partial u)/(\partial \mathbf{n})(x_0')$ at x_0' satisfies the condition (see Fig. C.5)*

$$\frac{\partial u}{\partial \mathbf{n}}(x_0') < 0 \, ,$$

unless the function $u(x)$ is a constant in D.

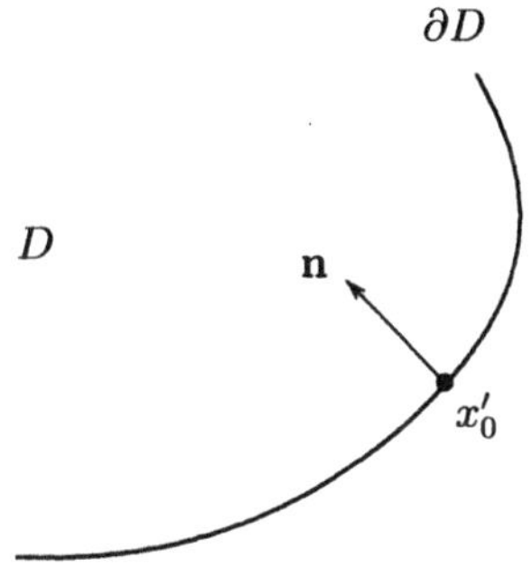

Fig. C.5.

Proof. By Theorem C.3, it suffices to consider the following case:

$$\begin{cases} u(x_0') = \max_{x \in \overline{D}} u(x) \geq 0 \, , \\ u(y) < u(x_0'), \quad y \in D \, . \end{cases} \tag{C.9}$$

The proof is divided into three steps.

Step 1: Now we assume, to the contrary, that

$$\frac{\partial u}{\partial \mathbf{n}}(x_0') = 0 . \tag{C.10}$$

We can find an open ball V contained in the domain D, centered at x_1, such that (see Fig. C.6)

(a) The point x_0' is on the boundary ∂V of V;
(b) $\mathbf{n} = s(x_1 - x_0')$ for some $s > 0$.

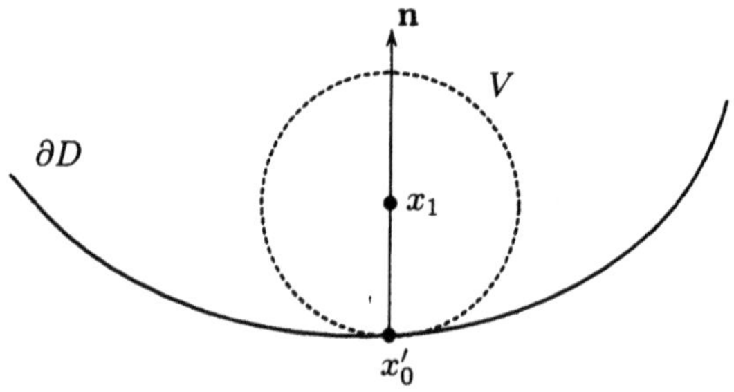

Fig. C.6.

Step 2: The next claim on the existence of "barriers" is an essential step in the proof of Lemma C.5, just as in the proof of Theorem C.3:

Claim C.6. *There exists a function $v(x) \in C^\infty(\overline{D})$ which satisfies the following five properties:*

(i) $v(x) > 0$ *in* V;
(ii) $v(x) < 0$ *on* $\overline{D} \setminus \overline{V}$;
(iii) $v(x) = 0$ *on* ∂V;
(iv) $W v(x_0') > 0$;
(v) $(\partial v / \partial \mathbf{n})(x_0') > 0$.

Proof. Near the boundary point x_0', we introduce local coordinate systems (x', x_N) such that $x' = (x_1, x_2, \ldots, x_{N-1})$ give local coordinates for the boundary ∂D and that (see Fig. C.7)

$$\begin{aligned} D &= \{(x', x_N) : x_N > 0\} , \\ \partial D &= \{(x', x_N) : x_N = 0\} , \\ x_0' &= (0, \ldots, 0, 0) , \\ x_1 &= (0, \ldots, 0, \rho) . \end{aligned}$$

Then we define a function $v(x)$ by the formula

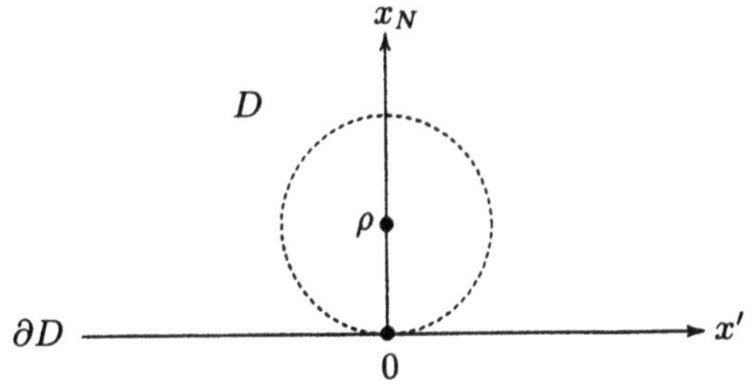

Fig. C.7.

$$v(x) = v(x', x_N) = \exp\left[-q(|x'|^2 + (x_N - \rho)^2)\right] - \exp\left[-q\rho^2\right] ,$$
$$\rho = |x_0' - x_1| ,$$

where q is a positive constant to be chosen later on. Then it is easy to see that the function $v(x)$ satisfies conditions (i) through (iii) and (v). Hence it suffices to show that $v(x)$ satisfies condition (iv) for q sufficiently large.

First, we estimate the function $Av(x_0')$ from below. To do this, it should be noticed that

$$v(x_0') = 0, \tag{C.11a}$$
$$\frac{\partial v}{\partial x_i}(x_0') = 0, \quad 1 \leq i \leq N-1, \tag{C.11b}$$
$$\frac{\partial v}{\partial x_N}(x_0') = 2\rho q \exp\left[-q\rho^2\right], \tag{C.11c}$$

and

$$\frac{\partial^2 v}{\partial x_i \partial x_j}(x_0') = -2q\delta_{ij} \exp\left[-q\rho^2\right], \quad 1 \leq i,\ j \leq N-1 ,$$
$$\frac{\partial^2 v}{\partial x_N^2}(x_0') = (4q^2\rho^2 - 2q) \exp\left[-q\rho^2\right] .$$

Hence we have

$$Av(x_0') = \left\{ 4a^{NN}(x_0')\rho^2 q^2 + 2\left(b^{NN}(x_0')\rho - \sum_{i=1}^{N} a^{ii}(x_0') \right) q \right\} \exp\left[-q\rho^2\right] .$$

Since the matrix (a^{ij}) is positive definite, we can estimate the function $Av(x_0')$ from below as follows:

$$Av(x_0') \geq \left(4a_0\rho^2 q^2 - Cq\right) \exp\left[-q\rho^2\right] , \tag{C.12}$$

where $C > 0$ is a constant independent of q.

Secondly, to estimate the function $Sv(x_0')$ we study the kernel $s(x_0', z)$: By conditions (C.9) and (C.10), it follows that

$$\frac{\partial u}{\partial x_i}(x_0') = 0, \quad 1 \leq i \leq N\,,$$
$$\frac{\partial^2 u}{\partial x_N^2}(x_0') \leq 0\,.$$

Hence we have

$$\begin{aligned} Au(x_0') &= \sum_{i,j=1}^{N} a^{ij}(x_0') \frac{\partial^2 u}{\partial x_i \partial x_j}(x_0') + c(x_0')u(x_0') \leq c(x_0')u(x_0') \\ &= a^{NN}(x_0') \frac{\partial^2 u}{\partial x_N^2}(x_0') + \sum_{i,j=1}^{N-1} a^{ij}(x_0') \frac{\partial^2 u}{\partial x_i \partial x_j}(x_0') + c(x_0')u(x_0') \\ &\leq 0\,, \end{aligned}$$

and also

$$Su(x_0') = \int_{\mathbf{R}^N \setminus \{0\}} s(x_0', z)\,(u(x_0' + z) - u(x_0'))\, m(dz) \leq 0\,.$$

This implies that

$$Au(x_0') = 0\,,$$
$$Su(x_0') = 0\,.$$

Indeed, it suffices to note that

$$0 \leq Wu(x_0') = Au(x_0') + Su(x_0') \leq 0\,.$$

Thus we obtain that

$$\begin{aligned} 0 &= Su(x_0') \\ &= \int_{\mathbf{R}^N \setminus \{0\}} s(x_0', z)\,(u(x_0' + z) - u(x_0'))\, m(dz) \\ &= \int_{x_0' + z \notin G} s(x_0', z)\,(u(x_0' + z) - u(x_0'))\, m(dz)\,, \end{aligned}$$

so that

$$s(x_0', z) = 0 \quad \text{if } x_0' + z \in \partial D \setminus G\,.$$

Here

$$G = \{x' \in \partial D : u(x') = \max_{x \in \overline{D}} u(x)\}\,.$$

Therefore, we can write the function $Sv(x_0')$ in the form

$$\begin{aligned} &Sv(x_0') \\ &= \int_{\substack{x_0' + z \in G \\ z \neq 0}} s(x_0', z) \left(v(x_0' + z) - v(x_0') - \sum_{j=1}^{N} z_j \frac{\partial v}{\partial x_j}(x_0') \right) m(dz)\,. \end{aligned}$$

For each $\varepsilon > 0$, we decompose the function $Sv(x_0')$ into the two terms

$$\begin{aligned} &Sv(x_0') \\ &= \int_{\substack{x_0'+z\in G \\ 0<|z|\leq\varepsilon}} s(x_0', z)\left(v(x_0'+z) - v(x_0') - \sum_{j=1}^{N} z_j \frac{\partial v}{\partial x_j}(x_0')\right) m(dz) \\ &+ \int_{\substack{x_0'+z\in G \\ |z|>\varepsilon}} s(x_0', z)\left(v(x_0'+z) - v(x_0') - \sum_{j=1}^{N} z_j \frac{\partial v}{\partial x_j}(x_0')\right) m(dz) \\ &:= S_1^{(\varepsilon)} v(x_0') + S_2^{(\varepsilon)} v(x_0') . \end{aligned}$$

Then, by using formulas (C.11) we can estimate the second term $S_2^{(\varepsilon)} v(x_0')$ as follows:

$$\begin{aligned} S_2^{(\varepsilon)} v(x_0') &= \int_{\substack{x_0'+z\in G \\ |z|>\varepsilon}} s(x_0', z)\Big\{ \left[\exp\left[-q(|x'|^2 + (x_N - \rho)^2)\right] - \exp\left[-q\rho^2\right]\right] \\ &\quad - 2\rho z_N q \exp\left[-q\rho^2\right]\Big\} m(dz) \\ &\leq \left\{ \int_{|z|>\varepsilon} m(dz) + 2\rho q \int_{|z|>\varepsilon} |z| m(dz) \right\} \exp\left[-q\rho^2\right] \\ &\leq \{\tau(\varepsilon) + 2\rho\delta(\varepsilon) q\} \exp\left[-q\rho^2\right] . \end{aligned} \tag{C.13}$$

Similarly, since we have, for $|z| \leq \varepsilon$,

$$x_0' + z \in G \Longleftrightarrow z_N = 0 ,$$

we can estimate the first term $S_1^{(\varepsilon)} v(x_0')$ as follows:

$$\begin{aligned} S_1^{(\varepsilon)} v(x_0') &= \int_{\substack{x_0'+z\in G \\ 0<|z|\leq\varepsilon}} s(x_0', z) v(x_0'+z) m(dz) \\ &= \int_{\substack{0<|z'|\leq\varepsilon \\ z_N=0}} s(0, z', 0) v(z', 0) m(dz) \\ &= \int_{\substack{0<|z'|\leq\varepsilon \\ z_N=0}} s(0, z', 0) \left[\exp\left[-q(|x'|^2 + \rho^2)\right] - \exp\left[-q\rho^2\right]\right] m(dz) \\ &\leq \left\{ \int_{\substack{0<|z'|\leq\varepsilon \\ z_N=0}} |z'|^2 m(dz) \right\} q \exp\left[-q\rho^2\right] \\ &\leq \sigma(\varepsilon) q \exp\left[-q\rho^2\right] . \end{aligned} \tag{C.14}$$

By combining estimates (C.13) and (C.14), we have proved that

$$|Sv(x_0')| \leq \{(2\rho\delta(\varepsilon) + \sigma(\varepsilon))\, q + \tau(\varepsilon)\} \exp\left[-q\rho^2\right] . \tag{C.15}$$

Therefore, it follows from estimates (C.12) and (C.15) that

$$Wv(x_0') = Av(x_0') + Sv(x_0') \geq Av(x_0') - |Sv(x_0')| > 0\,,$$

if q is sufficiently large.

The proof of Claim C.6 is complete. □

Step 3: If we introduce a function

$$u_\lambda(x) = u(x) + \lambda v(x)$$

for a positive constant λ, then we have, by Claim C.6, the following four assertions (see Fig. C.8):

(a) There exists a neighborhood V' of x_0' such that

$$Wv > 0 \quad \text{in } V' \cap \overline{D}\,,$$

since $Wv(x)$ is a continuous function on $\overline{D}$.

(b) $Wu_\lambda = Wu + \lambda Wv \geq \lambda Wv > 0$ in $V' \cap \overline{D}$.

(c) $u_\lambda = u + \lambda v \leq u \leq M$ on $\overline{D} \setminus V$, since $v \leq 0$ on $\overline{D} \setminus V$.

(d) $u_\lambda = u + \lambda v \leq M$ on $\overline{V} \setminus V'$ for λ sufficiently small, since $u < M$ on $\overline{V} \setminus V'$.

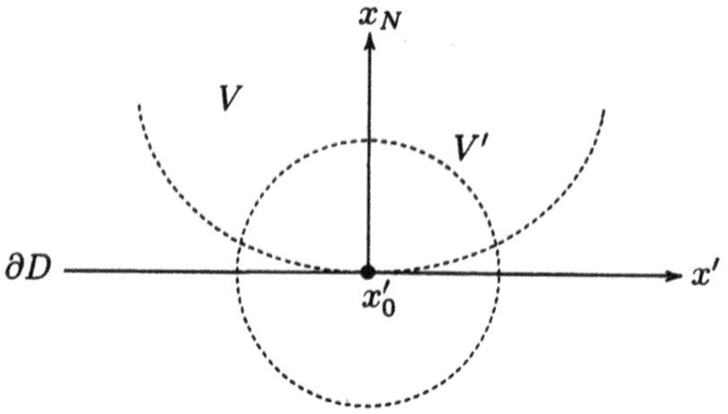

Fig. C.8.

Hence it follows from an application of part (ii) of Theorem C.1 that

$$u_\lambda \leq M \quad \text{in } V' \cap D\,,$$

so that

$$u_\lambda(y) = u(y) + \lambda v(y) \leq M = u(x_0') + \lambda v(x_0'), \quad y \in V' \cap D\,,$$

or equivalently

$$u(y) - u(x_0') \leq -\lambda (v(y) - v(x_0')), \quad y \in V' \cap D\,.$$

Therefore, we obtain that

$$\frac{\partial u}{\partial \mathbf{n}}(x_0') = -\lambda \frac{\partial v}{\partial \mathbf{n}}(x_0') < 0\,.$$

This contradicts hypothesis (C.10).

Now the proof of Lemma C.5 is complete. □

References

[Ad] Adams, R.A.: Sobolev spaces. Academic Press, New York (1975)

[Ag] Agmon, S.: Lectures on elliptic boundary value problems. Van Nostrand, Princeton (1965)

[ADN] Agmon, S., Douglis, A., Nirenberg, L.: Estimates near the boundary for solutions of elliptic partial differential equations satisfying general boundary conditions I. Comm. Pure Appl. Math., **12**, 623–727 (1959)

[AS] Aronszajn, N., Smith, K.T.: Theory of Bessel potentials I. Ann. Inst. Fourier (Grenoble), **11**, 385–475 (1961)

[BL] Bergh, J., Löfström, J.: Interpolation spaces, an introduction. Springer-Verlag, Berlin Heidelberg New York (1976)

[BG] Blumenthal, R.M., Getoor, R.K.: Markov processes and potential theory. Academic Press, New York (1968)

[BCP] Bony, J.-M., Courrège, P., Priouret, P.: Semigroupes de Feller sur une variété à bord compacte et problèmes aux limites intégro-différentiels du second ordre donnant lieu au principe du maximum. Ann. Inst. Fourier (Grenoble), **18**, 369–521 (1968)

[Bo] Bourdaud, G.: L^p-estimates for certain non-regular pseudo-differential operators. Comm. Partial Differential Equations, **7**, 1023–1033 (1982)

[Bt] Boutet de Monvel, L.: Boundary problems for pseudo-differential operators. Acta Math., **126**, 11–51 (1971)

[Ca] Calderón, A.P.: Lebesgue spaces of differentiable functions and distributions. Proc. Sym. Pure Math., **V**, 33–49, Amer. Math. Soc., Providence, Rhode Island (1961)

[Cn] Cancelier, C.: Problèmes aux limites pseudo-différentiels donnant lieu au principe du maximum. Comm. Partial Differential Equations, **11**, 1167–1726 (1986)

[CP] Chazarain, J., Piriou, A.: Introduction à la théorie des équations aux dérivées partielles linéaires., Gauthier-Villars, Paris (1981)

[CM] Coifman, R.R., Meyer, Y.: Au-delà des opérateurs pseudo-différentiels, Astérisque No. 57, Soc. Math. France, Paris (1978)

[Dy] Dynkin, E.B.: Markov processes I, II. Springer-Verlag, Berlin Göttingen Heidelberg (1965)

[DY] Dynkin, E.B., Yushkevich, A.A.: Markov processes, theorems and problems. Plenum Press, New York (1969)

[EN] Engel, K.-J. and Nagel, R.: One-parameter semigroups for linear evolution equations. Springer-Verlag, New York Berlin Heidelberg (2000)

[Es] Èskin, G.I.: Boundary value problems for pseudodifferential equations. Nauka, Moscow (1973), (in Russian), English translation. Amer. Math. Soc., Providence, Rhode Island (1981)

[EK] Ethier, S.N., Kurtz, T.G.: Markov processes, characterization and convergence. John Wiley & Sons, New York Chichester Brisbane Toronto Singapore (1986)

[Fe1] Feller, W.: The parabolic differential equations and the associated semigroups of transformations. Ann. of Math., **55**, 468–519 (1952)

[Fe2] Feller, W.: On second order differential equations. Ann. of Math., **61**, 90–105 (1955)

[Fo] Folland, G.B.: Introduction to partial differential equations. second edition, Princeton University Press, Princeton (1995)

[Fr1] Friedman, A.: Partial differential equations. Holt, Rinehart and Winston, New York (1969)

[Fr2] Friedman, A.: Foundations of modern analysis. Holt, Rinehart and Winston, New York (1970)

[Fu] Fujiwara, D.: On some homogeneous boundary value problems bounded below. J. Fac. Sci. Univ. Tokyo Sec. IA, **17**, 123–152 (1970)

[GB] Galakhov, E.I., Skubachevskiĭ, A.L.: On Feller semigroups generated by elliptic operators with integro-differential boundary conditions. J. Differential Equations, **176**, 315–355 (2001)

[GM] Garroni, M.G., Menaldi, J.L.: Green functions for second order integro-differential problems. Pitman Research Notes in Mathematics Series No. 275, Longman Scientific & Technical, Harlow (1992)

[GS] Gel'fand, I.M., Shilov,G.E.: Generalized functions I. Academic Press, New York London (1964)

[GT] Gilbarg, D., Trudinger, N.S.: Elliptic partial differential equations of second order. 1998 edition, Springer-Verlag, New York Berlin Heidelberg Tokyo (1998)

[GK] Gohberg, I.C., Kreĭn, M.G.: The basic propositions on defect numbers, root numbers and indices of linear operators. Uspehi Mat. Nauk., **12**, 43–118 (1957), (in Russian), English translation: Amer. Math. Soc. Transl., **13**, 185–264 (1960)

[Ho1] Hörmander, L.: Pseudodifferential operators and non-elliptic boundary problems. Ann. of Math., **83**, 129–209 (1966)

[Ho2] Hörmander, L.: Pseudo-differential operators and hypoelliptic equations. In: Proc. Sym. Pure Math., **X**, Singular integrals, A. P. Calderón (ed.), 138–183, Amer. Math. Soc., Providence, Rhode Island (1967)

[Ho3] Hörmander, L.: The analysis of linear partial differential operators III, Springer-Verlag, Berlin Heidelberg New York Tokyo (1985)

[IW] Ikeda, N., Watanabe, S.: Stochastic differential equations and diffusion processes. second edition, North-Holland/Kodansha, Amsterdam/Tokyo (1989)

[IM] Itô, K., McKean, H.P., Jr.: Diffusion processes and their sample paths, Springer-Verlag, Berlin Heidelberg New York (1965)

[Ka] Kannai, Y.: Hypoellipticity of certain degenerate elliptic boundary value problems. Trans. Amer. Math. Soc., **217**, 311–328 (1976)

[Kn] Knight, F.B.: Essentials of Brownian motion and diffusion. Amer. Math. Soc., Providence, Rhode Island (1981)

[Ko] Komatsu, T.: Markov processes associated with certain integro-differential operators. Osaka J. Math., **10**, 271–303 (1973)

[Ku] Kumano-go, H.: Pseudodifferential operators. MIT Press, Cambridge, Massachusetts (1981)

[La] Lamperti, J.: Stochastic processes. Springer-Verlag, New York Heidelberg Berlin (1977)

[LM] Lions, J.-L., Magenes, E.: Problèmes aux limites non-homogènes et applications 1, 2. Dunod, Paris (1968), English translation: Non-homogeneous boundary value problems and applications 1, 2. Springer-Verlag, Berlin Heidelberg New York (1972)

[Ma] Masuda, K.: Evolution equations. (in Japanese), Kinokuniya Shoten, Tokyo (1975)

[Mc] McLean, W.: Strongly elliptic systems and boundary integral equations. Cambridge University Press, Cambridge (2000)

[Me] Meyer, Y.: Remarques sur un théorème de J.-M. Bony. Suppl. Rend. Circ. Mat. Palermo, **II**-1, 1–20 (1981)

[Na] Nagase, M.: The L^p-boundedness of pseudo-differential operators with non-regular symbols. Comm. Partial Differential Equations **2**, 1045–1061 (1977)

[OR] Oleĭnik, O.A., Radkevič, E.V.: Second order equations with nonnegative characteristic form (in Russian), Itogi Nauki, Moscow (1971), English translation, Amer. Math. Soc., Providence, Rhode Island and Plenum Press, New York (1973)

[Pa] Pazy, A.: Semigroups of linear operators and applications to partial differential equations. Springer-Verlag, New York Berlin Heidelberg Tokyo (1983)

[PW] Protter, M.H., Weinberger, H.F.: Maximum principles in differential equations. Prentice-Hall, Englewood Cliffs, New Jersey (1967)

[RS] Rempel, S., Schulze, B.-W.: Index theory of elliptic boundary problems. Akademie-Verlag, Berlin (1982)

[RY] Revuz, D., Yor, M.: Continuous martingales and Brownian motion. third edition, Springer-Verlag, Berlin New York Heidelberg (1999)

[SU] Sato, K., Ueno, T.: Multi-dimensional diffusion and the Markov process on the boundary. J. Math. Kyoto Univ., **14**, 529–605 (1965)

[Sc] Schaefer, H.H.: Topological vector spaces, second edition, Springer-Verlag, New York Berlin Heidelberg (1991)

[Sr] Schrohe, E.: Fréchet algebras of pseudo-differential operators and boundary value problems. Birkhäuser, Basel Boston Stuttgart (to appear)

[Su] Schulze, B.-W.: Boundary value problems and singular pseudo-differential operators. John Wiley & Sons, Chichester (1998)

[Se1] Seeley, R.T.: Refinement of the functional calculus of Calderón and Zygmund. Proc. Nederl. Akad. Wetensch., Ser. A, **68**, 521–531 (1965)

[Se2] Seeley, R.T.: Singular integrals and boundary value problems. Amer. J. Math., **88**, 781–809 (1966)

[Sn1] Stein, E.M.: The characterization of functions arising as potentials II. Bull. Amer. Math. Soc., **68**, 577–582 (1962)

[Sn2] Stein, E.M.: Singular integrals and differentiability properties of functions. Princeton University Press, Princeton (1970)

[Sw] Stewart, H.B.: Generation of analytic semigroups by strongly elliptic operators under general boundary conditions. Trans. Amer. Math. Soc., **259**, 299–310 (1980)

[St] Stroock, D.W.: Diffusion processes associated with Lévy generators. Z. Wahrscheinlichkeitstheorie verw. Gebiete, **32**, 209–244 (1975)

[Tb] Taibleson, M.H.: On the theory of Lipschitz spaces of distributions on Euclidean n-space I. J. Math. Mech., **13**, 407–479 (1964)

[Ta1] Taira, K.: On some degenerate oblique derivative problems. J. Fac. Sci. Univ. Tokyo Sec.IA, **23**, 259–287 (1976)

[Ta2] Taira, K.: Diffusion processes and partial differential equations. Academic Press, San Diego New York London Tokyo (1988)

[Ta3] Taira, K.: Boundary value problems and Markov processes. Lecture Notes in Mathematics, **1499**, Springer-Verlag, Berlin Heidelberg New York Tokyo (1991)

[Ta4] Taira, K.: On the existence of Feller semigroups with boundary conditions. Memoirs Amer. Math. Soc. No. 475, (1992)

[Ta5] Taira, K.: Analytic semigroups and semilinear initial boundary value problems. London Mathematical Society Lecture Note Series, No. 223, Cambridge University Press, London New York (1995)

[Ta6] Taira, K.: Boundary value problems for elliptic integro-differential operators. Math. Z., **222**, 305–327 (1996)

[Ta7] Taira, K.: Brownian motion and index formulas for the de Rham complex. Mathematical Research series, No. 106, Wiley-VCH, Berlin (1998)

[TW] Takanobu S., Watanabe, S.: On the existence and uniqueness of diffusion processes with Wentzell's boundary conditions. J. Math. Kyoto Univ., **28**, 71–80 (1988)

[Tn] Tanabe, H.: Equations of evolution. Iwanami-Shoten, Tokyo (1975), (in Japanese), English translation. Pitman, London (1979)

[Ty] Taylor, M.: Pseudodifferential operators. Princeton University Press, Princeton (1981)

[Tv] Treves, F.: Topological vector spaces, distributions and kernels. Academic Press, New York London (1967)

[Tr] Triebel, H.: Theory of function spaces. Birkhäuser, Basel Boston Stuttgart (1983)

[Wa] Waldenfels, W.v.: Positive Halbgruppen auf einem n-dimensionalen Torus. Archiv der Math., **15**, 191–203 (1964)

[Wt] Watson, G.N.: A treatise on the theory of Bessel functions. second edition, Cambridge University Press, Cambridge (1944)

[We] Wentzell (Ventcel'), A.D.: On boundary conditions for multidimensional diffusion processes. (in Russian), Teoriya Veroyat. i ee Primen., **4**, 172–185 (1959), English translation: Theory Prob. and its Appl., **4**, 164–177 (1959)

[Yo] Yosida, K.: Functional analysis. sixth edition, Springer-Verlag, Berlin Heidelberg New York (1980)

Index of Symbols

Geometric Objects:

D (bounded domain), 2, 135, 245
∂D (boundary), 2, 135, 245
$\overline{D} = D \cup \partial D$ (closure), 2, 135
$M = \{x' \in \partial D : \mu(x') = 0\}$, 4
$M = \{x' \in \partial D : m(x') = 0\}$, 156
$\mathbf{n}$ (normal vector), 3, 138, 247, 319
$S = \mathbf{R}/2\pi\mathbf{Z}$ (unit circle), 15, 177
$T^*(\mathbf{R}^n)$ (cotangent bundle), 112
$T^*_{x'}(\partial D)$ (cotangent space), 138
Δ_ε (sector), 7
Δ_ω (sector), 38
$\Delta_\omega^{2\varepsilon}$ (sector), 38
$\Delta_{\overline{D}}$ (diagonal), 137
$\Delta_{\partial D}$ (diagonal), 139
Δ_M (diagonal), 113, 303
$\Delta_{\mathbf{R}^N}$ (diagonal), 137
Δ_Ω (diagonal), 71, 108
$\Sigma(\varepsilon)$, 8
$\Sigma_p(\varepsilon)$, 6
Σ_ω, 35
$\Sigma_\omega^\varepsilon$, 35
K (metric space), 48, 156
∂ (point at infinity), 49, 157
K_∂ (one point compactification), 49, 157
Ω (bounded domain), 97
$\partial\Omega$ (boundary), 97
$\overline{\Omega} = \Omega \cup \partial\Omega$ (closure), 97
M (double of Ω), 97
M (manifold), 97, 113, 121, 303
$T^*(M)$ (cotangent bundle), 122
$T^*_x(M)$ (cotangent space), 305
ω_n (area of unit sphere), 94

Spaces of Functions, Distributions, and Operators:

$B^{k-1/p,p}(\partial D)$ (Besov space), 4
$B^{1-1/p,p}_{L_0}(\partial D)$ (modified Besov space), 5
$B^{s,p}(\mathbf{R}^{n-1})$ (Besov space), 99
$B^s_{p,q}(\mathbf{R}^n)$ (Besov space), 274
$B^{s,p}(\partial\Omega)$ (Besov space), 101
$B^{s,p}_{\mathrm{loc}}(\partial\Omega)$ (localized Besov space), 101
$B^{s-1-1/p,p}_{B_0}(\partial\Omega)$ (modified Besov space), 128
$B_{\mathrm{loc}}(\Omega)$ (Borel measurable functions), 70
$C(\overline{D})$, 7
$C_0(\overline{D} \setminus M)$, 7, 157
$C(K)$, 55
$C_0(K)$, 55
$C^{1+\theta}_{L_0}(\partial D)$ (modified Hölder space), 6
$C_0(\Omega)$, 71
$C_0^2(\Omega)$, 71
$C^k(\Omega)$, $C^\infty(\Omega)$, 76
$C^k(\overline{\Omega})$, $C^\infty(\overline{\Omega})$, 77
$C^m(\Omega)$, $C^m(\overline{\Omega})$, 77
$C_0^m(\Omega)$, $C_0^\infty(\Omega)$ (test functions), 78
$C^\theta(\Omega)$, $C^{k+\theta}(\Omega)$ (Hölder spaces), 79
$C^\theta(\overline{\Omega})$, $C^{k+\theta}(\overline{\Omega})$ (Hölder spaces), 79
$\mathcal{D}'(\Omega)$ (distributions), 80
$\mathcal{E}'(\Omega)$ (compactly supported distributions), 85
$H^{k,p}(D)$ (Sobolev space), 4
$H^{s,p}(\mathbf{R}^n)$ (Sobolev space), 98

$H_p^s(\mathbf{R}^n)$ (Sobolev space), 274
$H^{s,p}(\Omega)$ (Sobolev space), 101
$H^{s,p}(M)$ (Sobolev space), 101
$H_{\mathrm{loc}}^{s,p}(\Omega)$ (localized Sobolev space), 101
$L_{\mathrm{loc}}^1(\mathbf{R}^n)$ (locally integrable functions), 81
$L^p(\mathbf{R}^n)$, $L^\infty(\mathbf{R}^n)$ (L^p spaces), 74, 75
$L_m^\infty(\mathbf{R}^n)$ (Sobolev space), 277
$\Lambda_r(\mathbf{R}^n) = B_{\infty,\infty}^r(\mathbf{R}^n)$ (Hölder space), 277
$W^{s,p}(\mathbf{R}^n)$ (Sobolev space), 99
$W^{1,\infty}(M)$ (Sobolev space), 304
$L_{\rho,\delta}^m(\Omega)$ (pseudo-differential operators), 108
$L_{\rho,\delta}^m(M)$ (pseudo-differential operators), 113
$L_{\mathrm{cl}}^m(\Omega)$ (classical pseudo-differential operators), 110
$L_{\mathrm{cl}}^m(M)$ (classical pseudo-differential operators), 114
$L^{-\infty}(\Omega)$ (regularizers), 110
$L^{-\infty}(M)$ (regularizers), 113
$\mathcal{S}(\mathbf{R}^n)$ (Schwartz space), 88
$\mathcal{S}'(\mathbf{R}^n)$ (tempered distributions), 89
$S_{\mathrm{cl}}^m(\Omega \times \mathbf{R}^N)$ (classical symbols), 103
$S_{\rho,\delta}^m(\Omega \times \mathbf{R}^N)$ (symbol class), 102
$S_{\rho,\delta}^\infty(\Omega \times \mathbf{R}^N)$ (symbol class), 105
$S^{-\infty}(\Omega \times \mathbf{R}^N)$, 102
$S_{1,\delta}^0(N, r)$ (non-regular symbols), 275

Multi-indices and (Pseudo-) Differential Operators:

$|\alpha|$, $\alpha!$, 73
∂^α, D^α, 74
A (elliptic differential operator or diffusion operator), 2, 9, 122, 136, 213, 245
A^* (adjoint), 111
A' (transpose), 111
A_Ω, 118
$\widetilde{\Lambda} = (A + \frac{\partial^2}{\partial y^2})$, 15, 177
$\widetilde{\Lambda}(\vartheta) = (A + e^{i\vartheta} \frac{\partial^2}{\partial y^2})$, 16, 193
$p(x, D)$ (operator defined by a symbol), 109
Δ (Laplacian), 93, 124, 170, 178, 194
Δ_0, Δ_j (Littlewood-Paley series), 272
$\sigma(A)$ (symbol), 110
$\sigma_A(x, \xi)$ (principal symbol), 110

Other Operators and Functionals:

A (densely defined, closed linear operator), 35, 253
A (infinitesimal generator), 60
A_p, 6, 191, 259
$\mathcal{A} = (A, B_0)$, 130
$\mathcal{A} = (A, L_0)$, 169
$\mathcal{A}(\lambda)$, 175
$\overline{A}$ (minimal closed extension), 219, 253
$\mathfrak{A}$ (infinitesimal generator), 24
$\mathfrak{A}$, 141, 235
$\mathfrak{A}_N$, 226
$\mathfrak{A}_\alpha$ (Yosida approximation), 30
$[A, \varphi] = A\varphi - \varphi A$ (commutator), 209
B (linear operator), 64
$\overline{B}$ (minimal closed extension), 64
B_0 (special boundary condition), 128
$\mathcal{D}(A)$ (domain of definition), 35, 61, 253
$\mathcal{D}(\mathfrak{A})$ (domain of definition), 24
$\mathcal{D} = \mathcal{D}(\overline{LH_\alpha})$ (common domain), 145, 223
G_α (resolvent), 27
G_α (Green operator), 68
G_N (Green operator), 14, 127
$G_\alpha = (\alpha I - \mathfrak{A})^{-1}$ (Green operator), 147, 236
G_α^0 (Green operator), 142, 214
$G_\alpha^N = (\alpha I - \mathfrak{A}_N)^{-1}$ (Green operator), 226
G_α^ν (Green operator), 159, 160
H_α (harmonic operator), 142, 214
L_0 (special Ventcel' boundary condition), 3, 169, 177, 193, 234, 246
L (general Ventcel' boundary condition), 138, 155, 213
L (general boundary condition), 146, 225
L_ν (transversal Ventcel' boundary condition), 156
L_N (Neumann boundary condition), 226
$\overline{LG_\alpha^0}$ (bounded extension), 145, 220
$\overline{LH_\alpha}$ (minimal closed extension), 145, 222

$\overline{L_N H_\alpha}$ (minimal closed extension), 281
$m(dz)$ (Radon measure), 10, 246, 311
$\mathcal{N}(A, s, p)$ (null space), 126
$\mathcal{N}(A - \lambda I, s, p)$ (null space), 176
$\mathcal{N}(A(\lambda))$ (null space), 176
$\mathcal{N}(\mathcal{T}(\lambda))$ (null space), 176
$\mathcal{N}(\widetilde{\Lambda}, s, p)$ (null space), 177
$\mathcal{N}(\widetilde{\Lambda}(\vartheta), s, p)$ (null space), 193
P (Poisson operator), 14, 126
$P(\lambda)$ (Poisson operator), 176
$\widetilde{P}$ (Poisson operator), 177
$\widetilde{P}(\vartheta)$ (Poisson operator), 193
$R(\lambda) = (A - \lambda I)^{-1}$ (resolvent), 35
S_r (Lévy operator), 136
S (Lévy operator), 9, 71, 245
T_t (semigroup), 23, 32
T_t (Feller semigroup), 60, 157
$T = B_0 P$, 130
$\mathcal{T}$, 131
$T = L_0 P$, 170
$\mathcal{T}$, 175
$T(\lambda) = L_0 P(\lambda)$, 176, 198
$\mathcal{T}(\lambda)$, 176
$T_p(\lambda)$, 198
$\widetilde{T} = L_0 \widetilde{P}$, 177
$\widetilde{\mathcal{T}}$, 179
$\widetilde{T}(\vartheta) = L_0 \widetilde{P}(\vartheta)$, 194
$\widetilde{T}_p(\vartheta)$, 199
U_t, U_z (semigroup), 35, 40
W (Waldenfels operator), 9, 136, 245, 311
$\overline{W}$ (minimal closed extension), 144
W_p, 12, 258
$\mathcal{W}$, 11, 257
$\overline{\mathcal{W}}$ (minimal closed extension), 22, 157, 262
$\mathfrak{W}$, 13, 201, 319
$r(x', y')$ (distribution kernel), 139
$s(x, dy)$ (Borel kernel), 71
$s(x, y)$ (distribution kernel), 137
$s(x, z)$ (integral kernel), 10, 246
$t(x, y)$ (distribution kernel), 139
$\sigma(x, y)$ (local unity function), 71, 137, 305
$\tau(x, y)$ (local unity function), 139
$\Pi = \frac{\partial P}{\partial \mathbf{n}}$, 130, 170
$\Pi_\alpha = \frac{\partial H_\alpha}{\partial \mathbf{n}}$, 148
$\Pi(\lambda) = \frac{\partial P(\lambda)}{\partial \mathbf{n}}$, 176
$\widetilde{\Pi} = \frac{\partial \widetilde{P}}{\partial \mathbf{n}}$, 178
$\widetilde{\Pi}(\vartheta) = \frac{\partial \widetilde{P}(\vartheta)}{\partial \mathbf{n}}$, 194
γ_0 (trace map), 101, 127
γ_1 (trace map), 127

Miscellaneous:

$a(x, \xi) \sim \sum a_j(x, \xi)$ (asymptotic expansion), 103
$(E, \|\cdot\|)$ (Banach space), 19
(K, ρ) (metric space), 48, 51
$\mathcal{B}$ (σ-algebra of Borel sets in K), 48
e^{tA} (exponential function), 22
$G_\alpha(x)$ (Bessel potential), 89
J^s (Bessel potential), 98, 274
J'^s (Bessel potential), 100
$L(E, E)$ (space of bounded operators on E), 21
$N(x)$ (Newtonian potential), 89
$R_j(x)$ (Riesz kernel), 90
$R_\alpha(x)$ (Riesz potential), 89
$\mathcal{D}(T)$ (domain of definition), 122
$\mathcal{N}(T)$ (null space), 122
$\mathcal{R}(T)$ (range), 122
$\operatorname{ind} T = \dim \mathcal{N}(T) - \operatorname{codim} \mathcal{R}(T)$ (index), 122
$P(B \mid \mathcal{G})$ (conditional probability), 48
$\mathcal{X} = (x_t, \mathcal{F}, \mathcal{F}_t, P_x)$ (Markov process), 50
$p(t, x, y)$ (probability density), 47
$p_t(x, E)$ (Markov transition function), 51, 136, 158
$p_t(x, dy)$ (transition function), 8, 51
∂ (terminal point), 51
$\zeta(\omega)$ (life time), 51
$\operatorname{sing\,supp} u$, 106
$\operatorname{supp} u$, 78, 85
$f * g$ (convolution), 76
$u * v$ (convolution), 86
$u \otimes v$ (tensor product), 85
$\mathcal{F}f$, $\hat{f}$ (Fourier transforms), 87, 90
$\mathcal{F}^* g$, $\check{g}$ (inverse Fourier transforms), 87, 90
$Y(x)$ (Heaviside function), 81
$\delta(x)$ (Dirac measure), 82
$\delta(x_n)$ (Dirac measure), 93
$\overline{u}$ (complex conjugate), 83
u^0 (extension of u), 93, 115
v.p. (valeur principale), 82

$\rho_\varepsilon(x)$ (mollifier), 80
χ_E (characteristic function), 52
$\|f\|_p$, $\|f\|_\infty$ (L^p norms), 74, 75
$\|f\|_\infty$ (maximum norm), 7
$\|f\|_\infty$ (supremum norm), 55
$\|T\|_\infty$ (operator norm), 21
$\|u\|_{s,p}$ (Sobolev norm), 99
$\|u\|_{j,p}$ (Sobolev seminorm), 192, 204
$|\varphi|_{\theta;D}$ (Hölder seminorm), 78
$|\varphi|_{s,p}$ (Besov norm), 100
$\|\sigma\|$ (norm of a symbol), 276

Notes: Symbols used only in a single section are generally not listed here.

Index

a priori estimate 133, 191
absorbing barrier Brownian motion 64
absorption 140
absorption phenomenon 3, 139
adjoint 111, 181, 200
adjoint operator 184
Agmon's method 15, 177, 192, 199
algebra of pseudo-differential operators 111
amplitude 106, 107
analytic 7, 8, 12, 13, 201, 243, 270
analytic semigroup 38
A_p-completely continuous 262
Ascoli-Arzelà theorem 251, 308
asymptotic expansion 103

Banach space valued function 19
Banach's closed graph theorem 169
Banach's closed range theorem 182
Banach-Steinhaus theorem 84
barrier 314, 320
Besov space 100, 274
Besov-space boundedness theorem 15, 171, 189
Bessel potential 89, 91, 92, 98, 100, 107, 111, 274
Bessel-potential space 99
bijective 169, 179
Borel kernel 71
Borel set 48, 49, 51
Borel-measurable 70
boundary condition 146, 225
boundary point lemma 150, 161, 174, 222, 227, 252, 257, 319
boundary value problem 121, 127, 135, 144, 218, 245
Boutet de Monvel calculus 118
Brownian motion 47, 53, 63
Brownian motion with constant drift 53, 63

C_0-function 56
Calderón–Zygmund integro-differential operator 93
Cauchy density 54
Cauchy process 54, 63
Cauchy's theorem 37, 40, 43
cemetery 51
Chapman-Kolmogorov equation 51
classical pseudo-differential operator 110, 113, 130, 148, 149, 151, 170, 176, 178, 194, 198, 199, 303
classical symbol 103
closable 64, 253
closed extension 11, 64, 144, 145, 219, 222, 253, 257, 270
closed graph theorem 169
closed operator 169
closed range theorem 182, 201
codimension 122
commutator 209, 266
compact operator 16, 183, 201, 251
compact support 78, 85, 91
compactification 157
complete symbol 110
completed π-topology tensor product 103
completely continuous 101, 262

complex interpolation space 296
composition of pseudo-differential operators 111
conditional expectation 48
conditional probability 48
conjugate exponent 75
conjugation 83
conservative 51, 55
contraction semigroup 23
contraction semigroup of class (C_0) 23
contractive 60
convex 9, 245
convolution 76, 86
cotangent bundle 112, 122
cotangent space 138, 305
criterion for hypoellipticity 114

degenerate elliptic differential operator 138, 150, 303
densely defined, closed operator 35, 131, 176, 179, 184, 191, 198, 199
density 97, 122, 139
diagonal 71, 108, 113, 137, 139, 151, 303
differential operator 83
differentiation 83
diffusion along the boundary 140
diffusion coefficient 3, 137
diffusion operator 3
diffusion process 53
Dini's theorem 232
Dirac measure 82, 89, 91, 93, 106
Dirichlet condition 3
Dirichlet problem 95, 121, 126, 141, 175, 177, 193, 214
distribution 80, 141, 144, 219
distribution kernel 92, 137, 139, 151, 303
distribution with compact support 85, 91
double layer potential 95
double of a manifold 97
drift 53, 63
drift coefficient 3, 137
drift vector filed 251
dual space 80, 81, 85, 89

elementary symbol 277
elliptic 104, 111, 114, 193, 304
elliptic boundary value problem 121, 135
elliptic differential operator 2, 122, 213, 306
elliptic integro-differential operator 9, 136, 245
elliptic pseudo-differential operator 111, 114
elliptic regularization 303
elliptic symbol 104
elliptic Waldenfels operator 311
equivalent 74
essentially bounded 75
existence and uniqueness theorem for the Dirichlet problem 141, 214
existence theorem 174
existence theorem for Feller semigroups 146, 213, 223
expectation 48
exponential function 22
extension 93, 115, 117–119

Feller function 56
Feller semigroup 8, 60, 135, 136, 157
Feller semigroup with reflecting barrier 226
first passage time 54
formulation of a boundary value problem 127
Fourier integral distribution 106
Fourier integral operator 107
Fourier inversion formula 89
Fourier transform 87, 90
Fredholm integral equation 15, 130
Fredholm operator 122, 131, 176, 179, 199
Fubini's theorem 37, 260
function rapidly decreasing at infinity 88
function space 97

Gagliardo-Nirenberg inequality 204, 205, 263
Gaussian kernel 24
general boundary value problem 144, 218
general existence theorem for Feller semigroups 146, 223
generalized Sobolev space 99, 274

generalized Young's inequality 75, 287
generation theorem for Feller semigroups 60, 141, 157
global regularity theorem for the Dirichlet problem 214
Green kernel 120
Green operator 14, 61, 127, 142, 147, 159, 160, 215, 226, 236
Green's representation formula 95

harmonic operator 142, 215
heat kernel 24
Heaviside function 81
Hille-Yosida theorem 29, 61
Hille-Yosida-Ray theorem 253
Hölder continuous 78
Hölder space 5, 79, 247, 277
Hölder's inequality 75
homogeneous principal symbol 111, 114
hypoelliptic 114, 183, 185

index 122, 175
inductive limit topology 78
infinitesimal generator 24, 60, 147, 160, 226, 236
integro-differential operator 9, 136, 245
interior normal 3, 138, 247
interior regularity theorem for the Dirichlet problem 214
interpolation 296, 309
invariance of pseudo-differential operators 111
inverse Fourier transform 87, 90
isomorphism 5, 6, 11, 187, 247

jump formula 94, 115
jump into the interior 140
jump on the boundary 140

kernel 92
kernel theorem 92

λ-dependent localization argument 16, 206, 264
Laplacian 93, 124, 170, 194
layer potential 94, 95, 115
Lebesgue convergence theorem 43, 284, 285
Leibniz formula 22, 83
Lévy operator 10, 137, 246
lifetime 51
Littlewood-Paley series 273
local unity function 71, 305
localized Besov space 101
localized Sobolev space 101
locally Hölder continuous 78
locally integrable function 81
L^p boundedness theorem 280, 286
L^p space 74

manifold 97, 113, 121, 150, 303
manifold with boundary 97, 121
Markov process 9, 50, 136, 158
Markov property 49, 52
Markov transition function 51, 136, 158
maximum norm 7, 136
maximum principle 16, 64, 173, 252, 253
Mazur's theorem 308
measurable 49
metric space 48, 51
minimal closed extension 11, 64, 144, 145, 219, 222, 253, 257, 270
modified Bessel function 90
mollifier 80, 216
moment condition 10, 246, 312
multi-index 73
Multiplication by functions 83

Neumann condition 226, 236
Neumann problem 14, 127, 129, 188
Neumann series 45
Newtonian potential 89, 91, 92, 122
non-negative 60, 151, 303
non-regular symbol 275
non-transversal 155
norm 21, 74, 75, 77, 79, 98–101, 276
norm continuous 21
norm differentiable 21
normal coordinate 97, 121
normal transition function 52
null space 122, 131, 176, 183, 185, 198

one-point compactification 157
operator 92
operator valued function 21

order of a Besov space 100
order of a Bessel potential 98
order of a distribution 81
order of a pseudo-differential operator 108
order of a Sobolev space 99
order of a symbol 102
oscillatory integral 106

parametrix 15, 111, 120, 171, 178, 189
Parseval formula 89
path 49
Peetre's definition of Besov spaces 273
Peetre's definition of Sobolev spaces 273
Peetre's Lemma 181, 200
Peetre's lemma 179
phase function 104, 106, 107
Plancherel theorem 91
plane-wave expansion 106
point at infinity 49, 157
Poisson integral formula 96
Poisson kernel 120, 123
Poisson operator 14, 126, 176, 177, 193
Poisson process 53, 62
positive Borel kernel 71
positive maximum principle 72
positively homogeneous 82, 102
potential 94, 95, 115, 119, 122, 123
principal part 103
principal symbol 72
principle of uniform boundedness 21
probability measure 48
probability space 48
projective topology tensor product 103
properly supported 109
pseudo-differential operator 108, 113, 115
pseudo-local property 108
pull-back 112
push-forward 112

Radon measure 10, 246, 312
random variable 48
range 122, 131, 176, 184, 199
rapidly decreasing 88
real interpolation space 309
reduction to the boundary 129
reflecting barrier Brownian motion 54, 63
reflection 140
reflection phenomenon 3, 226
reflexivity 84
regularity property 131, 170
regularity theorem 170
regularization 80, 87, 137
regularizer 109
regularizing effect 8
Rellich's theorem 101, 181, 183, 200, 201
residue theorem 37, 43
resolvent 6, 8, 12, 13, 27, 35, 197, 205, 258, 264
resolvent equation 143, 216
resolvent set 6, 8, 12, 13, 35, 197, 211, 258, 269
restriction 82
Riemannian metric 124, 170
Riesz kernel 90, 91
Riesz operator 92
Riesz potential 89, 91, 92
Robin condition 3

sample space 49
Schwartz kernel theorem 92
Schwartz space 88
Schwarz's inequality 75
second-order, elliptic differential operator 2, 122, 213, 306
semigroup 36
semigroup property 24, 60
seminorm 76, 77, 79, 88, 102
sheaf property 84
σ-algebra containing all open sets 48
σ-algebra of all Borel sets 51
single layer potential 94
singular Green operator 119
singular support 106
Sobolev space 99, 274, 304
Sobolev's imbedding theorem 203, 269
space of continuous functions 7, 55, 76, 156, 218
spectral parameter 15, 177, 192
state space 49
sticking 140
sticking barrier Brownian motion 54, 63

sticky barrier Brownian motion 63
stochastic process 49
strong Markov process 9, 136, 158
strong maximum principle 123, 314
strong topology 84
strongly continuous 19, 21, 60
strongly differentiable 20, 21
strongly integrable 20
support 85
supremum norm 55
surface potential 115, 123
symbol 102
symbol class 102

tempered distribution 89
tensor product 85
terminal point 51, 64
termination coefficient 3, 137
test function 78
trace 127
trace map 101, 128
trace operator 119
trace theorem 101, 128
trajectory 49
transition function 51
transition map 112
transmission property 116, 117, 119, 137, 139, 147, 149, 161
transpose 111
transversal 140, 147, 154, 159
transversality condition 150, 156
trap 54

uniform motion 52, 62
uniformly stochastically continuous 56
unique solvability theorem for pseudo-differential operators 150
uniqueness theorem 173

v.p. (valeur principale) 82
Ventcel' boundary condition 139, 213
viscosity 140
volume potential 116, 122

Waldenfels operator 10, 137, 246, 311
weak maximum principle 125, 149, 174, 215, 222, 232, 252, 312
weak* topology 84
weakly closed 308
weakly compact 308
Wiener measure 48

Yosida approximation 30, 61
Young's inequality 76, 283, 289, 297

Zygmund condition 276

Springer Monographs in Mathematics

This series publishes advanced monographs giving well-written presentations of the "state-of-the-art" in fields of mathematical research that have acquired the maturity needed for such a treatment. They are sufficiently self-contained to be accessible to more than just the intimate specialists of the subject, and sufficiently comprehensive to remain valuable references for many years. Besides the current state of knowledge in its field, an SMM volume should also describe its relevance to and interaction with neighbouring fields of mathematics, and give pointers to future directions of research.

Abhyankar, S.S. **Resolution of Singularities of Embedded Algebraic Surfaces** 2nd enlarged ed. 1998
Andrievskii, V.V.; Blatt, H.-P. **Discrepancy of Signed Measures and Polynomial Approximation** 2002
Ara, P.; Mathieu, M. **Local Multipliers of C*-Algebras** 2003
Armitage, D.H.; Gardiner, S.J. **Classical Potential Theory** 2001
Arnold, L. **Random Dynamical Systems** corr. 2nd printing 2003 (1st ed. 1998)
Aubin, T. **Some Nonlinear Problems in Riemannian Geometry** 1998
Auslender, A.; Teboulle M. **Asymptotic Cones and Functions in Optimization and Variational Inequalities** 2003
Bang-Jensen, J.; Gutin, G. **Digraphs** 2001
Baues, H.-J. **Combinatorial Foundation of Homology and Homotopy** 1999
Brown, K.S. **Buildings** 3rd printing 2000 (1st ed. 1998)
Cherry, W.; Ye, Z. **Nevanlinna's Theory of Value Distribution** 2001
Ching, W.K. **Iterative Methods for Queuing and Manufacturing Systems** 2001
Crabb, M.C.; James, I.M. **Fibrewise Homotopy Theory** 1998
Dineen, S. **Complex Analysis on Infinite Dimensional Spaces** 1999
Elstrodt, J.; Grunewald, F. Mennicke, J. **Groups Acting on Hyperbolic Space** 1998
Fadell, E.R.; Husseini, S.Y. **Geometry and Topology of Configuration Spaces** 2001
Fedorov, Y.N.; Kozlov, V.V. **A Memoir on Integrable Systems** 2001
Flenner, H.; O'Carroll, L. Vogel, W. **Joins and Intersections** 1999
Gelfand, S.I.; Manin, Y.I. **Methods of Homological Algebra** 2nd ed. 2003
Griess, R.L.Jr. **Twelve Sporadic Groups** 1998
Gras, G. **Class Field Theory** 2003
Ivrii, V. **Microlocal Analysis and Precise Spectral Asymptotics** 1998
Jech, T. **Set Theory** (3rd revised edition 2002)
Jorgenson, J.; Lang, S. **Spherical Inversion on SLn (R)** 2001
Kanamori, A.; **The Higher Infinite** (2nd edition 2003)
Khoshnevisan, D. **Multiparameter Processes** 2002
Koch, H. **Galois Theory of *p*-Extensions** 2002
Kozlov, V.; Maz'ya, V. **Differential Equations with Operator Coefficients** 1999
Landsman, N.P. **Mathematical Topics between Classical & Quantum Mechanics** 1998
Lebedev, L.P.; Vorovich, I.I. **Functional Analysis in Mechanics** 2002
Lemmermeyer, F. **Reciprocity Laws: From Euler to Eisenstein** 2000
Malle, G.; Matzat, B.H. **Inverse Galois Theory** 1999
Mardesic, S. **Strong Shape and Homology** 2000
Murdock, J. **Normal Forms and Unfoldings for Local Dynamical Systems** 2002
Narkiewicz, W. **The Development of Prime Number Theory** 2000
Parker, C.; Rowley, P. **Symplectic Amalgams** 2002
Peller, V. (Ed.) **Hankel Operators and Their Applications** 2003
Prestel, A.; Delzell, C.N. **Positive Polynomials** 2001
Puig, L. **Blocks of Finite Groups** 2002
Ranicki, A. **High-dimensional Knot Theory** 1998
Ribenboim, P. **The Theory of Classical Valuations** 1999
Rowe, E.G.P. **Geometrical Physics in Minkowski Spacetime** 2001
Rudyak, Y.B. **On Thom Spectra, Orientability and Cobordism** 1998
Ryan, R.A. **Introduction to Tensor Products of Banach Spaces** 2002

Saranen, J.; Vainikko, G. **Periodic Integral and Pseudodifferential Equations with Numerical Approximation** 2002
Schneider, P. **Nonarchimedean Functional Analysis** 2002
Serre, J-P. **Complex Semisimple Lie Algebras** 2001 (reprint of first ed. 1987)
Serre, J-P. **Galois Cohomology** corr. 2nd printing 2002 (1st ed. 1997)
Serre, J-P. **Local Algebra** 2000
Serre, J-P. **Trees** corr. 2nd printing 2003 (1st ed. 1980)
Smirnov, E. **Hausdorff Spectra in Functional Analysis** 2002
Springer, T.A. Veldkamp, F.D. **Octonions, Jordan Algebras, and Exceptional Groups** 2000
Sznitman, A.-S. **Brownian Motion, Obstacles and Random Media** 1998
Taira, K. **Semigroups, Boundary Value Problems and Markov Processes** 2003
Tits, J.; Weiss, R.M. **Moufang Polygons** 2002
Uchiyama, A. **Hardy Spaces on the Euclidean Space** 2001
Üstünel, A.-S.; Zakai, M. **Transformation of Measure on Wiener Space** 2000
Yang, Y. **Solitons in Field Theory and Nonlinear Analysis** 2001

www.ingramcontent.com/pod-product-compliance
Ingram Content Group UK Ltd.
Pitfield, Milton Keynes, MK11 3LW, UK
UKHW022230230426
12048UKWH00016BA/1168
* 9 7 8 3 6 6 2 0 9 8 5 8 5 *